McGRAW-HILL NETWORKING AND TELECOMMUNICATIONS

Build Your Own

Trulove — *Build Your Own Wireless LAN* (with projects)

Crash Course

Louis	*Broadband Crash Course*
Vacca	*I-Mode Crash Course*
Louis	*M-Commerce Crash Course*
Shepard	*Telecom Convergence, 2/e*
Shepard	*Telecom Crash Course*
Bedell	*Wireless Crash Course*
Kikta/Fisher/Courtney	*Wireless Internet Crash Course*

Demystified

Harte/Levine/Kikta	*3G Wireless Demystified*
LaRocca	*802.11 Demystified*
Muller	*Bluetooth Demystified*
Evans	*CEBus Demystified*
Bayer	*Computer Telephony Demystified*
Hershey	*Cryptography Demystified*
Taylor	*DVD Demystified*
Bates	*GPRS Demystified*
Symes	*MPEG-4 Demystified*
Camarillo	*SIP Demystified*
Shepard	*SONET / SDH Demystified*
Topic	*Streaming Media Demystified*
Symes	*Video Compression Demystified*
Shepard	*Videoconferencing Demystified*
Bhola	*Wireless LANs Demystified*

Developer Guides

Vacca	*I-Mode Crash Course*
Guthery	*Mobile Application Development with SMS*
Richard	*Service and Device Discovery: Protocols and Programming*

Professional Telecom

Smith/Collins	*3G Wireless Networks*
Bates	*Broadband Telecom Handbook, 2/e*
Collins	*Carrier Grade Voice over IP*
Harte	*Delivering xDSL*
Held	*Deploying Optical Networking Components*
Minoli/Johnson/Minoli	*Ethernet-Based Metro Area Networks*
Benner	*Fibre Channel for SANs*

Bates	*GPRS*
Sulkin	*Implementing the IP-PBX*
Lee	*Lee's Essentials of Wireless*
Bates	*Optical Switching and Networking Handbook*
Wetteroth	*OSI Reference Model for Telecommunications*
Russell	*Signaling System #7, 4/e*
Minoli/Johnson/Minoli	*SONET-Based Metro Area Networks*
Nagar	*Telecom Service Rollouts*
Louis	*Telecommunications Internetworking*
Russell	*Telecommunications Protocols, 2/e*
Minoli	*Voice over MPLS*
Karim/Sarraf	*W-CDMA and cdma2000 for 3G Mobile Networks*
Bates	*Wireless Broadband Handbook*
Faigen	*Wireless Data for the Enterprise*

Reference

Muller	*Desktop Encyclopedia of Telecommunications, 3/e*
Botto	*Encyclopedia of Wireless Telecommunications*
Clayton	*McGraw-Hill Illustrated Telecom Dictionary, 3/e*
Radcom	*Telecom Protocol Finder*
Pecar	*Telecommunications Factbook, 2/e*
Russell	*Telecommunications Pocket Reference*
Kobb	*Wireless Spectrum Finder*
Smith	*Wireless Telecom FAQs*

Security

Nichols	*Wireless Security*

Telecom Engineering

Smith/Gervelis	*Cellular System Design and Optimization*
Rohde/Whitaker	*Communications Receivers, 3/e*
Sayre	*Complete Wireless Design*
OSA	*Fiber Optics Handbook*
Lee	*Mobile Cellular Telecommunications, 2/e*
Bates	*Optimizing Voice in ATM / IP Mobile Networks*
Roddy	*Satellite Communications, 3/e*
Simon	*Spread Spectrum Communications Handbook*
Snyder	*Wireless Telecommunications Networking with ANSI-41, 2/e*

BICSI

Network Design Basics for Cabling Professionals
Networking Technologies for Cabling Professionals
Residential Network Cabling
Telecommunications Cabling Installation

CARRIER GRADE VOICE OVER IP

Carrier Grade Voice Over IP

Daniel Collins

McGraw-Hill

New York   Chicago   San Francisco   Lisbon
London   Madrid   Mexico City   Milan   New Delhi
San Juan   Seoul   Singapore   Sydney   Toronto

The McGraw·Hill Companies

Cataloging-in-Publication Data is on file with the Library of Congress.

7 8 9 0   DOC/DOC   0 9 8 7 6

ISBN 0-07-140634-4

The sponsoring editor for this book was Scott Grillo and the production supervisor was Pamela Pelton. It was set in Century Schoolbook by MacAllister Publishing Services, LLC.

Printed and bound by RR Donnelley.

 This book is printed on recycled, acid-free paper containing a minimum of 50 percent recycled de-inked fiber.

McGraw-Hill books are available at special quantity discounts to use as premiums and sales promotions, or for use in corporate training programs. For more information, please write to the Director of Special Sales, Professional Publishing, McGraw-Hill, Two Penn Plaza, New York, NY 10121-2298. Or contact your local bookstore.

To Ann, my wife and friend.

CONTENTS

Contents

PREFACE

I wrote the preface to the first edition of this book in mid-2000. At that time, the United States was still enjoying a long period of economic growth, and the telecommunications industry was one of the greatest beneficiaries of the expansion. New companies were using new technologies, such as *Voice over IP* (VoIP), to compete with traditional telecommunications companies. The result was a wealth of choices for the customer. All was well, and the preface to the first edition of *Carrier Grade over VoIP* painted a very rosy future. Two years later, it seems that my predictions (like those of many others) might have been a little premature.

As I write this preface, we are experiencing an economy that is making a faltering recovery from recession. Although some industries are recovering from the downturn, the telecommunications industry has suffered more than any other, and recovery is not yet in sight. Many of the new telecommunications companies that emerged in the late 1990s no longer exist. In hindsight, one could say that my positive outlook was proven wrong. I agree, but only to a limited extent.

Although the growth of telecommunications has not maintained its breakneck pace, growth is continuing, and much of the growth involves VoIP systems. There are more VoIP-based products and networks in service today than ever before. More multimedia applications exist that merge voice and data. In addition to the new carriers, the established carriers have embraced VoIP technology. A good deal of today's voice traffic is already carried as VoIP, and the percentage is increasing. I remain convinced that the future for VoIP is very bright.

One reason why VoIP is here to stay is the fact that true carrier-grade solutions exist—solutions that provide superior quality, reliability, and scalability. More important, these solutions cost less to purchase and operate than traditional systems. In addition, VoIP systems can enable true multimedia services—something that circuit switching cannot offer. By incorporating advances made in standards groups such as the *Internet Engineering Task Force* (IETF), today's VoIP systems offer everything that circuit-switched systems can offer and more.

The advantages of VoIP systems do not mean that we can expect a sudden wholesale replacement of installed circuit-switched systems. Instead, we can expect traditional systems to remain with us for some time, but eventually they will be phased out. Wherever there is a need for new capacity, new features, or multimedia capabilities, VoIP systems will be deployed. Given the steadily increasing demand for voice service and the emerging demand for true multimedia, VoIP deployment will continue to grow. Eventually long, VoIP systems will be more prevalent than circuit-switching systems. Perhaps that situation will not materialize as quickly as we once thought, but it will happen. We are already moving in that direction.

DANIEL COLLINS

CHAPTER 1

Introduction

The two terms that make up the title of this book, "carrier-grade" and "*Voice over IP* (VoIP)," mean many things to many people, and each term may have several meanings depending on the context. Furthermore, many people might consider the two terms to be mutually exclusive, making the phase carrier-grade VoIP an oxymoron. That used to be the case, but those days are over. The excitement that VoIP has generated and the resources that have been applied to developing technical solutions for VoIP have allowed it to become a serious alternative for voice communications. That is, VoIP is a serious alternative to the circuit-switched telephony that we have known for decades. Not only can VoIP technology now offer straightforward telephony services, but it can offer a lot more besides, and it can do it with the same *quality of service* (QoS) that we have become familiar with in traditional telephony.

The overall objectives of this book are to examine the technical solutions that make VoIP a reality and to illustrate how today's VoIP solutions can offer everything that traditional circuit switching provides, plus a lot more. This first chapter introduces the reader to VoIP in general and looks at some of the reasons why it is such a hot topic. Thereafter, a brief overview of technical challenges and solutions is presented. This chapter can be considered a glimpse of the topics that are covered in more depth in later chapters. Firstly, however, we need to understand what exactly is meant when we talk about terms such as VoIP and carrier-grade.

What Is Meant by Carrier-Grade?

Perhaps the easiest way to describe the term carrier-grade is to think of the last time you picked up your residential phone and failed to get a dial tone. With the exception of extraordinary circumstances, such as an earthquake, most people cannot remember such a time. Basically, the phone always works.

Carrier-grade means extremely high availability. In fact, the requirement for availability in commercial telephony networks is 99.999 percent. In other words, a carrier-grade network should be operational at least 99.999 percent of the time. This corresponds to a down time of no more than about five minutes per year and is known as *five nines reliability*.

Carrier-grade means high capacity, the capability to support hundreds of thousands, perhaps millions, of subscribers and similar numbers of simultaneous calls. For example, at the end of 2001, the Verizon network supported over 70 million voice access lines.[1] Meanwhile, the AT&T network carries about 300 million voice calls a day.[2]

Carrier-grade means that when you dial a number, you get through to the number you dialed. It means that when you finish dialing, the phone at the other end starts ringing within two to three seconds. It means that when someone answers and conversation takes place, the speech quality is very high, without any perceptible echo, noticeable delay, or annoying noise on the line.

These are tough standards to meet. In terms of the network, these requirements translate to systems that are fully redundant, in some cases self-healing, highly scalable, and manageable. They mean compliance with numerous technical specifications to ensure interoperability with other networks. They also imply a highly skilled network maintenance organization, on duty around the clock, and ready to respond to a network problem in an instant.

What Is Meant by VoIP?

VoIP is simply the transport of voice traffic using the *Internet Protocol* (IP), hardly a surprising definition. However, it is important to note that VoIP does not automatically imply voice over the Internet.

The Internet is a collection of interconnected networks, all using IP. The connections between these networks are used by anyone and everyone for a wide range of applications, from e-mail to file transfer to *electronic commerce* (e-commerce). As we shall see later, one of the greatest challenges to VoIP is voice quality and one of the keys to acceptable voice quality is bandwidth. If we are to ensure that sufficient bandwidth is available to enable high-quality voice, then we need to control and prioritize access to the available bandwidth. Currently, that does not happen on the Internet. In fact, each user of the Internet is at the mercy of all other users. One person transferring a huge file may cause other users' transactions to proceed more slowly. As a result, voice quality over the Internet today may vary from acceptable to atrocious.

---

[1]Verizon Annual Report, 2001.

[2]www.att.com/network.

All is not lost, however. The Internet is changing rapidly in terms of its size, the number of users, and the technology that it uses. As technology changes and as more and more bandwidth is made available, it is possible that high-quality voice over the Internet may become the norm rather than the exception. However, that day is still not at hand. Consequently, while the term carrier-grade VoIP may not be an oxymoron, the term carrier-grade voice over the Internet may well be. Internet telephony may be considered a subset of, or a special case of, VoIP (aka IP telephony). However, within this book, when we refer to VoIP, we generally exclude Internet telephony. We shall see that high voice quality in IP networks requires the use of managed networks, QoS solutions, and *service-level agreements* (SLAs) between providers. Although the Internet might include all those things in the future, today's Internet does not.

In the meantime, a significant opportunity is available to deploy VoIP in networks where access and bandwidth are better managed. One such environment is that of next-generation Telcos, those companies building telephone networks using VoIP from the outset and posing a challenge to traditional carriers.

A Little about IP

IP is a packet-based protocol. This means that traffic is broken into small packets that are sent individually to their destination. In the absence of special technical solutions, the route that each packet takes to its destination is determined independently at each network node on a packet-by-packet basis. IP is not the only packet-based protocol in existence. However, it is by far the most successful. The explosive growth of the Internet proves this.

IP itself provides no guarantees. For example, it is possible for different packets to take different routes from the origination point to the destination point, leading to the possibility of packets arriving at a destination in a different order than originally sent. Not only can packets arrive out of sequence, but some packets might not arrive at all or might be severely delayed.

To combat these shortcomings, other protocols have been developed to operate in conjunction with IP to ensure that packets are delivered to their ultimate destination in the correct sequence and without loss. The most notable such protocol is the *Transmission Control Protocol* (TCP). TCP includes functions for the retransmission of packets that may have been lost or delayed and the assembly of packets in the correct order at the destination end. TCP is so widespread that the term TCP/IP is almost as common as IP and is sometimes used synonymously with IP. Although the mechanisms used by TCP are appropriate and successful for data transfer (such as file transfers and e-mail), they are not appropriate for the delivery of voice traffic.

Most data traffic is asynchronous and extremely error sensitive. For example, it hardly matters if an e-mail message takes 10 seconds or 30 seconds to reach its destination, but it is critical that every bit is received correctly. On the other hand, voice traffic is synchronous in nature and a little more tolerant of errors. In speech, when someone speaks, the listener should hear it practically immediately, although it is not as critical that every millisecond of speech is heard.

Given that IP provides no guarantees regarding the efficient transport of data packets, one wonders why IP would even be considered as a means for transporting voice, particularly with the stringent delay requirements that voice imposes. One also wonders how VoIP can be made to match the quality, reliability, and scalability of traditional networks. After all, if VoIP is to be a successful competitor to traditional telephony technology, then it must meet all the requirements met by traditional telephony, it must offer new and attractive capabilities beyond traditional telephony, and it must do so at a lower cost.

Why VoIP?

It is probably best to break this question into two parts. The first is, why worry about carrying voice at all when a large and lucrative market for data services exists? The second question is, assuming that a carrier intends to compete in the voice market, why should IP be considered the transport mechanism?

Why Carry Voice?

IP and the Internet have led to many new and exciting services. Whereas in the past it might have taken weeks to access certain types of information, it is now possible to find out practically anything in an instant. It is now possible to communicate at the touch of a button and to share information between colleagues, friends, and family. Many companies use IP technology to share information between their business systems and those of their suppliers and customers in order to reduce ordering times and streamline production. All of this means that people are using IP technology in new and exciting ways. The technology has led to new revenue opportunities and to the creation of companies and enterprises to capitalize on those opportunities. However, the revenues generated as a result of those opportunities are miniscule compared to the revenues generated by telephone companies who carry voice. Voice is still the killer application.

To illustrate this point, let us look at some real examples. In 2001, Amazon.com had net sales of over $3 billion but still failed to make a profit.[3] Yahoo!, one of the pillars of the Internet, generated net revenues of over $700 million and lost money.[4] By comparison, BellSouth generated net income of over $2.5 billion from revenues of over $24 billion, with about half the revenues coming from local wireline voice services.[5] Meanwhile, Verizon generated an adjusted net income of over $8 billion from revenues of over $67 billion, again with a large portion of the revenue from regular telephony.[6] Voice is big business.

Why Use IP for Voice?

Traditional telephony carriers use circuit switching for carrying voice traffic. Given that these carriers generate large revenues and profits, why should anyone consider using a packet technology such as IP to transport voice instead of using circuit switching? Circuit switching was designed for voice from the outset and does what it is supposed to do very well. So why try to fix something that is clearly not broken?

The fact that circuit switching was designed for voice is both its strength and its weakness. There is no doubt that circuit switching carries voice very well, although it could be argued that it is an expensive solution. Circuit switching is not particularly suitable for much else, however. These days, people want to talk as much as ever, but they also want to communicate in myriad other ways—e-mail, instant messaging, video, and so on. Circuit switching really does not qualify as a suitable technology for this new world

---

[3]Amazon.com Annual Report, 2001.

[4]Yahoo!, Inc. Annual Report, 2001.

[5]Bellsouth Annual Report, 2001.

[6]Verizon Annual Report, 2001.

of multimedia communications. Furthermore, circuit-switching equipment has a number of characteristics that are less than ideal and that negatively affect vendors and operators of that equipment. IP offers not only another technology choice, but also a new way of doing business, as described shortly.

IP is an attractive choice for voice transport for several reasons. These include

- Lower equipment cost
- Lower operating expense
- Integration of voice and data applications
- Potentially lower bandwidth requirements
- The widespread availability of IP

Lower Equipment Cost

Although thousands of standards and technical specifications for circuit-switched telephony have been made, the systems themselves are generally proprietary in nature. A switch from a given supplier will often use proprietary hardware, and a proprietary operating system and the applications will use proprietary software. A given switch supplier may well design and produce everything, including processors, memory cards, cabling, power supplies, the operating system, and application software, as well as perhaps even the racks to hold the equipment. The opportunities for third parties to develop new software applications on these systems are extremely limited. The systems generally also require extensive training to operate and manage.

Consequently, when an operator chooses to implement systems from a given vendor, it is not unusual to choose just a single vendor for the whole network, at least at the beginning. If the operator wants to replace operating equipment with products from a new vendor, the cost and effort involved are huge. The result of all this is that the equipment vendor, once chosen, is in a position to generate revenue not only from the sale of the equipment, but from training, support, and feature development. In this environment, proprietary systems are the order of the day.

This is very much like the mainframe computer business of several years ago. In those days, computers were large monolithic systems. The system vendor was responsible for all the components of the system, plus installation, training, and support. The choice of vendor was limited and once a system from a particular vendor was installed, it was extremely difficult to change to another vendor. Since those days, the computer world has changed drastically from the mainframe model. Although many mainframe systems are still in place, many of the big vendors of mainframe systems no longer exist. Those who still exist do so in a very different form.

The IP world is very different from the monolithic systems of mainframe computers and circuit-switching technology. Though some specialized hardware systems exist, much of the hardware is standard mass-produced computer equipment. This offers a greater choice to the purchaser and the opportunity to benefit from the large volumes that are produced. It is common for the operating system to be less tightly coupled to the hardware and for the application software to be quite separate again. This separation enables a greater range of choices for the purchaser of the system and the ability to contract with a separate company to implement unique features. Furthermore, IP systems tend to use a distributed client-server architecture rather than large monolithic systems, which means that it is easy to start small and grow as demand dictates. In addition, this type of architecture means that certain companies make only portions of the network solution, allowing the customer to pick those companies that are the best in the different areas to create a solution that is optimum in all respects. Figure 1-1 gives a pictorial comparison between the mainframe approach of circuit-switched systems and that of packet-switching technologies. In the words of Cisco CEO John Chambers, "When a horizontal business model meets a vertical business model, horizontal wins every time."

Not only are IP architectures more open and competition-friendly, but the same applies to IP standards. They are more open and flexible than telephony standards, allowing for the implementation of unique features, so that a provider can offer new features quicker and be better able to customize service offerings. New features can be developed and deployed in a

Figure 1-1

Architecture and business model comparison

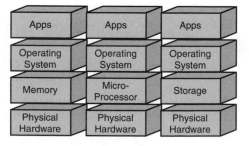
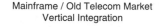
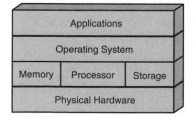

few months, which is a big difference from the traditional telephony world where new features can take 18 months to 2 years.

One could argue that an *intelligent network* (IN) architecture in telephony also offers great flexibility and rapid feature development. Although that is what IN has promised from the beginning, it does not match the openness and flexibility of IP solutions. With the exception of a few highly successful services, such as toll-free calling, IN has not had the enormous success that was originally envisaged. Of course, those IN services that are successful happen to be extremely important and also need to be provided by an IP-based solution. As we shall see in Chapter 7, "VoIP and SS7," VoIP networks can interwork with *Signaling System 7* (SS7) and take advantage of IN services that are built on SS7. Hence, IP can take advantage of what is already there and can do more besides.

All of the foregoing points mean greater competition, greater flexibility on the part of the network operator, and significantly lower cost. Furthermore, the packet-based systems obey Moore's Law, which states that processing power doubles roughly every 18 months, which is a much faster pace of development than we see with traditional circuit-switched systems. Figure 1-2 shows a comparison between the two. We can clearly see that circuit-switching equipment lags behind packet and frame technologies in the speed of development.

Of course, all the major manufacturers of traditional telephony switches now also produce VoIP solutions, but a quick search on the Internet will reveal a large number of new companies also producing VoIP products. The range of choices is large, the equipment cost is lower than that of circuit-switching products, and the pace of development is faster.

Voice/Data Integration and Advanced Services

IP is the standard for data transactions—everything from e-mail to Web browsing to e-commerce. When we combine these capabilities with the transport of voice, it is easy to imagine advanced features that can be based on the integration of the two. One example is an application where the receiver of an e-mail can call the sender simply by clicking a header within the e-mail. Another example is a button on a web page that a user can click to speak to a customer service representative for assistance in navigating the site or conducting a transaction, as depicted in Figure 1-3. When the user clicks that button, it is possible to direct the call to the most appropriate customer service representative based on what the user was attempting to do just before deciding to seek help.

Figure 1-2
Cost performance
comparison

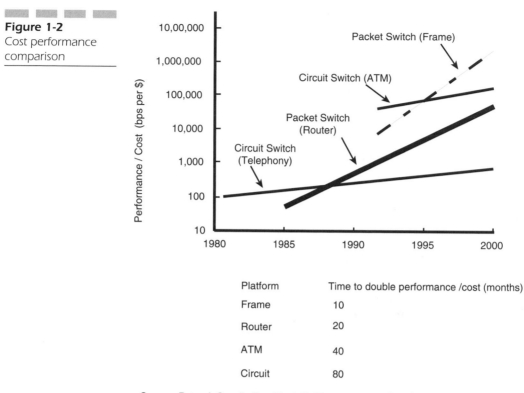

Platform	Time to double performance /cost (months)
Frame	10
Router	20
ATM	40
Circuit	80

Source: Peter J. Sevcik, President, NetForecast, peter@netforecast.com
"Why Circuit Switching is Doomed," Business Communications Review, Sep. 1997

Web collaboration is a service whereby one user can navigate the Web and another user at another location is presented with the same web page. The two users may talk to each other at the same time, as depicted in Figure 1-4.

Yet another example is video conferencing and all that it entails. Not only does this include person-to-person video telephony, but it can also include shared whiteboard sessions. Design teams can come together to discuss issues and sketch out solutions while sitting hundreds or thousands of miles from each other. With IP multicasting, virtual auditorium services are available, something of great interest for distance learning in particular.

These types of applications are not just wishful thinking. Many applications exist today, and more are coming. Such applications include IP-based *private branch exchanges* (PBXs), IP-based call centers, and IP-based voice mail, as well as straightforward telephony. IP-based telephony devices exist today, and they are far more feature-rich than the standard 12-button key-

Figure 1-3
Click-to-talk
application

Figure 1-4
Example application:
web collaboration

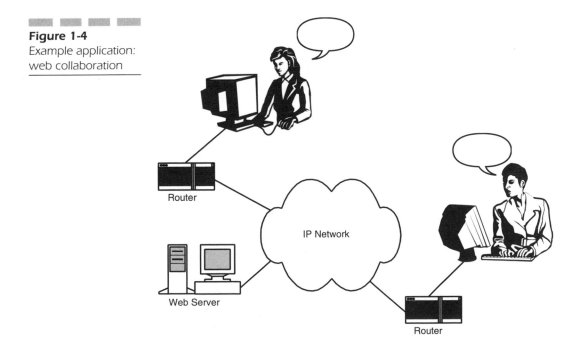

pad that we have known for years. Many include web browsers, address book applications, e-mail clients, and so on. The user interface is graphical and intuitive, and we no longer have to worry about obscure "star codes" to activate features. Figure 1-5 shows an example of two IP-based telephony devices. Contrast these examples with the traditional telephone keypad and ask yourself which user interface you would prefer.

Potentially Lower Bandwidth Requirements

Circuit-switched telephone networks transport voice at a rate of 64 Kbps. This is based on the Nyquist Theorem, which states that it is necessary to sample an analog signal at twice the maximum frequency of the signal in order to fully capture the signal. Typical human speech has a maximum frequency of somewhat less than 4,000 Hz. Therefore, in digitizing human speech, a telephone system takes 8,000 samples per second, as directed by

Figure 1-5
New versus old phones

Pingtel

Congruency, Inc.

the Nyquist Theorem, and each sample is represented with 8 bits. This leads to a bandwidth requirement of 64 Kbps in each direction for a standard telephone call. This voice-coding scheme is standardized in *International Telecommunication Union* (ITU) Recommendation G.711.

Although G.711 is the standard for traditional telephony, many other coding schemes are available. These use more sophisticated coding algorithms, which enable speech to be transmitted at a range of different speeds, such as 32, 16, 8, 6.3, or 5.3 Kbps. Furthermore, some of these coding techniques utilize silence suppression such that traffic is passed over the network only when something is being said. Given that most conversations involve one person listening and one person talking at a given time, the bandwidth saved by not transmitting silence can be significant. Various speech-coding schemes are discussed in detail in Chapter 3, "Speech-Coding Techniques," and all of those described can be used in VoIP systems. Therefore, VoIP can offer significant advantages over circuit-switched telephony from a bandwidth-efficiency point of view.

It should be noted, however, that the availability of these coding schemes is not something unique to VoIP. Strictly speaking, these various coders could also be deployed in traditional telephony networks. However, they are not generally implemented in circuit-switched networks, as it would require an enormous investment to do so. Currently, it is possible to place a call from any telephone to any other telephone in the world. To take advantage of more advanced coding schemes would require that these coding techniques be implemented in practically every telephone switch in the world. Alternatively, it would be necessary that existing networks support the capability for the two ends of a conversation to negotiate the coding scheme to use. At the bare minimum, it would require that calls be routed from the origination to the destination based on the coding choice of the originator so that, if different coders are used at either end, the call could be passed though some interworking device. None of these capabilities are supported in traditional telephony networks. As a result, traditional telephony uses G.711 and is unlikely ever to change.

On the other hand, VoIP solutions are far more recent than traditional telephony. They generally support more efficient coding schemes than G.711 and the standards written for VoIP allow for the two ends of a call to negotiate which coding scheme to use. Thus, VoIP is in a position to significantly reduce bandwidth requirements—to as little as one-eighth of that used in the circuit-switched world. Given that transmission capacity can account for a large percentage of a carrier's initial investment and a very large portion of a carrier's operational costs, such bandwidth savings can mean a big difference to the bottom line.

Although it is true that bandwidth-efficient coding schemes can reduce overall bandwidth requirements, we shall see in Chapter 9, "Designing a VoIP Network," that IP and associated protocols add some overhead beyond the bandwidth of the coded voice itself. Therefore, coding voice at, say, 32 Kbps, does not necessarily mean a 50 percent bandwidth reduction compared to 64 Kbps G.711 coding. Beware of anyone who suggests such a simple comparison. The exact bandwidth saving compared to traditional telephony is dependent on a number of factors and requires some calculation. These calculations are described in Chapter 9.

Not only can transmission costs be reduced through the use of more efficient voice-coding schemes, but the fundamental architecture of VoIP systems lends itself to more transmission-efficient network designs. Recall that traditional circuit-switching networks tend to be composed of large monolithic systems. These systems lend themselves to centralized network architectures, with both call control signaling and bearer traffic being routed through the same centralized machines. VoIP systems, however, are designed to be distributed. Although call control signaling can be centralized in a limited number of centralized machines, the bearer traffic does not have to be routed through the same machines. Rather, the bearer traffic can be routed more directly from source to destination, leading to shorter transmission distances and lower cost.

The Widespread Availability of IP

IP is practically everywhere. Every personal computer produced today supports IP. It is used in corporate *local area networks* (LANs) and *wide area networks* (WANs), as well as in dial-up Internet access. IP applications now even reside within hand-held computers and various wireless devices. As a result, IP expertise is widespread and application development companies are numerous. This ubiquitous presence alone makes IP a suitable choice for transporting voice, or any other digital media stream for that matter.

Of course, VoIP is not the only packet-based solution available to a commercial carrier. In fact, *Voice over Frame Relay* (VoFR) and *Voice over Asynchronous Transfer Mode* (VoATM) are powerful alternatives. One of the great disadvantages of such solutions, however, is that they do not have the same ubiquitous presence as IP. Although many users currently use IP, whether they know it or not, the same cannot be said for Frame Relay or ATM. IP is already at the desktop and it is hard to imagine that the same will ever apply to ATM or Frame Relay.

To some extent, IP at the desktop might not matter to a next-generation Telco if the main objective is just to capture market share in the long-distance business. Such a business could be easily served by other technologies besides IP. To ignore IP, however, would be to decline possible opportunities offered by voice and data integration. It would also limit choices and opportunities for advanced applications, and it would reduce the possible benefits from the myriad advances being made almost daily in the IP community. That is not to say that ATM and Frame Relay do not have a role to play. They certainly do, perhaps in the core of the network as an underlying network transport for IP, but not at the user device.

The VoIP Market

It is clear from the previous descriptions that VoIP can offer cost savings, new services, and a new way of doing business, both for carriers and network operators. It is also clear that voice is an important service, but not the only one. Rather, it is part of a suite of services that, when packaged together, can offer exciting new capabilities. All of this translates into serious revenue opportunities. Although the business is currently only in its infancy and much of the revenue is yet to be seen, the future is certainly bright. Figure 1-6 gives one projection of the market for Voice over Packet.

Figure 1-6
Voice over Packet
revenue projection

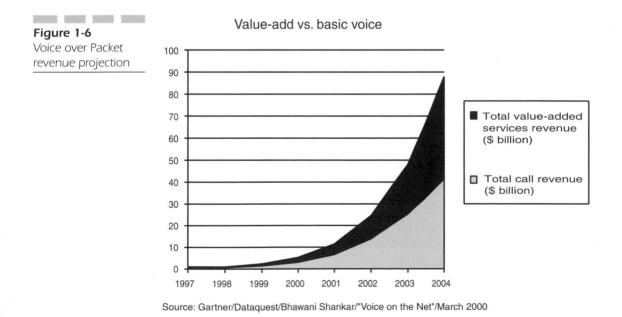

Value-add vs. basic voice

- ■ Total value-added services revenue ($ billion)
- □ Total call revenue ($ billion)

Source: Gartner/Dataquest/Bhawani Shankar/"Voice on the Net"/March 2000

Figure 1-7

Projected breakdown
of VoIP-related
revenue

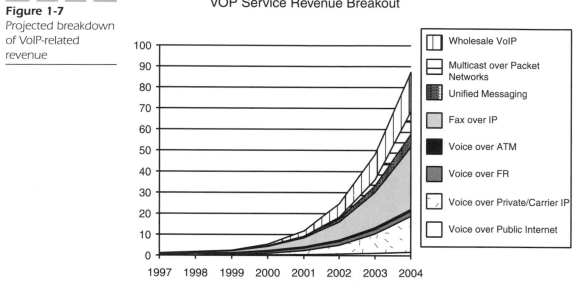

VOP Service Revenue Breakout

Source: Gartner/Dataquest/Bhawani Shankar/"Voice on the Net"/March 2000

It is interesting to note that Figure 1-6 refers to Voice over Packet, as opposed to VoIP. One would be curious to know what the breakdown might be, in particular how IP fares against other technologies such as ATM or Frame Relay. Figure 1-7 provides that breakdown, which shows that IP is expected to be dominant, and that ATM and Frame Relay are expected to comprise a relatively small percentage of the market.

VoIP Challenges

To offer a credible alternative to traditional circuit-switched telephony, VoIP must offer the same reliability and voice quality. In other words, the five nines availability requirement must be met and speech must be of toll quality. Toll quality generally means a *Mean Opinion Score* (MOS) of 4.0 or better on the following scale, as specified in *International Telecommunication Union Telecommunications Standardization Sector* (ITU-T) Recommendation P.800:

- 5—Excellent
- 4—Good

▪ 3—Fair

▪ 2—Poor

▪ 1—Bad

Speech Quality

Perhaps the most important issue in VoIP is ensuring high speech quality. Anyone who has made calls over the Internet can attest to the fact that the quality is variable at best. In order for VoIP to be a commercial challenge to circuit-switched technology, the voice quality must be at least as good as experienced in today's telephony networks and should not vary. The reason why this is such a challenge for VoIP networks is the fact that IP was not originally designed to carry voice or similar real-time interactive media.

IP was designed from the outset for data. Among the characteristics of most data traffic is that it is asynchronous, which means that it can tolerate delay. Data traffic is also extremely sensitive to packet loss, which requires mechanisms for the recipient of data packets to acknowledge a receipt, such that the sender can retransmit packets if an acknowledgment is not returned in a timely manner. Voice, however, is very sensitive to delay but is more tolerant of packet loss. Clearly, voice traffic imposes requirements that are quite different from those imposed by data traffic.

Delay   In many ways, real-time communications such as voice are almost the exact opposite of data, at least from a requirements perspective. To begin with, voice is more tolerant of packet loss, provided that lost packets are kept to a small percentage of the total (fewer than five percent). Voice, however, is extremely intolerant of delay. Anyone who has ever made an international call over a satellite knows how annoying delay can be. In a satellite call, it takes about 120 milliseconds for the signal to travel from the earth to the satellite and another 120 to come back down. If we add in some additional delay due to processing and terrestrial transmission, we have a one-way delay from A to B of 250 to 300 milliseconds. This means a round-trip delay of 500 to 600 milliseconds.

It is the round-trip time that is important. Consider, for example, a conversation between A and B. A is speaking and B decides to interrupt. However, because of a delay, what B is hearing is already old by the time he hears it. Furthermore, when B interrupts, it takes time for A to hear the interruption. Therefore, it seems to B that A has ignored the interruption. Therefore, B stops. Now, however, A finally hears the interruption and stops, so there is silence. Then they may both try to start again and both back off again. The whole experience can be very frustrating if the round-trip delay is much more than about 300 milliseconds.

ITU-T Recommendation G.114 states that the round-trip delay should be 300 milliseconds or less for telephony. Contrast this with the delay requirements for transmission of an e-mail. It hardly matters if an e-mail takes 30 seconds or 3 minutes to reach the recipient, provided that it all arrives exactly as sent.

Jitter   Just as important as the actual delay is delay variation, also known as jitter. If some delay occurs in a given conversation, then it is possible for the parties to adjust to the delay fairly quickly, provided that the delay is not too large. However, it is difficult to adjust to delay if it keeps varying. Therefore, another strong requirement is to ensure that jitter is minimized so that whatever delay exists, it is at least kept constant. This can be achieved through the use of jitter buffers, where speech packets are buffered so that they can be played out to the receiver in a steady fashion. The downside of these buffers, however, is the fact that they add to the overall delay.

Jitter is a concern only for real-time communications such as voice. It is not an issue for the types of traffic traditionally carried over IP networks. It is caused in IP networks in two ways. First, packets can take different routes from sender to receiver and consequently experience different delays. Second, a given packet in a voice conversation can experience longer queuing times than the previous packet—even if they both traveled the same route. This is due to the fact that the network resources, and particularly queues within routers, might be used by other traffic as well as voice traffic.

In contrast to the situation in IP networks, jitter is not an issue in circuit-switched networks, because an open, dedicated pipe exists from the sender to the receiver for the duration of the conversation. Thus, all the speech follows the same path and does not have to share resources with any other traffic. What goes in one side comes out the other side after a fixed delay. That delay is generally tiny—at least in the absence of satellite links.

Packet Loss   Another factor in high-quality voice is that little or none of the speech should be lost between the speaker and listener. Obviously, it is important to ensure that speech should be received as transmitted. It is possible, however, that queues might overflow in network nodes between sender and receiver, resulting in the loss of packets. Although traditional retransmission mechanisms are used in the case of lost data packets, those same mechanisms cannot be applied to real-time communications (such as voice). First, it takes some time for the two ends to determine that a packet is missing. Second, it takes more time to retransmit the missing packet.

For example, if packets 1 through 5 and 7 through 9 have been received, then the receiving end has a choice. It can either present what has been received or hold packets 7 through 9 until packet 6 is received and then play packets 6 through 9. The second of these two approaches simply does not work, because the delay would be intolerable. Therefore, if a packet is missing, then the end systems must carry on without it. If, however, a packet turns up out of sequence, then there is the risk that it will be presented to the recipient out of sequence (another no-no).

In a circuit-switched network, all speech in a given conversation follows the same path and is received in the order in which it is transmitted. If something is lost, then that loss is the result of a fault rather than an inherent characteristic of the system. If IP networks are required to carry voice with high quality, then they must emulate the performance of circuit-switched networks for voice traffic. In other words, mechanisms are required to ensure that packet loss is minimized.

Managing Access and Prioritizing Traffic

One of the great advantages that VoIP offers is the capability to use a single network to support a wide range of applications, including data, voice, and video. A single multifunctional network, as opposed to multiple discreet networks, means lower capital cost and lower operating cost. A single network does, however, introduce certain challenges. Because of the fact that voice and data have different quality requirements, as described earlier, it is important that they are not treated the same way within the network. For example, it is critical to make sure voice packets are not waiting in a queue somewhere while a large file transfer is occupying network bandwidth.

The fundamental difference between different types of traffic translates to a number of network requirements. First, it is necessary that a voice call not be connected if not enough network resources are available to handle the call. The last thing that we want is to answer the phone and find that conversation is not possible because not enough bandwidth is available. Therefore, it is necessary to ensure up front that sufficient bandwidth exists before any phone rings.

Secondly, different types of traffic must be handled in the network in different ways. It is necessary to prioritize certain types of traffic to ensure that the most critical traffic is least effected by network congestion. In practical terms, if a network becomes heavily loaded, e-mail traffic should feel the effects before synchronous traffic (such as voice).

Managing and prioritizing traffic as well as minimizing packet loss, delay, and jitter all fall into the realm of QoS. This is an area of development within the IP community that has seen a huge effort in recent times. Chapter 8, "Quality of Service," describes a number of solutions for QoS in IP networks.

Speech-Coding Techniques

Yet another factor in high-quality voice is the choice of speech-coding technique. It is all very well to reduce bandwidth, minimize delay, and minimize packet loss. However, that is not much good if the speech still sounds tinny or synthetic. Consequently, we need to select a solution that reduces bandwidth while maintaining natural-sounding speech. In general, coding techniques are such that speech quality degrades as bandwidth reduces. However, the relationship is not linear, so it is possible to significantly reduce bandwidth at the cost of only a small degradation in quality. In general, the objective is to offer, or at least approach, toll-quality voice, where toll quality relates to an MOS of 4.0 or better on the ITU-T Recommendation P.800 scale listed earlier in the chapter. Table 1-1 shows MOS values for several common coding schemes.

Although concepts like "good" and "fair" do not sound very scientific, they are not quite as nebulous as they might first appear. The ratings can be considered a summation of various quality parameters such as delay, jitter, echo, noise, crosstalk, intelligibility, and overall lack of distortion. The man-

Table 1-1

Common speech coder bit rates and MOSs

Speech coder	Bit rate (Kbps)	MOS
G.711	64	4.3
G.726	32	4.0
G.723 (celp)	6.3	3.8
G.728	16	3.9
G.729	8	4.0

ner in which various coders are tested and the scoring methodology behind it are well standardized to the extent that the MOS scale is the most common standard for voice-quality measurement today.

Nonetheless, MOS values are still subjective in nature. Therefore, regardless of how well tests are carried out, the subjective character of the evaluation means that one might find slightly different ratings for a given coding scheme. The variations may depend on the details of the test environment and the individual test participants. Therefore, anyone who intends to deploy VoIP should perform his or her own analysis of the available speech coders in a test environment that closely matches the expected network conditions. Chapter 3 provides details regarding voice-coding technologies.

Network Reliability and Scalability

Not only should a voice sound clear and crisp and be free from noticeable delay, but the service should be always available. Recall that the phone system almost never fails; contrast that with office computer networks, some of which seem to fail on a regular basis. If VoIP is to be a commercial challenge to established telephony networks, then the network must be rock solid. In other words, the five nines availability requirement must be met.

Unlike earlier systems, today's VoIP solutions do offer reliability and resilience. Moreover, the solutions are designed to enable redundancy and load-sharing within the network. Therefore, individual network nodes are less likely to fail, and if a node does fail, then another node in the network can assume the load of the failed node. Of course, networks must be designed to take advantage of these capabilities. Building redundancy into a network, however, means that the network will be more expensive. Therefore, a balance must be struck between network cost and network quality. Finding the right balance is the responsibility of the network architect.

Not only must a network be reliable, but it must also be scalable. In other words, it must be possible to increase the capacity of the networks to handle millions of simultaneous calls. The early VoIP systems did not scale to such capacities. Nowadays, however, VoIP systems can handle enormous demands. Some of today's VoIP gateway products, for example, can handle many gigabits per second of voice traffic with equipment that occupies just a single 19-inch shelf.

An advantage of VoIP architectures is that they are relatively easy to start on a small scale and then expand as traffic demand increases. Standardized protocols enable small network nodes to be controlled by the same call-processing systems as used for large network nodes. Therefore, an expansion of the network capacity can be done in small or large steps as needs dictate.

Overview of the Following Chapters

This chapter has provided a brief introduction to VoIP and highlighted some of its attractions and challenges. The remainder of this book is devoted to addressing these issues in further detail.

We begin these detailed discussions in Chapter 2, "Transporting Voice by Using IP," with a review of IP networking in general. This will help the reader to understand what IP offers, why it is a best-effort protocol, and why carrying real-time traffic (such as voice) over IP has significant challenges. We focus particularly on those protocols that run on top of IP and help to make such real-time transport a possibility. Specifically, we review the *Real-Time Transport Protocol* (RTP), the protocol that is used for carrying voice packets over IP.

Chapter 3 focuses on voice-coding techniques. We have already mentioned that IP as a transport protocol can take advantage of sophisticated and efficient voice-coding schemes. Numerous coding schemes exist, however, and choosing the right coding scheme for a particular network or application is not necessarily a simple matter. When choosing a coding scheme, we must strike a balance between the conflicting goals of low bandwidth and high speech quality, which means that we must understand the advantages and disadvantages of all the coding schemes available.

Chapter 4, "H.323," discusses the H.323 architecture and protocol suite. H.323 has been the standard for VoIP for several years. Even though VoIP technology is relatively new, it has already seen significant deployment—both in private VoIP networks and over the Internet. Initially, however, VoIP implementations used proprietary technology, which did not easily enable communication between users of different equipment. The emergence of H.323 has helped to change that situation. As a result, H.323 is the most widely deployed VoIP technology at the moment. Standardized alternatives to H.323 are now available and some experts believe that H.323 will not be the future standard of choice. Nonetheless, H.323 has a large embedded base and is more mature than other technologies. H.323 is likely to play a major role in VoIP networks for some time to come.

Chapter 5, "The Session Initiation Protocol (SIP)," discusses the rising star of VoIP technology. Though quite new and not yet widely deployed, SIP is hailed by many as the standard of the future for VoIP. Simple in design and easy to implement, the simplicity of SIP is one of its greatest advantages. However, simplicity does not mean that flexibility is sacrificed. Instead, SIP is extremely flexible and supports a range of advanced features. Many in the VoIP industry are convinced that SIP will overtake H.323 in the marketplace.

Interworking with other networks, such as the *Public Switched Telephone Network* (PSTN), is a major concern in the deployment of VoIP networks. Such interworking requires the use of gateways. Hand in hand with SIP are two protocols that deal with the control of those gateways: the *Media Gateway Control Protocol* (MGCP) and MEGACO.[7] The two protocols address the same issue. MGCP is the standard of today but is to be replaced in the near future with MEGACO (aka ITU-T H.248). The main advantage of each is that they enable a widely distributed VoIP network architecture, whereby call control can be centralized and gateways can be distributed close to the sources and sinks of traffic. This is known as the softswitch architecture. The softswitch architecture, MGCP and MEGACO, are described in Chapter 6, "Media Gateway Control and the Softswitch Architecture."

Of critical importance to a communications system is the signaling system that it uses. In fact, H.323, SIP, MGCP, and MEGACO are all signaling protocols, while the state of the art in PSTN signaling is SS7. SS7 is the signaling system that has made call setup as quick as it is today and it provides the foundation for numerous services such as caller-ID, toll-free calling, calling card services, and mobile telephony. These are all services that need to be supported just as well in a VoIP network as in a traditional network. Moreover, traffic must be able to flow between new VoIP networks and established SS7-based networks. Thus, VoIP networks must be able to interwork with SS7 networks. Chapter 7 "VoIP and SS7" describes the technology used to enable such interworking.

Chapter 8 deals with QoS. We have already described the quality challenges that a VoIP network must face to meet the stringent performance requirements that define a carrier-grade network. A great deal of work has

---

[7]MEGACO also stands for "Media Gateway Control Protocol," but it is a different protocol from MGCP.

been done in this area and various solutions are discussed in Chapter 8. These solutions can enable VoIP to reach the goal of being the new technology of choice for voice transport.

Understanding technologies and protocols is all very well. At the end of the day, however, those technologies and protocols are simply a set of tools. To design and deploy a carrier-grade VoIP network, we must understand how to use those tools to best effect. We need to address issues of network dimensioning, traffic engineering, and traffic routing. We need to understand how we can build redundancy and diversity into a network without losing sight of the trade-off between network quality and network cost. Chapter 9 "Designing a Voice over IP Network" describes these issues in detail.

Transporting Voice by Using IP

The Internet is a collection of networks, both public and private. The private networks include the *local area networks* (LANs) and *wide area networks* (WANs) of various institutions, corporations, businesses, and government agencies. The public networks include those of the numerous *Internet service providers* (ISPs). Although the private networks may communicate internally using various communications protocols, the public networks communicate using the *Internet Protocol* (IP). In order for one private network to communicate with another via a public network, then each private network must have a gateway to the public network and that gateway must speak the language of that network—IP.

We illustrate this situation in Figure 2-1, where a number of private networks are shown, connected to a number of routers. The various entities within a given private network may communicate internally in any way they want, but when communicating with other networks via the Internet, they do so with IP. Routers provide the connectivity between various networks and form the backbone of the Internet.

Figure 2-1

Interconnecting
networks

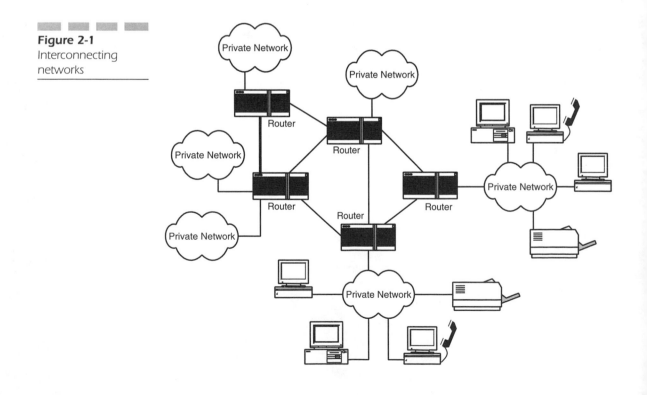

The foregoing does not mean to imply that private networks like LANs and WANs do not use IP. The vast majority of them do. The reasons are numerous, including the fact that IP is embraced by all major developers of computer systems, and it supports a vast array of applications. Moreover, compared to some other types of networks, it can be relatively easy to implement an IP-based network.

IP and some of the related protocols would require a separate book to describe in any detail. In fact, numerous books are available. The main intention in this chapter is only to provide a high-level overview of IP in general and of some of the aspects of particular significance to the transport of voice using IP. This will allow a better understanding of the architectures and protocols described in subsequent chapters. A part of this chapter is dedicated to a brief explanation of the process for developing an Internet standard. Given that at the time of writing this book, many new specifications or enhancements for *Voice over IP* (VoIP) are being finalized or enhanced, it is important to have a high-level understanding of the standardization process.

Overview of the IP Protocol Suite

Strictly speaking, IP is a routing protocol for the passing of data packets. Other protocols invoke IP for the purposes of getting those data packets from origination to destination. Thus, IP must work with the cooperation of

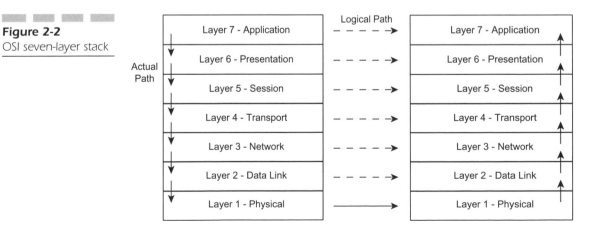

Figure 2-2
OSI seven-layer stack

higher-layer protocols in order for any application or service to operate. Equally, IP does not incorporate any physical transmission facilities, so it must work in cooperation with lower-layer transmission systems.

When we use the term layer, we generally refer to the *Open System Interconnection* (OSI) seven-layer model. Figure 2-2 shows the OSI model. Each side of a communications transaction involves a number of layers. The top layer represents the useable information to be passed to the other side. However, in order to pass this information, it must be packaged appropriately, it must be routed correctly, and it must traverse some physical medium.

Consider, for example, a simple conversation between person A and person B. Person A has a thought to be communicated to person B. In order to communicate that thought, person A must articulate the thought in language using words and phrases. Those are communicated to the mouth using nerve impulses from the brain. The mouth constructs sounds, which incorporate that language. Those sounds are transmitted through the air using vibrations. At person B, the vibrations are collected and the sounds extracted. These are passed to the brain using nerve impulses. At the brain, the original thought or concept is extracted from the language used. Thus, the thought has passed from the mind of person A to the mind of person B, but it has used the services of other functions in order to do so.

The OSI model includes seven layers as follows:

- The *physical layer* includes the physical media used to carry the information, such as cables and connectors. It also includes the coding and modulation schemes used to represent the actual 1's and 0's.

- The *data link layer* enables the transport of information over a single link. This includes the packaging of the information into frames and may also include functions for error detection and/or correction and retransmission.

- The *network layer* provides functions for the routing of traffic through a network from the source to the destination. Since a source of traffic might not be connected directly to the required destination, the information needs to pass through intermediate points, where the information can be forwarded in the right direction until it hits the correct destination.

- The *transport layer* provides functions that ensure the error-free, omission-free, in-sequence delivery of data. To support multiple streams of information from the source to destination, it provides mechanisms for identifying each individual stream and ensures that each stream is passed upward to the correct application.

- The *session layer* deals with the commencement (logging in) and completion (logging out) of a session between applications. It establishes the type of dialog to be established, such as one way at a time or both ways at the same time. In the case of one way at a time, it keeps track of whose turn is next.

- The *presentation level* specifies the language to be used between applications. This includes the encoding to be used for the various pieces of information, such as whether certain pieces of information are coded as ASCII or binary.

- The *application layer* provides an interface to the user or at least to the user application. For example, file transfer programs and web browsers belong to the application layer.

Figure 2-3 shows a correlation between the IP protocol suite and the OSI model. We can see that IP itself resides at layer three. Above IP is a choice of protocols: the *Transmission Control Protocol* (TCP) and the *User Datagram Protocol* (UDP).

TCP ensures a reliable, error-free delivery of data packets. When multiple packets are being sent, it ensures that all of them arrive at the

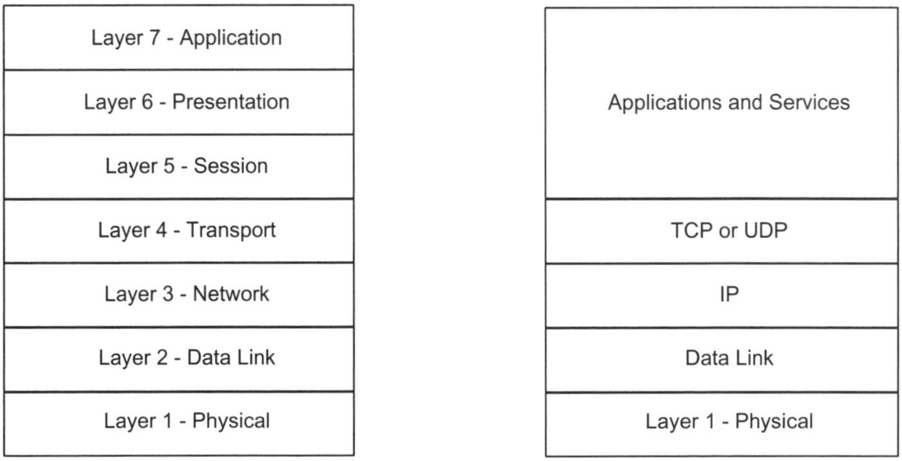

Figure 2-3
The IP suite and the
OSI stack

Layer 7 - Application	Applications and Services
Layer 6 - Presentation	
Layer 5 - Session	
Layer 4 - Transport	TCP or UDP
Layer 3 - Network	IP
Layer 2 - Data Link	Data Link
Layer 1 - Physical	Layer 1 - Physical

Figure 2-4
Packet structure

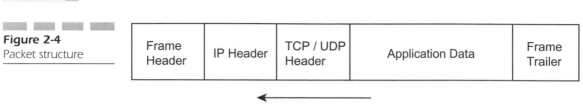

Frame Header	IP Header	TCP / UDP Header	Application Data	Frame Trailer

←

Transmission direction

destination and are presented in the correct order to the application. UDP is a simpler protocol aimed mainly at one-shot transactions, where a packet is sent and a response received. It does not include sequencing and retransmission functions. Above TCP or UDP, we find a number of different applications and services. The application or service itself determines whether TCP or UDP is used.

Taking these facts into account, Figure 2-4 shows how an IP packet would appear when being passed from one node in a network to another. The application data itself is passed down to the transport layer, where a TCP or UDP header is applied. This combination is passed to the network layer where an IP header is applied. The packet is then passed down to the datalink layer, where it is packaged into a frame structure for transmission. At the destination end, each layer extracts its own header information and passes the remainder to the layer above until the destination application receives just the original data.

Internet Standards and the Standards Process

As we move forward in our discussion, we will make reference to numerous Internet specifications and standards. Before doing so, it is worth discussing Internet standards in general, including naming conventions and the standardization process itself. This information will help to put some of the specifications in a better context.

The Internet Society

The Internet Society is a nonprofit organization with the overall objective of keeping the Internet alive and growing. Its formal mission statement is

"To assure the open development, evolution, and use of the Internet for the benefit of all people throughout the world."

To meet this goal, the Internet Society, among other things, undertakes the following efforts:

- Supports the development and dissemination of Internet standards.
- Supports research and development related to the Internet and internetworking.
- Assists developing countries in implementing and evolving Internet infrastructure and use.
- Forms liaisons with other organizations, government bodies, and the public for education, collaboration, and coordination related to Internet development.

A great many of the solutions described in this book have been developed under the auspices of the Internet Society. It is the Internet Society that is the copyright holder for those specifications. A copy of the copyright statement is included with all standards developed under the auspices of the Internet Society.

The Internet Architecture Board (IAB)

The IAB is the technical advisory group of the Internet Society. It provides technical guidance to the Internet Society, oversees the Internet standards process, and provides a liaison with other organizations on behalf of the Internet Society in technical matters.

The Internet Engineering Task Force (IETF)

The IETF comprises a huge number of volunteers, who cooperate in the development of Internet standards. These individuals come from equipment vendors, network operators, and research institutions and generally have significant technical expertise. They perform the detailed technical work of developing Internet standards.

The work of the IETF is mainly done within working groups, formed to address specific areas or topics (such as security or routing). Most of the work is done by individuals in the working groups who share their work with others in the group through the use of mailing lists. The IETF as a whole meets three times a year.

The Internet Engineering Steering Group (IESG)

The IESG is responsible for management of the IETF's activities. In particular, it is the IESG that approves whether a particular specification becomes an official standard or can even advance within the standards process.

The Internet Assigned Numbers Authority (IANA)

The IANA is responsible for the administration of unique numbers and parameters used in Internet standards. Whenever a new standard proposes to utilize parameters that have specific values and specific meanings associated with those values, then those parameters must be registered with the IANA.

The Internet Standards Process

The process for developing an Internet standard is documented in RFC 2026. The following is meant simply as an overview of what is described in RFC 2026.

An Internet standard begins its life as an Internet draft, which is just the early version of a specification that can be updated, replaced, or made obsolete by another document at any time. It is stored in the IETF's Internet Drafts directory, where it can be viewed by anyone who has web access. If an Internet draft is not revised within six months or has not been recommended for publication as a *Request for Comments* (RFC), it is removed from the directory and officially ceases to exist.

Any reference made to an Internet draft must be done with great care and with emphasis that the draft is a work in progress and that it can change at any time without warning.

Once an Internet draft is considered sufficiently complete, it may be published as an RFC, with an RFC number. Publication as an RFC does not mean that the specification is a standard, however. In fact, several steps need to be taken along the "standards track" before a draft becomes a standard. The first step is where the RFC becomes a proposed standard. A proposed standard is a specification that is stable, complete, and well understood and that has garnered significant interest within the Internet community. It is not a strict requirement that the specification be implemented and demonstrated before becoming a proposed standard, but such implementation and demonstration may be required by the IESG in the case of particularly critical protocols.

The next step is where the RFC becomes a draft standard. To achieve this status, there must have been at least two independent successful implementations of the specification and interoperability must have been demonstrated. If any portion of the specification has not met the requirement for independent and interoperable implementations, then such a portion must be removed from the specification before it can be granted draft standard status. In this way, a high level of confidence will support the details of the specification.

An RFC becomes a standard once the IESG is satisfied that the specification is stable and mature and can be readily and successfully implemented on a large scale. This step requires that the specification be mature and that there be significant operational experience in using it. Once an RFC becomes a standard, it is given a *standard* (STD) number. The specification retains its original RFC number, however.

Not all RFCs are standards, nor do all RFCs document technical specifications. For example, some document *best current practices* (BCPs), which generally outline processes, policies, or operational considerations related to the Internet or related to the Internet community itself. Others are known as *applicability statements*, which describe how a given specification may be utilized to achieve a particular goal or how different specifications can work together. STD 1 lists the various RFCs and indicates whether they are standards, draft standards, proposed standards, or applicability statements. The list is updated periodically.

The Internet Protocol (IP)

IP itself is specified in RFC 791, with amendments according to RFCs 950, 919, and 920. RFCs 1122 and 1123 give requirements for Internet hosts in their use of IP, and RFC 1812 documents requirements for IP routers (at least for IP version 4).

IP is a protocol for routing packets through a network from origination to destination. A packet of data is equipped with an IP header, which contains information about the originator and the destination address. The data packet with the IP header is known as an IP datagram. The information in the header is used by routers to route the packet to its destination.

IP itself makes no guarantees that a given packet will be delivered, let alone that multiple packets will be delivered in sequence and without omission. IP is known as a "best-effort" protocol. This means that a packet should be delivered under normal circumstances, but that a packet may be lost along the way due to transmission errors, congestion in buffers or transmission facilities, link failures, and so on.

The IP Header

The IP header appears as in Figure 2-5. Although all the fields in the header are there for a reason, a number are of particular importance. The first of these is the version. The currently deployed version of IP is 4 and the header shown in Figure 2-5 applies to IP version 4. The next version to be deployed is version 6.

The Identification, Flags, and Fragment Offset fields are used for cases where a datagram needs to be split into fragments in order to traverse a link, particularly in the case where the size of the datagram is greater than the maximum that can be handled by a given link. The Identification field is used to identify datagram fragments that belong together. The Flags field indicates whether a datagram may be fragmented or not, and in the case of

Figure 2-5
IP datagram format

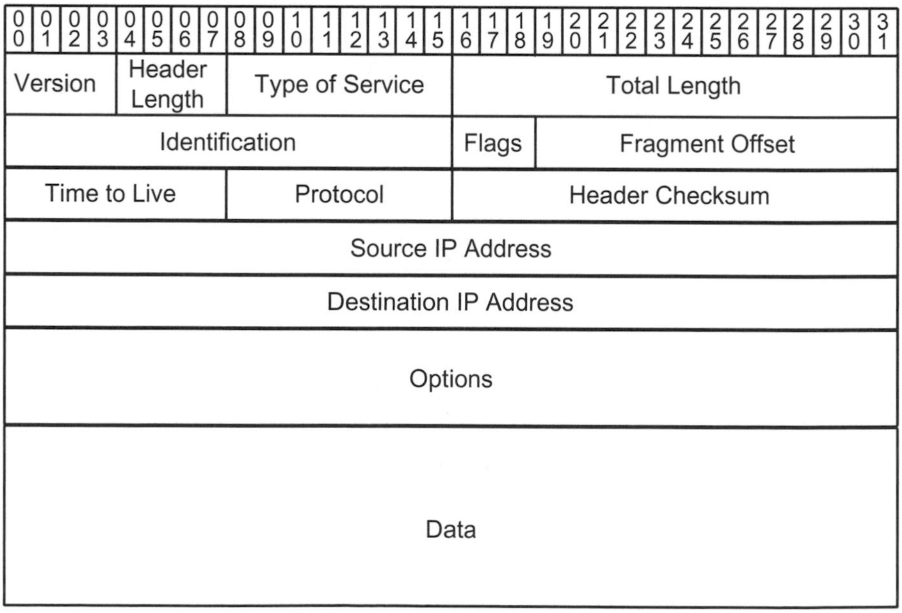

| 0 0 0 0 0 0 0 0 | 0 0 1 1 1 1 1 1 | 1 1 1 1 2 2 2 2 | 2 2 2 2 2 2 3 3 |
0 1 2 3 4 5 6 7	8 9 0 1 2 3 4 5	6 7 8 9 0 1 2 3	4 5 6 7 8 9 0 1
Version	Header Length	Type of Service	Total Length
Identification		Flags	Fragment Offset
Time to Live		Protocol	Header Checksum
Source IP Address			
Destination IP Address			
Options			
Data			

a fragmented datagram, it is used to indicate whether a particular fragment is the last fragment. The fragment offset is a number describing where the fragment belongs in the overall datagram, thereby enabling the destination to put the different pieces together correctly.

The *Time to Live* (TTL) is an indication of how long a datagram should exist in the network before being discarded. Although this field was originally meant to represent a number of seconds, it actually represents a number of hops. It is set by the originating node and is decremented by one at each subsequent node. If the datagram does not reach its destination by the time the TTL reaches zero, then the datagram is discarded.

The Protocol field indicates the higher-layer protocol or application to which this datagram belongs. For example, at the destination node, the value of this field indicates whether this is a TCP or UDP datagram or something else. The protocol value for TCP is 6. The protocol value for UDP is 17.

The source and destination IP addresses in IP version 4 are 32 bits (4 octets) long. An IP address is represented in dotted decimal notation as XXX.XXX.XXX.XXX, with the maximum possible value being 255.255.255.255.

IP Routing

The main function of IP is the routing of data from source to destination. This routing is based upon the destination address contained in the IP header and is the principal function of routers.

Routers can contain a range of different interfaces—Ethernet, Frame Relay, *Asynchronous Transfer Mode* (ATM), and so on. Each interface will provide a connection to another router or perhaps to a LAN hub or even to a device such as a PC. The function of the router is to determine the best outgoing interface to use for passing on a given IP datagram, that is, to determine the next hop. The decision is made based on the content of the routing tables, which basically match destination IP addresses with interfaces on which the datagrams should be sent.

Of course, routers do not have every possible IP address loaded within them. Instead, a match is made between the first portion of the IP address and a given interface. When we say the first portion of the IP address, how much of the address does that mean? The answer varies. For example, one table entry might specify that any address starting with 182.16.16 should be routed on interface A, whereas another table entry might specify that any address starting with 182.18 should be routed on interface B. In the first case, the IP route mask is 255.255.255.0, while in the second case the IP route mask is 255.255.0.0.

The IP route mask is basically a string of 1's, followed by a string of 0's. A logical AND is performed between the mask and the destination address in the packet to determine if it should be routed over a particular interface. If, after the AND operation, the address of the packet to be routed matches the address in the routing table, then a match occurs and the corresponding interface should be used. For example, if a datagram has a destination address of 182.16.16.7 and a logical AND is performed with mask 255.255.255.0, then the result is 182.16.16.0. This result corresponds to the table entry previously mentioned and the corresponding interface out of the router should be used. If two separate table entries begin with the same prefix, then when a destination address is being compared with the table entries, the entry chosen is the one that provides the longest match with the address in question.

Populating Routing Tables   Of course, tables in routers are somewhat more sophisticated than the foregoing suggests, but the basic description still applies and serves the purpose of giving an indication of the large amount of data that might need to be populated in a given router (and, more importantly, kept up to date).

The main issue with routing tables is how to populate them with the correct information in the first place and keep the information current in an environment where new individuals and companies connect to the Internet every day and where new routers are constantly being added. Another issue arises when several paths could lead to a given destination—which is the best one to take?

One could certainly determine from an external source which routing information needs to be populated in routing tables and then add the information manually. However, it would be far better if a given router could somehow automatically keep its own tables up to date with the latest and best routing information. Not only would that take care of the addition of new nodes in the network, but it would also take care of situations where a link to a router is lost and tables need to be reconfigured to send traffic via another router.

Fortunately, mechanisms exist whereby routing tables can be populated automatically. The most common such protocol in large networks is known as *Open Shortest Path First* (OSPF), which is termed a link state protocol. OSPF uses a concept known as an *autonomous system* (AS). An AS is a group of routers that share routing information between them. The AS is further divided into areas, where an area is defined as a set of networks and where a network is defined by an IP address and mask. For example, the address 182.16.0.0 with mask 255.255.0.0 could represent a network. A router will belong to an area if it has direct connectivity to one of the networks in the area.

A special area, Area 0, is known as the backbone area. This area comprises routers that connect the different areas together. As shown in Figure 2-6, certain border routers will exist in a given area and also in the backbone area to enable routing between different areas.

Within an area, every router will have an identical database describing the topology of the area. The topology describes what is connected to what, that is, the best way to get from one point to another. Whenever the topology changes, such as an interface (link) failure, this information is transmitted from the router attached to the failed interface to all other routers in the same area. Thus, each router within an area will always know the shortest path to a given destination in the area. If the router needs to send packets out of the area, then it needs only to know the best border router to use. It is provided with this information by the border router itself,

Figure 2-6
OSPF areas

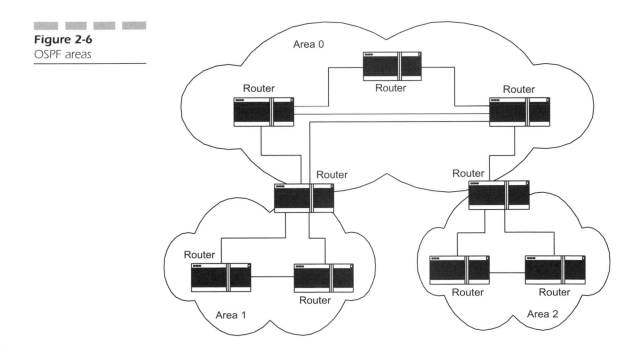

which passes back information regarding the backbone to the routers in its own area and which also passes information about its own area to other routers in the backbone.

Usually associated with OSPF is the *Border Gateway Protocol* (BGP). Recall that an AS is a collection of networks whose routers communicate routing information using a single protocol such as OSPF. The obvious question is how routing information is passed from one AS to another so that a router in one AS knows how to get to an IP address that is outside the AS. BGP is designed to take care of the issue.

Within an AS, a BGP system has visibility of all the networks within the AS. The BGP system, which could be a host or a router, is connected to the BGP system of a neighboring AS and tells the neighboring BGP system about the networks contained in its AS. That system relays the information to the BGP systems of its neighboring ASs and so on. The important part is that each BGP update sent from BGP peer to BGP peer for a given network lists all the ASs that need to be crossed to reach the network in question. Thus, if AS number 6 receives information about the path to network N1, then the information could list AS numbers 5, 4, and 3 as the route that needs to be taken. If AS number 6 is to pass the information on to AS number 7, then the route will specify AS numbers 6, 5, 4, and 3. If a given BGP system receives an update from a neighbor and finds that its own AS number is already in the information that it receives, then a loop has occurred, in which case the information is discarded. The current version of BGP is BGP-4. It is specified in RFC 1771.

The Transmission Control Protocol (TCP)

TCP resides in the transport layer, above the IP layer. Its primary function is to ensure that all packets are delivered to the destination application in sequence, without omissions and without errors. Recall that IP is a best-effort protocol and that packets could be lost in transit from source to destination. The primary function of TCP is to overcome the lack of reliability that is inherent in IP through an end-to-end confirmation that packets have been received. In the case that packets have not been received, then TCP ensures that retransmission occurs. TCP also includes flow control, so that an application at one end does not overwhelm a slower application at the other end. Applications that use the services of TCP include web browsing, e-mail, and file transfer sessions. TCP is specified in RFC 793.

At a high level, TCP works by breaking up a data stream into chunks called segments. Attached to each segment is a TCP header, which includes a sequence number for the segment in question. The segment and header are then sent down the stack to IP, where an IP header is attached and the datagram is dispatched to the destination IP address. At the destination, the segment is passed from IP to TCP. TCP checks the header for errors and, if all is well, sends back an acknowledgement to the originating TCP. Back at the originating end, TCP knows that the data has reached the far end. On the other hand, if the originating end does not receive an acknowledgement within a given timeout period, the segment is transmitted again.

The TCP Header

The TCP header has the format shown in Figure 2-7. The first fields we notice are the source and destination port numbers.

TCP Port Numbers   Contrary to what the name implies, a port number does not represent any type of physical channel or hardware. Instead, a port number is a means of identifying a specific instance of a given application.

Figure 2-7
TCP header format

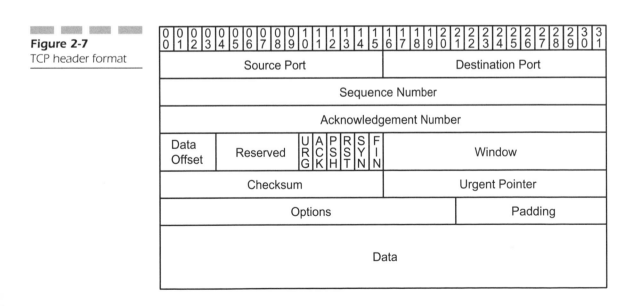

Imagine, for example, a computer where three things are happening at the same time—browsing the web, e-mail, and file transfer. Since each of these uses TCP, how does the computer distinguish between data received for one application versus data received for another? That is where port numbers come into play. Whenever a given application wants to begin a TCP session, a unique port number will be assigned for that particular session. TCP will pass that port number to the far end in any TCP messages relating to that session.

So the near end knows its own port number, but how does it know the port number being used by the far end in advance of any communication with the far end? There are two ways to know this. The easiest is through the use of well-known port numbers. These are port numbers from 0 to 1,023 reserved by the IANA for certain applications. For example, port number 23 is used for TELNET and port number 25 is used for the *Simple Mail Transfer Protocol* (SMTP, used for e-mail).

If a client wants to access a service at a server, and that service is allocated a well-known port number, then that port number will be used as the destination port number. Of course, for a given server, many clients may be trying to access the same type of service. The server can distinguish between the various clients since TCP and IP together include both the destination address and port number, as well as the source address and port number. The combination of all four will be unique. The combination of an IP address and a port number is called a *socket address*.

Sequence and Acknowledgment Numbers The next field we encounter is the sequence number. Although, as the name suggests, a sequence number is used to identify individual segments in a data stream, the sequence number does not actually number the segments. In other words, if a given segment has sequence number 100, then the next sequence will not have sequence number 101. The reason is because the sequence number is actually used to count data octets. The sequence number is a number that is applicable to the first data octet in the segment. Thus, if a given segment has a sequence number of 100 and contains 150 octets of data, then the next segment will have a sequence number of 250.

The acknowledgement number indicates the next segment number that the receiver is expecting. If, for example, a segment has been received with a sequence number of 100 and if the segment contains 150 octets of data, then the acknowledgement returned will contain an acknowledgement number of 250, indicating that the next segment expected should have a sequence number of 250.

Other Header Fields The data offset, also known as the header length, indicates the length of the TCP header in 32-bit words. This field also indicates where the actual user data begins, hence the name given to the field.

The TCP header also contains a number of flags as follows:

- **URG** This is set to 1 if urgent data is included, in which case the urgent pointer field is of significance.
- **ACK** This is set to 1 to indicate an acknowledgement. This flag will be set in all segments sent except for the very first segment sent when establishing communication.
- **PSH** This indicates a push function. This will be set in cases where the data should be delivered promptly to the recipient.
- **RST** Meaning reset, this is used to indicate an error and to abort a session.
- **SYN** Meaning synchronize, this is used in the initial messages when setting up a data transfer.
- **FIN** Meaning finish, this is used to close a session gracefully.

The next field is the Window field. This is used to indicate the amount of buffer space available for receiving data. Data that is received is stored in a buffer until retrieved by the application. When TCP receives data and sends an acknowledgement, it uses the window field to indicate the amount of empty buffer space available at that instant. This enables the sending side to ensure that it does not send more data in the next segment(s) than the receiving end can handle.

The checksum field contains the 1's complement of the 1's complement sum of all 16-bit words in the TCP header, the data, and a pseudoheader, as shown in Figure 2-8. Of course, the pseudoheader is not part of the TCP header, but actually comprises components of the IP header. The checksum field is used to verify that the header and data have been received without error.

Figure 2-8
TCP pseudoheader

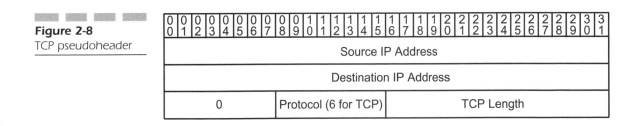

The urgent pointer contains an offset from the current segment number to the first segment after the urgent data, and it indicates the length of the urgent data. This pointer is useful to enable critical information to be sent to the user application as soon as possible.

The only real option defined in RFC 793 is the maximum segment size. This indicates the maximum data size that the receiving system is capable of handling.

TCP Connections

TCP involves the exchange of data streams between a client and a server. This exchange is accomplished through the establishment of a connection. The application at a client decides that it needs to exchange some data with an application at a server. The client application sends data to TCP, where the data is buffered. TCP constructs a TCP segment, including the source and destination port numbers, a sequence number (which is a random number, such as 2,000), the window size, and optionally the maximum segment size. TCP sets the SYN flag to 1 and sends the segment to the far end, as shown in Figure 2-9.

The server receives the segment and responds with a segment of its own. This segment includes the window size, the maximum segment size, a random sequence number of its own (such as 4,000), and an acknowledgement number, which is one greater than the received sequence number (2,001 in our example). The segment has the SYN and ACK flags set.

The client issues another segment containing a sequence number equal to the original sequence number plus 1 (that is, the same as the acknowledgment number received, which is 2,001 in our example). The acknowledgment number in the segment is 1 plus the sequence number received from the server (such as 4,001). The client also sets the ACK flag.

At this point, the TCP connection is open, and both ends are ready to exchange media. If the client were to now send a segment, it would contain a sequence number of 2,001 and an acknowledgment number of 4,001.

During a connection, a server or client will acknowledge the segments received. It is not necessary, however, that every received segment be individually acknowledged. Nor is it necessary for a sender to wait until a given segment is acknowledged before sending another segment. It is only necessary that a given segment is acknowledged before a timeout occurs at the sender of that segment.

For example, let us assume that a client sends 3 segments with segment numbers 100, 200, and 300. Let us also assume that the third of these seg-

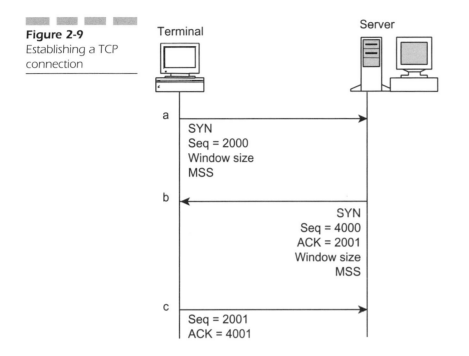

Figure 2-9
Establishing a TCP
connection

Terminal

Server

a

SYN
Seq = 2000
Window size
MSS

b

SYN
Seq = 4000
ACK = 2001
Window size
MSS

c

Seq = 2001
ACK = 4001

MSS = Maximum Segment Size

ments contains 100 octets of data. It is sufficient for the server to send a single acknowledgement for all 3, in which case the acknowledgment number would be 400, indicating the next sequence number expected. This acknowledgement also indicates that all three segments have been safely received.

The client will have a timer running for each segment that has not been acknowledged. In our example, let's suppose that the first of the three segments was lost. In that case, the server would *not* return an ACK, with an acknowledgment number of 400. In fact, let's assume that the server returns nothing. Upon the expiration of the timer at the client, the first segment will be retransmitted. The server could then respond that it has received all 3 segments by sending an ACK, with an acknowledgement number of 400.

Closing a connection is initiated by the sending of a segment with the FIN flag set. Upon receipt of such a segment at the receiver, an ACK will be returned, followed by a FIN. This second FIN (in the reverse direction to the first) will be acknowledged and the connection closure is complete.

The User Datagram Protocol (UDP)

UDP performs a very simple function. At the sending end, it simply passes individual pieces of data from an application to IP to be routed to the far end. At the receiving end, it simply passes incoming pieces of data from IP to the appropriate application.

Unlike TCP, UDP provides no acknowledgment functionality. Therefore, no guarantee exists that anything sent from an application that uses UDP will actually get to the desired destination. UDP is inherently unreliable. Why, then, should there be any interest in using UDP instead of using a more reliable protocol, such as TCP? The answer is that many applications require a quick one-shot transmission of a piece of data or a simple request/response. An example is the *Domain Name System* (DNS), where a client sends a domain name to a domain name server, which responds with an IP address. The establishment and teardown of a TCP connection just for the sake of such a simple transaction would be wasteful and cumbersome.

What happens if a UDP packet is lost? If the application expects a reply to a particular request and does not get one within a reasonable period of time, then the application should retransmit the request. Of course, this could then lead to the far end receiving two requests. In such situations, it would be a smart thing for the application itself to include a request identifier within the data to enable duplicates to be recognized.

Figure 2-10 shows the format of the UDP header. As is the case for TCP, the header contains source and destination port numbers. The source port number is optional and should be used when the sending application expects a response. If not used, then a value of zero is inserted. One should note that a TCP port number of a given value and a UDP port number of the same value represent different ports. Thus, if two applications are running at the same time on a given host (one using UDP and one using TCP), both applications can use the same port number without risk of data being sent to the wrong application.

In the UDP header, we also notice a length field that gives the total length of the datagram, including the header and the user data. The checksum gives the 16-bit 1's complement sum of the header, the data, and a

Figure 2-10
UDP header format

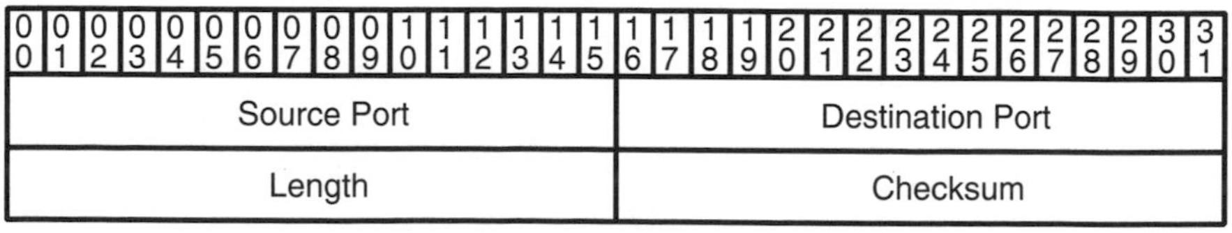

0 0 0 0 0 0 0 0 0 0 1 1 1 1 1 1 1 1 1 1 2 2 2 2 2 2 2 2 2 2 3 3	
0 1 2 3 4 5 6 7 8 9 0 1 2 3 4 5 6 7 8 9 0 1 2 3 4 5 6 7 8 9 0 1	
Source Port	Destination Port
Length	Checksum

pseudoheader. The pseudoheader is not actually part of the UDP header or data at all, but it includes the IP header fields, Source IP Address, Destination IP Address, and UDP Length and Protocol, as shown in Figure 2-11. The checksum is simply a means of checking that the received data is error-free.

Voice over UDP, not TCP

When voice is to be carried on IP, it is UDP that is used rather than TCP. Why should UDP, an inherently unreliable protocol, be chosen over TCP, a protocol than ensures ordered delivery without loss of data? The answer is related to the characteristics of speech itself. In a conversation, the occasional loss of 1 or 2 packets of voice, while not very desirable, is certainly not a catastrophe, since voice traffic will generally use small packets (10 to 40 milliseconds in duration). Obviously, modern voice-coding techniques operate better if no packets are lost, but the coding and decoding algorithms do have the capability to recover from losses, such that the occasional lost packet does not cause major degradation.

On the other hand, voice is very delay sensitive. Anyone who has made a phone call over a satellite connection, where the round-trip time may be many hundreds of milliseconds, knows how annoying a delay can be. Unfortunately, the connection setup routine in TCP and TCP's acknowledgment routines introduce delays, which should be avoided. Even worse, in the case of lost packets, TCP will cause retransmission and thereby introduce even more delay. Tolerating some packet loss is far better than introducing a significant delay in voice transmission.

So how much packet loss is tolerable? Packet loss of about 5 percent is generally acceptable, provided that the lost packets are fairly evenly spaced. A problem can arise, however, if groups of successive packets are lost. This problem can be alleviated if resource management and reservation techniques are used. These techniques generally involve the allocation of a suitable amount of resources (bandwidth and buffer space) before a call

Figure 2-11
UDP pseudoheader

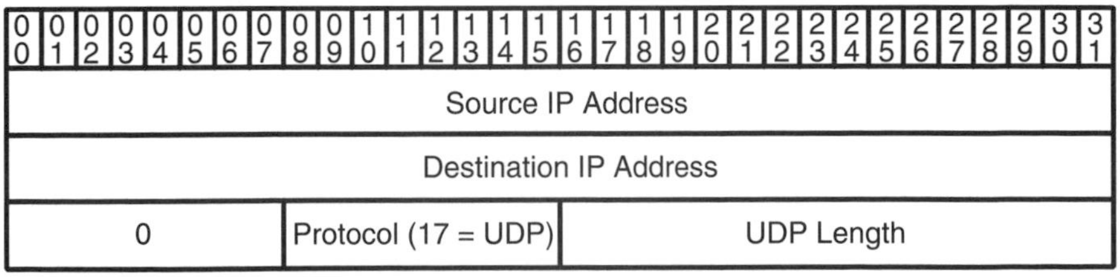

| 0 0 0 0 0 0 0 0 0 0 1 1 1 1 1 1 1 1 1 1 2 2 2 2 2 2 2 2 2 2 3 3 |
0 1 2 3 4 5 6 7 8 9 0 1 2 3 4 5 6 7 8 9 0 1 2 3 4 5 6 7 8 9 0 1
Source IP Address
Destination IP Address
0 · Protocol (17 = UDP) · UDP Length

is established. We discuss these issues in some detail in Chapter 8, "Quality of Service." Remember that we focus here on VoIP as an alternative to traditional circuit-switched voice. We are not necessarily addressing voice over the Internet. Although it might be unreasonable to expect controlled access and resource reservation on a grand scale on the Internet, such an expectation is far more reasonable in a managed IP network.

What if packets arrive at a destination in the wrong order? If the application uses UDP, packets will be passed to the application above regardless of the order of arrival. In fact, UDP has no concept of packet order. However, most packets should arrive in sequence. The reason is because all packets in a given session usually take the same route from source to destination. Of course, it is possible that different packets will take different routes, but this will usually occur only if a link fails or if some other topology change occurs in the network. Although these events do happen, the chances of such an event occurring during a particular session are relatively small. Furthermore, *quality of service* (QoS) management techniques can involve establishing a set path through the network for a given session, thus further reducing the probability of an out-of-sequence arrival of packets. Again, these techniques apply to managed networks and not necessarily to the Internet as a whole.

The foregoing arguments are more related to the problems of carrying voice using TCP than they are to the suitability of UDP for voice. In fact, UDP was not designed with voice traffic in mind, and voice traffic over UDP is not exactly a marriage made in heaven. Voice traffic over UDP just happens to be a better match than voice traffic over TCP. Nevertheless, it would be nice if some of the shortcomings of UDP could be overcome without having to resort to TCP. Fortunately, protocols exist that can help mitigate against some of the inherent weaknesses in UDP when used for real-time applications, such as voice or video. We discuss these protocols in the following sections of this chapter.

The Real-Time Transport Protocol (RTP)

RTP is specified in RFC 1889. The official title for this RFC is "RTP: A Transport Protocol for Real-Time Applications." The reason for the somewhat lengthy title is that the RFC actually describes two protocols: RTP and the *RTP Control Protocol* (RTCP). Between the two, they provide the network transport services required for the support of real-time applications, such as voice and video.

As previously discussed, UDP does nothing in terms of avoiding packet loss or even ensuring ordered delivery. RTP, which operates on top of UDP, helps address some of these functions. For example, RTP packets include a sequence number, so that the application using RTP can at least detect the occurrence of lost packets and can ensure that received packets are presented to the user in the correct order. RTP packets also include a timestamp that corresponds to the time at which the packet was sampled from its source media stream. The destination application can use this time stamp to ensure synchronized play-out to the destination user and to calculate delay and jitter. Note that RTP itself does not do anything about these issues; rather, it simply provides additional information to a higher-layer application so that the application can make reasonable decisions about how the packet of data or voice should be best handled.

As mentioned, RTP has a companion protocol, RTCP. This protocol provides a number of messages that are exchanged between session users and that provide feedback regarding the quality of the session. The type of information includes such details as the numbers of lost RTP packets, delays, and interarrival jitter. Although the actual voice packets themselves are carried within RTP packets, RTCP packets are used for the transfer of this quality feedback.

Under normal circumstances, whenever an RTP session is opened, an RTCP session is also opened. In other words, when a UDP port number is assigned to an RTP session for the transfer of media packets, a separate port number is assigned for RTCP messages. An RTP port number should always be even and the corresponding RTCP port number should be the next highest number (odd). RTP and RTCP may use any UDP port pair between 1025 and 65,535. However, the ports of 5,004 and 5,005 have been allocated as default ports when port numbers are not explicitly allocated.

The use of RTCP with RTP is not mandatory. Some VoIP implementations include RTCP as an option that can be turned off by the system operator. Turning off RTCP, however, eliminates the quality feedback mechanisms that RTCP provides. For that reason, I recommend that RTCP always be enabled.

RTP Payload Formats

RTP carries the actual digitally encoded voice by taking one or more digitally encoded voice samples and attaching an RTP header so that we have

RTP packets comprising an RTP header and a payload of the voice samples. These RTP packets are sent to UDP, where a UDP header is attached. The combination is then sent to IP, where an IP header is attached and the resultant IP datagram is routed to the destination. At the destination, the various headers are used to pass the packet up the stack to the appropriate application.

Given that many different voice and video-coding standards exist, RTP must include a mechanism for the receiving end to know which coding standard is being used, so that the payload data can be correctly interpreted. RTP does this by including a payload type identifier in the RTP header. The interpretation of the payload type number is specified in RFC 1890, "RTP Profile for Audio and Video Conferences with Minimal Control," which is a companion specification to the RTP specification itself. RFC 1890 allocates a payload type number to various coding schemes and provides high-level descriptions of the coding techniques themselves. The RFC also provides references to the formal specifications for the coding schemes.

Since the initial publication of RFC 1890 as a proposed standard, a number of new coding schemes have become available. For example, although RFC 1890 describes the *Global System for Mobile Communications* (GSM) full-rate coder, it makes no reference to the GSM *enhanced full-rate* (EFR) coder. Therefore, recognizing that a number of new coding schemes have become available, a new Internet draft has been prepared, providing an updated RTP profile. This draft is effectively a revised version of RFC 1890 and includes a number of new coding schemes. At the time of writing this book, this Internet draft has the same title as RFC 1890 and it is planned to supersede RFC 1890 as part of the advancement of RFC 1890 from proposed standard to draft standard status. As with all Internet drafts, it should be read with the understanding that it is a work in progress and subject to change at any time.

Table 2-1 and Table 2-2 list the various RTP audio and video payload types respectively defined in the new profile draft.

As can be seen from Table 2-1 and Table 2-2, some coding schemes are given a specific payload type number. These are known as static payload types. Other payload types are marked *Dyn*, meaning dynamic, and are not allocated any specific payload type number.

The main reason for different types of payload types is historical. RTP has existed for several years. When it was originally developed, separate signaling standards for real-time applications did not exist. Therefore, it was not possible for a sending application to let a receiving application know in advance (and out of band) what type of coding was to be used in a given session. Therefore, specific payload type numbers were allocated to a number of

Table 2-1

Payload types for
audio encodings

Payload type	Encoding name	Media type	Clock rate (Hz)	Channels
0	PCMU	Audio	8,000	1
1	Reserved	Audio		
2	G726-32	Audio	8,000	1
3	GSM	Audio	8,000	1
4	G723	Audio	8,000	1
5	DVI4	Audio	8,000	1
6	DVI4	Audio	16,000	1
7	LPC	Audio	8,000	1
8	PCMA	Audio	8,000	1
9	G722	Audio	8,000	1
10	L16	Audio	44,100	2
11	L16	Audio	44,100	1
12	QCELP	Audio	8,000	1
13	Reserved			1
14	MPA	Audio	90,000	1
15	G728	Audio	8,000	1
16	DVI4	Audio	11,025	1
17	DVI4	Audio	22,050	1
18	G729	Audio	8,000	1
19	Reserved			1
20	Unassigned			1
21	Unassigned			1
22	Unassigned			1
23	Unassigned			1
dyn	G726-40	Audio	8,000	1
dyn	G726-24	Audio	8,000	1
dyn	G726-16	Audio	8,000	1
dyn	G729D	Audio	8,000	1
dyn	G729E	Audio	8,000	1
dyn	GSM-EFR	Audio	8,000	1
dyn	L8	Audio	Variable	Variable
dyn	RED	Audio	Conditional	1
dyn	VDVI	Audio	Variable	1

Table 2-2

Payload types for
video encodings

Payload type	Encoding name	Media type	Clock rate (Hz)
24	Unassigned	Video	
25	CelB	Video	90,000
26	JPEG	Video	90,000
27	Unassigned	Video	
28	nv	Video	90,000
29	Unassigned	Video	
30	Unassigned	Video	
31	H261	Video	90,000
32	MPV	Video	90,000
33	MP2T	Audio/video	90,000
34	H263	Video	90,000
35–71	Unassigned		
72–76	Reserved	N/A	N/A
77–95	Unassigned		
96–127	Dynamic		90,000
Dyn	BT656	Video	90,000
Dyn	H263-1998	Video	90,000

popular coding schemes. When an RTP packet would arrive at a receiver, the payload type number would unambiguously indicate the coding scheme chosen by the sender so that the receiver would know what actions to take to decode the data. That was fine, provided the chosen coding scheme was supported at the receiver. Unfortunately, there was no way for a sender to find out in advance which coding schemes a particular receiver could handle.

In more recent times, separate signaling systems have been developed to enable the negotiation of media streams as part of the call setup process and in advance of actually sending any voice or video. For example, the *Session Initiation Protocol* (SIP) and the *Session Description Protocol* (SDP), both described in Chapter 5, "The Session Initiation Protocol (SIP)," support such capabilities. Therefore, the two ends of a conversation can now agree in advance on the type of coding scheme to be used. As part of that agreement, they can also agree on the payload type number to be applied to that coding scheme for that call.

For example, let us assume that party A wants to send voice coded according to some coding scheme XXX and party B is able to handle that coding scheme. The parties can agree at the start of the call that RTP packets carrying voice that is coded according to XXX shall have a payload type of 120. When packets arrive with payload type 120, party B knows exactly how to interpret the data. After the call is over, then payload type 120 could be used to indicate some other coding scheme in a subsequent call.

The RTP audio/video profile allocates the payload type numbers 96 to 127 to be used for dynamic payload types. Furthermore, some of the newer coding schemes referenced by the profile specification are not given a fixed (static) payload type, but are considered dynamic. The use of dynamic payload types is very useful. It enables many new coding schemes to be developed in the future without running the risk of exhausting the available payload type number space (a maximum of 127).

Note that the encoding name is also of significance. Whenever a new coding scheme is to be used with RTP, particularly if it is to be a dynamic payload type, then the encoding name should be specified in a separate payload specification document and the encoding name should be registered with the IANA. The reason for this is the fact that external signaling systems such as SDP make use of the encoding name to unambiguously refer to a particular payload specification.

Of the dynamic payload types, one is a little different from all the rest and deserves special mention. This is the *Redundant* (RED) payload type and is specified in RFC 2198. Normally, a single RTP packet contains one or more samples of coded voice or video, with each sample in the packet coded in the same way. If the packet is lost, the receiving application must cope with the loss of data as best it can, such as by replaying the content of a previous packet to fill the void in time. The idea with the redundant payload type is for a single packet to contain one or more voice samples as normal, plus one or more copies of previous samples already sent to the receiver. In other words, each packet contains the current sampled speech, plus a backup copy of the previously sampled speech. If a given packet is lost, then a copy of the lost speech sample arrives in the next packet. Note that the backup copy may use a different coding scheme (more bandwidth-efficient perhaps) than is used for the primary copy. If the backup copy uses a lower-bandwidth coding scheme, then it provides for redundancy to cope with packet loss, but without the need for excessive extra bandwidth.

The RTP Header

As mentioned, RTP carries the actual encoded voice. What this means is that one or more digitally coded samples make up the RTP payload. An RTP header is attached to this payload and the packet is sent to UDP.

The RTP header includes information necessary for the destination application to reconstruct the original voice sample (at least to as close a match as is enabled by the coding scheme used). The RTP header is shown in Figure 2-12.

The meanings of the various header fields are as follows:

- **Version (V)** This field should be set to the value 2, which is the current version of RTP.

- **Padding (P)** This field is comprised of one bit and indicates whether the packet contains one or more padding octets at the end of the payload. Because of the fact that the payload needs to align with a 32-bit boundary, it is possible that the end of the payload contains padding to align with the boundary. If such padding exists, then the P bit is set and the last octet of the padding contains a count of how many padding octets are included. Of course, such padding octets should be ignored.

- **Extension (X)** This field is comprised of one bit and indicates whether the fixed header, as shown in Figure 2-12, contains a header extension. If this bit is set, then the header is followed by exactly one header extension, as described later and with the format shown in Figure 2-13.

- **CSRC Count (CC)** The CC uses a four-bit field indicating the number of contributing source identifiers included in the header and takes a value of 0 to 15. Refer to "Contributing Source Identifiers" described later.

Figure 2-12
RTP header format

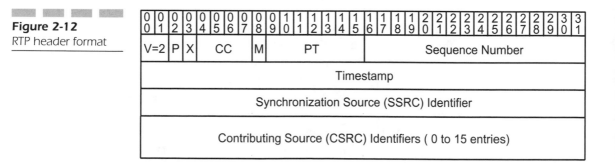

0 0 0 0 0 0 0 0	0 0 1 1 1 1 1 1	1 1 1 1 2 2 2 2	2 2 2 2 2 2 3 3
0 1 2 3 4 5 6 7	8 9 0 1 2 3 4 5	6 7 8 9 0 1 2 3	4 5 6 7 8 9 0 1
V=2 P X CC M	PT	Sequence Number	
Timestamp			
Synchronization Source (SSRC) Identifier			
Contributing Source (CSRC) Identifiers (0 to 15 entries)			

Figure 2-13
RTP header extension

```
 0 0 0 0 0 0 0 0 0 0 1 1 1 1 1 1 1 1 1 1 2 2 2 2 2 2 2 2 2 2 3 3
 0 1 2 3 4 5 6 7 8 9 0 1 2 3 4 5 6 7 8 9 0 1 2 3 4 5 6 7 8 9 0 1
```

Profile-specific information	Length
Header extension	

- **Marker (M)** The marker is a one-bit field, the interpretation of which is dependent upon the payload being carried. RFC 1889 does not mandate any specific use of this bit, but RFC 1890 (audio-video profile) does. RFC 1890 states that, for an application that does not send packets during periods of silence, this bit should be set in the first packet after a period of silence (the first packet of a talkspurt). Applications that do not support silence suppression should not set this bit.

- **Payload Type (PT)** This field comprises seven bits and indicates the format of the RTP payload. The content of the PT field is interpreted according to Table 2-1 or Table 2-2. In general, a single RTP packet will contain media coded according to only one payload format. The exception to this rule is the RED profile, where several payload formats may be included. In the case of RED, the payload itself contains information as to how the primary and redundant samples are coded and how they are arranged within the payload. Specifically, the first several octets of the payload contain information about the primary and redundant samples, including the coding schemes used, sample lengths, and timestamp data. The exact number of octets used to convey this information depends on the number of primary and redundant samples contained in the RTP packet. Refer to RFC 2198.

- **Sequence Number** This is a 16-bit field. It is set to a random number by the sender at the beginning of a session and is incremented by one for each successive RTP packet sent. It enables the receiver to detect packet loss and/or packets arriving in the wrong order.

- **Timestamp** This is a 32-bit field indicating the instant at which the first sample in the payload was generated. The sampling instant must be derived from a clock that increases monotonically and linearly in time so that far-end applications may play out the packets in a synchronized manner and so that jitter calculations may be performed. The resolution of the clock needs to be adequate to support synchronized playout. The clock frequency is dependent on the format of the payload data. For static payload formats, the applicable clock frequency is defined in the RTP profile. For example, the frequency for typical voice-coding schemes is 8,000 Hz.

The increase in the timestamp value from packet to packet will depend on the number of samples contained in a packet. The timestamp is tied to the number of sampling instants that occur from one packet to the next. For example, if 1 packet contains 10 voice samples and a timestamp value of 1, then the next packet should have a timestamp value of 11. Assuming that the sampling occurs at a rate of 8,000 Hz (every 0.125 milliseconds), then the difference of 10 in the timestamp indicates a difference in time of 1.25 milliseconds. If no packets are sent during periods of silence, then the next RTP packet may have a timestamp significantly greater than the previous RTP packet. The difference will reflect the number of samples contained in the first packet and the duration of silence. The initial value of the timestamp is a random number, chosen by the sending application.

- **Synchronization Source (SSRC)**   This is a 32-bit field indicating the synchronization source, which is the entity responsible for setting the sequence number and timestamp values and is normally the sender of the RTP packet. The identifier is chosen randomly by the sender so as not to be dependent upon network addresses. It is meant to be globally unique within an RTP session. In the case when the RTP stream is coming from a mixer, then the SSRC identifies the mixer, not the original source of media. Refer to the discussion of mixers and translators later in this chapter.

- **Contributing Source (CSRC)**   This is a 32-bit field containing an SSRC value for a contributor to a session. The field is used when the RTP stream comes from a mixer and is used to identify the original sources of media behind the mixer. There may be 0 to 15 CSRC entries in a single RTP header. Refer to the discussion of mixers and translators later in this chapter.

RTP Header Extensions   The RTP header is designed to accommodate the common requirements of most, if not all, media streams. Certain payload formats may, however, require additional information. This information may be contained within the payload itself, such as by specifying in the payload profile that the first n octets of the payload have some specific meaning. The RED profile uses such an approach. Alternatively, an application can apply an RTP header extension.

The existence of a header extension will be indicated by setting the X bit in the RTP header to 1. If the X bit is set to 1, then a header extension of the form shown in Figure 2-13 will exist between the CSRC fields and the actual data payload. Note that RTP only requires that the header extension should contain a length indicator in a specific location. Beyond that requirement, RTP does not specify what type of data is to be included in a header extension, or even how long the extension may be. Provided that a length indication exists, it is possible for the boundary between the header extension and the payload to be clearly identified.

Mixers and Translators

Mixers are applications that enable multiple media streams from different sources to be combined into one overall RTP stream. A good example is an audio conference, as depicted in Figure 2-14. The example shows 4 participants, and each is sending and receiving audio at 64 Kbps. If the bandwidth available to a given participant is limited to 64 Kbps in each direction, then that participant does not have sufficient bandwidth to send or receive individual RTP streams to or from each of the other participants. In the example, the mixer takes care of this problem by combining the individual streams into a single stream that runs at 64 Kbps. A given participant does not hear the other three participants individually, but hears the combined audio of everyone.

Another example would be where one of the participants is connected to the conference via a slow-speed connection, while the other participants have a high-speed connection. In such a case, the mixer could also change the data format on the slower link by applying a lower-bandwidth coding scheme. In terms of RTP header values, the RTP packets received by a given participant would contain an SSRC value specific to the mixer and three CSRC values corresponding to the three other participants. The mixer sets the timestamp value.

Figure 2-14
An RTP mixer

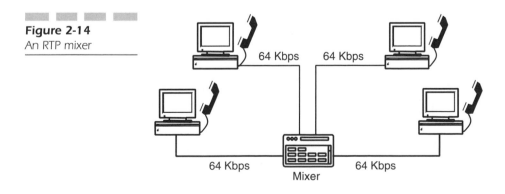

A translator is used to manage communications between entities that do not support the same media formats or bit rates. An example is shown in Figure 2-15, where participant A can support only 64 Kbps G.711 mu-law encoded voice (payload type 0), while participant B can support only 32 Kbps G.726 *adaptive differential pulse code modulation* (ADPCM) (payload type 2). In terms of RTP headers, the RTP packets received by participant A would have an SSRC value that identifies participant B, not a value identifying the translator. The difference between a mixer and a translator is that a mixer combines several streams into a single stream perhaps with some translation of the payload format, whereas a translator performs some translation of the media format, but does not combine media streams.

The advantage of a mixer over a translator is that, in the case of several streams, extra bandwidth does not need to be allocated for each stream. The disadvantage is that the receiver of the RTP stream cannot mute the stream from one source while continuing to listen to the stream from another.

The RTP Control Protocol (RTCP)

As mentioned previously, RTP comes with a companion control protocol, RTCP. This protocol enables the periodic exchange of control information between session participants, with the main goal of providing quality-related feedback. This feedback can be used to detect and potentially correct distribution problems. By using RTCP and IP multicast, a third party (such as a network operator who is not a session participant per se) can monitor session quality and detect network problems.

RTCP defines five different types of RTCP packets:

- *Sender Report* (SR) is used by active session participants to relay transmission and reception statistics.

- *Receiver Report* (RR) is used to send reception statistics from those participants that receive but do not send media.

Figure 2-15
An RTP translator

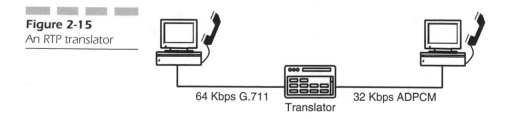

64 Kbps G.711   32 Kbps ADPCM
            Translator

- *Source Description* (SDES) contains one or more descriptions related to a particular session participant. In particular, the SDES must contain a *canonical name* (CNAME), which is used to identify session participants. The CNAME is separate from the SSRC because the SSRC might change in case of a host reset. Furthermore, within a given session, it is possible for a given source to generate multiple RTP streams, as would be the case if both audio and video were being transmitted. In such a case, the two streams would have different SSRC values, but would have the same CNAME value. At a receiver, the CNAME would be used to link the two streams in order to provide a synchronized playout. Within a given session, the CNAME must be unique. Other information may also be within the SDES packet, such as a regular name, e-mail address, or phone number for the participant in question.

- BYE indicates the end of participation in a session.

- APP stands for *application-specific functions*. The APP packet enables RTCP to send packets that convey information specific to a particular media type or application. RTCP does not specify the detailed contents of an APP packet.

Though the previous RTCP packet types are defined, they are never sent individually from one session participant to another. In reality, two or more RTCP packets will be combined into a compound packet. A reason for generating compound packets is that sender and/or receiver reports should be sent as often as practical to allow better statistical resolution. Another reason is that new receivers in a session must receive the identity (CNAME) of various media sources very quickly to allow a correlation between media sources and the received media. In fact, the RTCP specification states that all RTCP packets will be sent as part of a compound packet and that every packet must contain a report packet (SR/RR) and an SDES packet. When a SDES packet needs to be sent, a report packet (SR/RR) is sent even if there is no data to report. In such a case, the compound packet includes an empty RR packet in front of the SDES packet.

An RTCP compound packet would appear as in Figure 2-16. In this particular example, the sender of the compound packet happens to be leaving the session, as indicated by the inclusion of the BYE packet. Note the presence of the optional encryption prefix. If a compound packet is to be encrypted, then a 32-bit encryption prefix must be included in front of the compound packet. The encryption prefix is a random number chosen by the sender of the compound packet. The default encryption algorithm for RTCP is the *Data Encryption Standard* (DES).

Figure 2-16

Example of an RTCP
compound packet

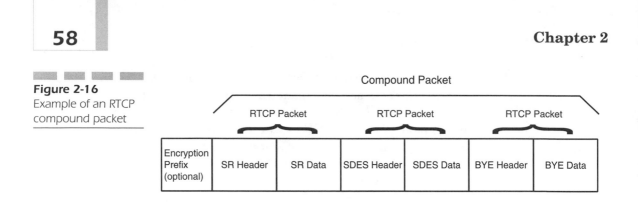

RTCP Sender Report (SR)

The SR is used by session participants who actively send RTP packets as well as receive them. The format of the SR is as shown in Figure 2-17. The SR has three distinct sections: header information, sender information, and a number of receiver report blocks. Optionally, the SR can also have a profile-specific extension.

It can be seen that the RTCP packet format bears quite a resemblance to the format of an RTP packet. As is the case for an RTP packet, we first encounter the version field (V), which has the value 2 for the current version of RTP and RTCP. We also encounter a padding bit (P), which is used to indicate if padding octets are at the end of the packet. If the P bit is set, then the final octet of the packet will contain a count of the number of padding octets. Those octets should be ignored.

The first difference between an RTP packet and an RTCP packet is noticeable in the next field (the RC field), which indicates how many reception report blocks are contained in this report. The RC is a 5-bit field, which means that a maximum of 31 receiver report blocks may be included in a single RTCP sender report packet. What happens in a large conference when a given participant needs to provide receiver report data for more than 31 other session participants? Such as situation is one in which the compounding of RTCP packets is useful, since it is perfectly valid to have a sender report followed by a receiver report in the same RTCP compound packet.

The next field we encounter is the Payload Type field, which has the value 200 for a sender report. This field is followed by a length indicator.

The sender information contains the SSRC of the sender plus some timing information and statistics regarding the number of packets and octets sent by the receiver. Specifically, the sender information includes the following fields:

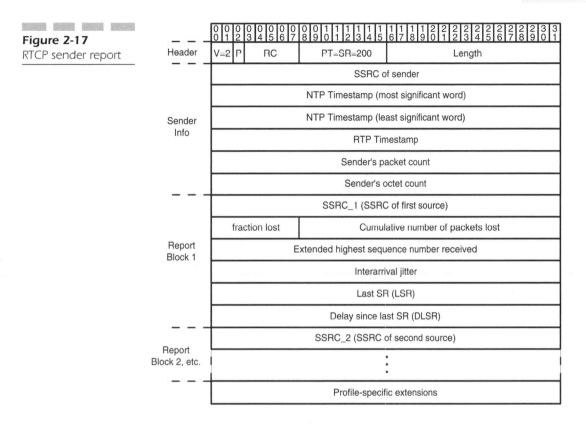

Figure 2-17
RTCP sender report

■ **SSRC of Sender**

■ **NTP Timestamp**   Standing for the Network Time Protocol
Timestamp, which is a 64-bit field, indicating the time elapsed in
seconds since 00:00 on January 1, 1900 (GMT). The most significant 32
bits represent the number of seconds while the 32 least significant bits
represent the fractions of a second, enabling a precision of about 200
picoseconds. Using 32 bits to represent the number of seconds since
1900 might suggest that we will run into difficulty in the year 2036,
when the 32 bits used to reflect the number of seconds since January 1,
1900 will wrap around zero. In reality, however, there will not be a
problem. For RTP and RTCP purposes, the timestamp is used for
relative time measurements. In other words, what matters is the
difference between NTP timestamps rather than the absolute value of
a single NTP timestamp. Assuming that consecutive RTCP reports are
not decades apart, there will not be a problem. The NTP protocol
version 3 is specified in RFC 1305.

■ **RTP Timestamp**   This is a timestamp corresponding to the NTP timestamp but that uses the same units and with the same offset as used for RTP timestamps in RTP packets. Recall that RTP timestamps in RTP packets begin at a random value and increase with each packet sent in units that are dependent on the payload format. Including both the RTP timestamp and the NTP timestamp in the same report enables receivers of the report to better synchronize with the sender of the report, particularly if they share the same timing source.

■ **Sender's Packet Count**   This is the total number of RTP packets transmitted by the sender from the start of the session up to the time this report was issued. Thus, the packet count is cumulative. The packet count is reset within a given session only if the sender's SSRC value has changed.

■ **Sender's Octet Count**   This field shows the total number of RTP payload octets transmitted by the sender since the beginning of the session. As is the case for the packet count, the octet count is reset within a session only if the sender's SSRC value changes.

One or more RR blocks follow the sender information. These provide other session participants with information as to how successfully the RTP packets that they have previously sent have been received. For each RR block, the following fields are included:

■ **SSRC_n**   The source identifier of the session participant to which the data in this particular RR block pertains.

■ **Fraction Lost**   An 8-bit field indicating what fraction of packets have been lost since the last report issued by this participant. The value is represented as a fixed point number, with the decimal point to the left of the field. The field has 8 bits, with a maximum value of 256. If, for example, the field contained the value 32, it would imply that the fraction lost is 32/256 (12.5 percent).

■ **Cumulative Number of Packets Lost**   The total number of packets from the source in question that have been lost since the beginning of the RTP session.

■ **Extended Highest Sequence Number Received**   The sequence number of the last RTP packet received from the source. The low 16 bits contain the last sequence number received. The high 16 bits indicate the number of sequence number cycles, used in the case where the sequence numbers from a particular source cycle through zero one or more times.

- **Interarrival Jitter**   An estimate of the variance in RTP packet arrival. See the section on calculating jitter later in this chapter.

- **Last SR Timestamp (LSR)**   The middle 32 bits of the 64-bit NTP timestamp used in the last sender report received from the source in question. This field is a means of indicating to the source whether sender reports issued by that source are being received.

- **Delay Since Last SR (DLSR)**   The duration, expressed in units of 1/65,536 seconds, between the reception of the last sender report from the source and the issuance of this receiver report block.

RTCP does not define the profile-specific extensions. If they are to be used with a given type of media (a given RTP payload type), then the profile definition of the payload in question should include a specification of the format and usage of any extensions.

RTCP Receiver Report (RR)

The RR is issued by a session participant who receives RTP packets but does not send, or has not yet sent, RTP packets of its own to other session participants. The RR has the format shown in Figure 2-18. The report

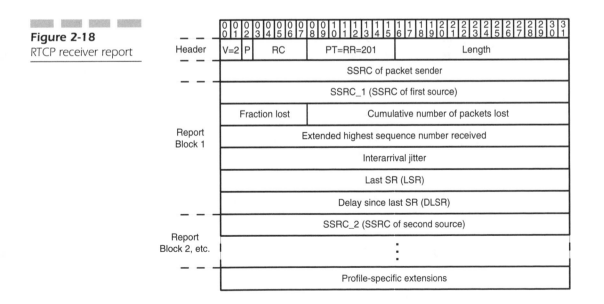

Figure 2-18
RTCP receiver report

packet is almost identical to the sender report with the exception that the payload type has the value 201 and the sender-specific information is not included.

RTCP Source Description Packet (SDES)

The SDES packet provides identification and other information regarding session participants. The SDES packet must exist in every RTCP compound packet. The general format of the SDES packet is shown in Figure 2-19.

The SDES packet is comprised of two main parts: a header and some chunks of information. The header contains a length field, a payload type value field with a value of 202, and a *source count* (SC) field. The SC field occupies five bits and indicates the number of chunks in the packet.

Each chunk contains an SSRC or CSRC value, followed by one or more identifiers and pieces of information that are applicable to that SSRC or CSRC. These pieces of information are known as SDES items and include such items as an e-mail address, phone number, and name (e.g., Jane Doe), among others. All of these SDES items and their structures are defined in RFC 1889. Of the possible SDES items, only one is mandatory—the *canonical name* (CNAME).

The CNAME is meant to be a unique identifier for a session participant and does not change within a given session. Normally, the SSRC value does not change either. It is possible, however, for a given participant's SSRC value to change during a session we find that two participants have chosen the same value of SSRC or if a host resets during a session. In order to ensure that the CNAME is unique, it should have the form user@host,

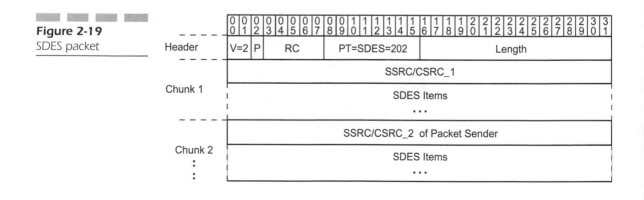

Figure 2-19
SDES packet

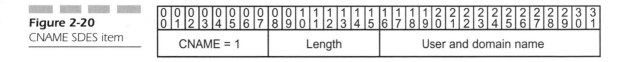

Figure 2-20
CNAME SDES item

0 0 0 0 0 0 0 0 0 0 1 1 1 1 1 1 1 1 1 1 2 2 2 2 2 2 2 2 2 2 3 3
0 1 2 3 4 5 6 7 8 9 0 1 2 3 4 5 6 7 8 9 0 1 2 3 4 5 6 7 8 9 0 1

CNAME = 1	Length	User and domain name

where host is a fully qualified domain name. The format of the CNAME parameter is shown in Figure 2-20.

RTCP BYE Packet

The BYE packet is used to indicate that one or more media sources are no longer active. It may also include a text string indicating a reason for leaving the session.

The BYE packet has the format shown in Figure 2-21. The SC field indicates the number of source identifiers included in the packet, that is, the number of SSRC or CRSC entries. After the list of SSRC/CSRC values, an optional text string may be provided, indicating the reason for leaving. In such a case, the text string is preceded by a length field indicating the number of octets in the text string.

Application-Defined RTCP Packet

An analysis of the various RTP and RTCP packet types reveals that RTP and RTCP are written so that packets can be extended with application-specific data. In keeping with this flexible approach, RTCP includes an application-defined packet, with the format shown in Figure 2-22.

Obviously, this type of packet is used for some nonstandardized applications; hence, the format of the application data is not specified. Some identifier is necessary, however, to enable the packet recipient to understand the data; the payload type value of 204 is not sufficient. If, for example, an experimental application were to require two new types of RTCP packets, each carrying different information, then there must be a means to differentiate between the two types.

At this point, the Name field comes into play. This field contains an ASCII string specified by the creator (designer) of the new application and is used to imply a particular coding and significance for the application

Figure 2-21
RTCP BYE packet

Figure 2-21
RTCP BYE packet

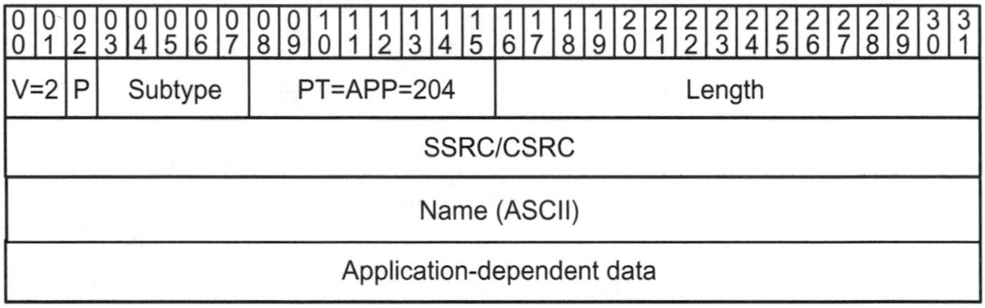

Figure 2-22
RTCP application-defined packet

data. The subtype field is also available for the application creator to indicate the use of any options within the particular application.

Calculating Round-Trip Time

One of the functions enabled by the sender and receiver reports of RTCP is the calculation of round-trip time. This quantity refers to the time that would be taken for a packet to be sent from a particular source to a particular destination and back again. It is a useful metric when measuring voice quality and is calculated from the data in RTCP reports as follows.

Let us assume that participant A issues report A at time T1. This report is received by participant B at time T2. Subsequently, participant B issues report B at time T3 and this report is received by A at time T4. Participant A can calculate the round-trip time as simply

$$T4 - T3 + T2 - T1 \text{ or}$$

$$T4 - (T3 - T2) - T1$$

The data within the RTCP reports either explicitly provide the values for T1, T2, T3, and T4 or provide participant A with sufficient information to determine the values. For example, report B in our example contains an NTP timestamp for when the report is issued (T3). Report B also contains the LSR timestamp, which points participant A to the timestamp included in report A and hence to the value T1. The DLSR timestamp in report B contains the delay between reception by B of report A and the sending from B of report B (T3–T2). Finally, participant A takes note of the instant that report B is received (T4).

Calculating Jitter

As mentioned previously, delay is significant in the perceived voice quality and should be minimized. Delay, however, cannot be eliminated completely. Given that some delay will always exist, the delay should be constant or close to constant. In other words, if a given packet takes 50 milliseconds to get from its source to its destination, then the next packet should take about the same time. Unfortunately, this delay is not always constant, and the variation in delay is known as jitter.

Jitter is defined as the mean deviation of the difference in packet spacing at the receiver compared to the packet spacing at the sender for a pair of packets. This value is equivalent to the deviation in transit time for a pair of packets.

If Si is the RTP timestamp from packet i, and Ri is the time of arrival in RTP timestamp units for packet i, then for the two packets, i and j, the deviation in transit time, D, is given by

$$D(i, j) = (Rj - Ri) - (Sj - Si) = (Rj - Sj) - (Ri - Si)$$

The interarrival jitter is calculated continuously as each data packet, i, is received from a given source SSRC_n, using the difference, D, for that packet and the previous packet, i − 1, in order of arrival according to the formula.

$$J(i) = J(i - 1) + (\,|\,D(i - 1, i)\,|\,-J(i - 1))/16$$

The current value of J is included in sender and receiver reports.

Timing of RTCP Packets

RTCP provides useful feedback regarding the quality of an RTP session. RTCP enables participants and/or network operators to obtain information about delay, jitter, and packet loss and to take corrective action where possible to improve quality. Of course, the better the information, the better the decisions that can be made—and, of course, the quality of the information is linked to how often RTCP reports are issued. Therefore, one would assume that RTCP reports should be sent as often as possible. RTCP packets consume bandwidth, however, and if a large number of RTCP reports are being sent, then these reports themselves contribute to delay, jitter, and RTP packet loss. In other words, too many RTCP packets can make the quality worse. Therefore, we need to strike a balance between the need to provide sufficient feedback and the need to minimize the bandwidth consumed by RTCP packets.

Determining how often to send RTCP packets involves keeping the following considerations in mind. The control traffic should be limited to a small and known fraction of the session bandwidth. The RTCP portion of total bandwidth should be small so that the primary function of the transport protocol to carry data is not impaired. The RTCP portion should be known by all participants, so that the control traffic can be included in the bandwidth specification given to a resource reservation protocol and so that each participant can independently calculate its share. In fact, the total bandwidth consumed by RTCP should be set to 5 percent of the total bandwidth for the session.

RFC 1889 provides an algorithm for calculating the interval between RTCP packets. The algorithm is designed to meet the foregoing criteria. It is also designed to ensure that RTCP can scale from small to large sessions. The algorithm includes the following main characteristics:

- Senders are collectively allowed at least 25 percent of the control traffic bandwidth so that in sessions with a small number of participants, newly arriving participants quickly receive the CNAME for sending participants.

- The calculated interval, which is the mean interval between RTCP packets, is to be greater than a minimum of 5 seconds.

- The interval between RTCP packets varies between 0.5 and 1.5 times the calculated interval. This helps to avoid unintended synchronization where all participants send RTCP packets at the same instant, hence clogging the network.

- A dynamic estimate of the average compound RTCP packet size is calculated, including all those received and sent, to automatically adapt to changes in the amount of control information carried. In other words, if RTCP compound packets start becoming bigger, perhaps because of more participants, then the rate at which the RTCP packets are sent needs to be adapted to ensure that excessive bandwidth is not consumed.

IP Multicast

Many situations occur where a given IP datagram needs to be sent to multiple hosts. Perhaps the example most applicable to VoIP is a conference, where voice streams from one or more participants need to be distributed to many listeners. Sending packets to each destination individually is possible, but a far better solution is to send the same packet to a single address that is associated with all listeners. This type of solution is known as *multicasting*. The following section provides a brief overview of the concept of multicasting.

Multicasting involves the definition of multicast groups and a given group is assigned a multicast address. All members of the group retain their own IP addresses, but also receive datagrams sent to the multicast address.

In order to support multicasting, a number of requirements must be met. There must be an address space allocated for multicast addresses so that they are not treated as ordinary addresses. Hosts that want to join a particular group must be able to do so and must inform local routers of their membership in a group. Routers must be able to recognize when one or more attached hosts belong to a particular group, as identified by a particular multicast address. Routers must support the capability to forward multicast datagrams correctly. Finally, routing protocols such as OSPF must support the propagation of routing information for multicast addresses. Furthermore, routing tables should be set up so that the minimum number of datagrams is sent.

For example, let's assume that a particular host joins a group and that the host is connected to router A. Let's also assume that other members of the group are connected to router B. Finally, let's assume that some host attached to router C wants to send a datagram to the group. In that case, there needs to be a means for router C to be aware that datagrams for the group's multicast address must be sent to both router A and router B.

As a second example, let's assume that multicast router D exists in the path between router C and routers A and B. In that case, router C should not send two datagrams to routers A and B via router D. Instead, router C should forward a single copy of the datagram to router D. Router D should then branch out the datagram to routers A and B. In other words, the number of copies of a multicast datagram should be kept to a minimum.

The *IP version 4* (IPv4) address space 224.0.0.0 to 239.255.255.255 is allocated for multicast addresses. Therefore, if a router receives a datagram for such an address, it must be able to route it appropriately. Of these addresses, a number are allocated for special purposes. Examples include the following:

- **224.0.0.1** All hosts on a local subnet
- **224.0.0.2** All routers on a local subnet
- **224.0.0.5** All routers supporting OSPF
- **224.0.0.9** All routers supporting *Routing Information Protocol* (RIP) version 2

Hosts in a particular group use the *Internet Group Message Protocol* (IGMP) to advertise their membership in a group to routers so that multicast datagrams may be forwarded to the appropriate hosts. IGMP version 1 is specified in RFC 1112, while IGMP version 2 is specified in RFC 2236.

Support for multicasting within routing protocols has also been developed. For example, RFC 1584 specifies extensions to OSPF to accommodate the propagation of multicast routing information.

IP Version 6 (IPv6)

The foregoing discussion regarding IP and higher-layer protocols has been based upon IPv4, which is the version of the protocol that is currently deployed throughout the world. The next generation of IP, however, *IP version 6* (IPv6), is starting to be deployed. Accordingly, a brief overview of IPv6 is appropriate here.

Among the biggest drivers for the creation of IPv6 has been the explosive growth of the Internet and IP-based networks in general. It is clear that such growth cannot be sustained indefinitely with an address space of only 32 bits. Furthermore, IP networks are being used more and more for real-time and interactive applications that were not strongly considered in the initial design of IP. IPv6 is designed to address these issues and provide greater flexibility for the support of future applications that are as yet unknown.

Compared to v4, IPv6 offers the following enhancements:

- **Expanded address space**   Each address is allocated 128 bits instead of 32.
- **Simplified header format**   This enables easier processing of IP datagrams.
- **Improved support for headers and extensions**   This enables greater flexibility for the introduction of new options.
- **Flow-labeling capabilities**   This enables the identification of traffic flows (and therefore better support at the IP level) for real-time applications.
- **Authentication and privacy**   Support for authentication, data integrity, and data confidentiality are included at the IP level rather than through separate protocols or mechanisms above IP.

IPv6 Header

The IPv6 header in Figure 2-23 clearly shows that the header—at least in the basic form shown in Figure 2-23—is less complex than the v4 header. The fields are as follows:

- **Version**   The version number is 6.
- **Traffic Class**   This field of the IPv6 header corresponds to the *Type of Service* (TOS) field of the IPv4 header. The field comprises 8 bits and is used to enable certain types of traffic to be differentiated from others— and potentially to be given a high priority. For example, real-time applications may be given a higher priority than other applications, such as e-mail. In fact, we will see in Chapter 8 how this field is used to ensure better QoS for real-time applications.
- **Flow Label**   This is a 20-bit field, which may be used by a source to label sequences of packets that belong to a single flow. A flow is a sequence of packets sent from a particular source to a particular destination for which the source desires special handling by the intervening routers. For example, a given VoIP stream from one source to another could be considered a flow. The nature of the required special handling for a particular flow might be conveyed to the routers by a control protocol, such as a resource reservation protocol, or by information within the flow's packets themselves. A given flow is identified by a unique combination of source address and a nonzero flow label.

Figure 2-23
IPv6 Header

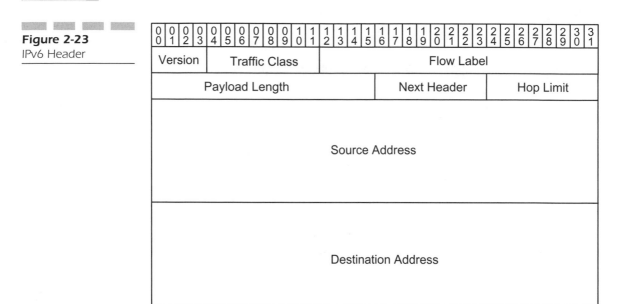

- **Payload Length** This is a 16-bit unsigned integer indicating the length of the payload in octets. Note that IPv6 enables header extensions to be included after the fixed header shown in Figure 2-23. As far as the length field is concerned, such extensions are part of the payload.

- **Next Header** This field comprises 8 bits, and it serves the same purpose and uses the same values as the Protocol field in the IPv4 header. Thus, the field is used to indicate the next higher layer protocol to which the payload data should be passed at the destination node. The Next Header field is also used to indicate the existence of IPv6 header extensions, as explained later in this chapter.

- **Hop Limit** This field is an 8-bit unsigned integer that indicates the maximum number of hops that an IP packet should experience before being discarded. The value in the Hop Limit field is decremented by 1 at each node that forwards the packet, and the packet is discarded if the value reaches 0. This field is analogous to the TTL field in the IPv4 header.

- **Source and Destination Addresses** Each of these addresses is a 128-bit IPv6 address. The IPv6 addressing structure is described in the following section.

IPv6 Addresses

As previously described, IPv4 addresses are generally represented by a dotted decimal notation with a maximum value of 255.255.255.255. IPv6 addresses are significantly different. They have the format XXXX:XXXX:XXXX:XXXX:XXXX:XXXX:XXXX:XXXX, where each X is a hexadecimal character. Thus, the address consists of eight fields, with each field comprising four hexadecimal characters. The maximum possible value for an IPv6 address is therefore FFFF:FFFF:FFFF:FFFF:FFFF:FFFF:FFFF:FFFF. Clearly, this value is several orders of magnitude greater than the maximum IPv4 address.

In terms of how we represent an IPv6 address, it is not necessary that leading zeros be included in a given field, but an entry must exist for every field. Therefore, for example, 1511:1:0:0:0:FA22:45:11 is a valid address. The one exception to this rule regarding leading zeros is the case where several contiguous fields have the value zero. In such a case, the symbol "::" can be used to represent a number of contiguous fields with zero values. For example, the address 1511:1:0:0:0:FA22:45:11 could be represented as 1511:1::FA22:45:11. This notation can be applied also in the case where the first several fields contain zero values; the address 0:0:0:0:AA11:50:22:F77 could be represented as: ::AA11:50:22:F77. If the "::" notation appears once in the address, then it is possible to determine the correct content of each of the eight fields of the address. However, that is no longer possible if the "::" notation appears more than once. Therefore, the notation is not allowed to appear more than once in an address. The formatting and usage of IPv6 addresses is specified in RFC 2373.

One important point to note is that the use of IPv6 should not generally require significant changes to upper-layer protocols. An important exception, however, is the use of pseudoheaders when calculating TCP and UDP checksums. Recall that the checksum calculation in UDP and TCP includes a pseudoheader, which incorporates certain fields of the IP header, including the source and destination addresses. Since these are significantly different in IPv6 compared to IPv4, it is important that TCP and UDP checksum calculations be suitably adapted when IPv6 is used as the routing protocol instead of IPv4.

IPv6 Special Addresses   IPv6 identifies a number of special addresses. These include the following:

■ **The all-zeros address**   Meaning an unspecified address, this address is sometimes used when a node does not yet know its address.

This can be represented as either 0:0:0:0:0:0:0:0 or simply :: when using the short notation. The all-zeros address must not be used as a destination address in an IPv6 packet.

- **The loopback address**   This is used by a node to send an IPv6 packet to itself. This is an address with the last field set to 1, that is, 0:0:0:0:0:0:0:1 or simply ::1 when using the short notation. This address can be used to route a packet on a virtual internal interface, but it should never be used as a destination address in a packet that is sent outside a physical node.

- **IPv6 address with embedded IPv4 address (type 1)**   This is an IPv6 address where the first 96 bits are 0 and the last 32 bits contain an IPv4 address. This type of address is termed an IPv4-compatible IPv6 address and it is used by IPv6 hosts and routers that tunnel IPv6 packets through an IPv4 infrastructure. Such an address can be represented in a manner that indicates the embedded IPv4 address in the familiar dotted decimal format. For example, the address 0:0:0:0:0:0:145.180.15.15, or simply ::145.180.15.15, would be used to represent an IPv6 address with the embedded IPv4 address 145.180.15.15.

- **IPv6 address with embedded IPv4 address (type 2)**   This is an IPv6 address where the first 80 bits are 0, the next 16 bits are set to FFFF, and the last 32 bits are set to an IPv4 address. This type of address is used to give an IPv6 address representation to a pure IPv4 address, which would apply to nodes that do not support IPv6 at all. Using the same IPv4 address as in the previous example, the IPv6 representation would be 0:0:0:0:0:FFFF:145.180.15.15 or simply ::FFFF:145.180.15.15.

IPv6 Header Extensions

IPv6 enables extension headers to be placed between the fixed header and the actual data payload. With one exception, these header fields are not acted upon by intermediate nodes and are only considered at the destination node. Recall that the IPv6 header contains a field called Next Header, which is used to indicate the type of payload carried in the IP datagram (and therefore to what upper-layer protocol the packet should be passed at the destination node). For example, if the value in the header is 6, the protocol identifier for TCP, then the payload is a TCP packet and the overall datagram would appear as shown in Figure 2-24.

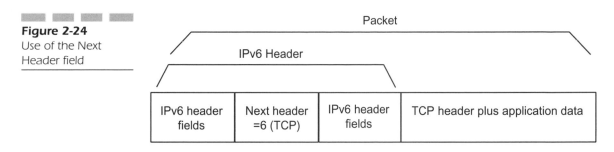

Figure 2-24
Use of the Next
Header field

The Next Header field, however, can also be used to indicate the existence of a header extension, which may be used to carry certain optional information. If a header extension exists, then the Next Header value will indicate the type of extension. Furthermore, each extension will have its own Next Header field, which allows for several contiguous header extensions to be included between the IP header and the upper-layer data. The value of Next Header in the IP header will point to the first extension, the value of Next Header in the first extension will point to the second extension, and so on. Finally, the value of Next Header in the last extension will indicate the upper-layer protocol being used. Figure 2-25 shows an example in which the three extensions Hop-by-Hop, Destination Options, and Routing are used with a TCP packet.

Note that with a few exceptions IPv6 itself does not usually specify the details of the various extension headers—at least not the specifics of what optional information and what corresponding functions are to be invoked. Rather, the header extension mechanisms specified by IPv6 offer a means to support applications and requirements that might be as yet unknown.

Hop-by-Hop Extension   In general, header extensions are not analyzed by nodes along the delivery path and are meant to be acted upon only by the destination node. The one exception to this rule is the Hop-by-Hop extension, which must be examined and processed by every node along the way. Because of the fact that it must be examined at each node, the Hop-by-Hop extension, if present, must immediately follow the IP header.

The Hop-by-Hop extension is of variable length, with the format shown in Figure 2-26. The Next Header value is used to indicate the existence of a subsequent header extension or the upper-layer protocol to which the packet payload applies. The Length field gives the length of this header extension in units of 8 octets, not including the first 8 octets. The individual options themselves, whatever they might signify, are of the format *type-length-value* (TLV), where the type field is 8 bits long, the length field is 8

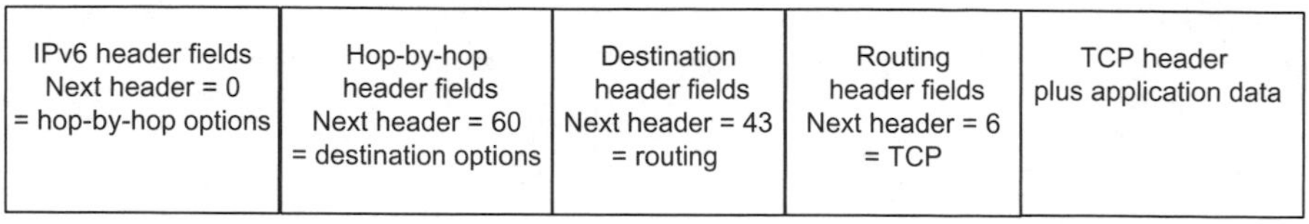

Figure 2-25
Concatenated
header extensions

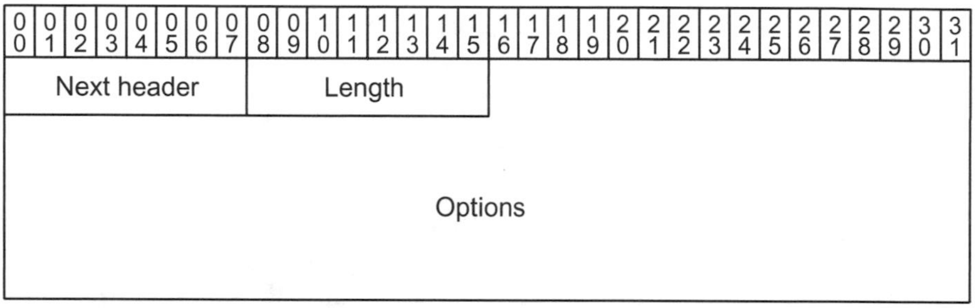

IPv6 header fields Next header = 0 = hop-by-hop options	Hop-by-hop header fields Next header = 60 = destination options	Destination header fields Next header = 43 = routing	Routing header fields Next header = 6 = TCP	TCP header plus application data

Figure 2-26
IPv6 Hop-by-Hop
header extension

| 0 0 0 0 0 0 0 0 | 0 0 1 1 1 1 1 1 | 1 1 1 1 2 2 2 2 | 2 2 2 2 2 2 3 3 |
0 1 2 3 4 5 6 7	8 9 0 1 2 3 4 5	6 7 8 9 0 1 2 3	4 5 6 7 8 9 0 1
Next header		Length	
Options			

bits long, and the data is a variable number of octets according to the length field. This arrangement is shown in Figure 2-27.

Within the Options Type field, the first three bits are of special significance. The first two bits indicate how the node should react if it does not understand the option in question, as follows:

- **00** Skip this option and continue processing the header.
- **01** Discard the packet.
- **10** Discard the packet and send an ICMP Parameter Problem, Code 2 message to the originator of the packet.
- **11** Discard the packet and send an ICMP Parameter Problem, Code 2 message to the originator of the packet, but only if the destination address in the IP header is not a multicast address.

The third bit indicates whether the option data may be changed enroute. If set to 0, then it cannot be changed. If set to 1, then it can change.

The remainder of the Options field is used to carry whatever options the sender of the datagram requires. The IPv6 specification itself does not specify what those options might be.

Destination Options Extension The Destination Options extension has the same format as the Hop-by-Hop extension in that it is of variable length and contains a number of options in the TLV format. These options are to be examined only by the destination node. The Destination Options

Figure 2-27
Format of options

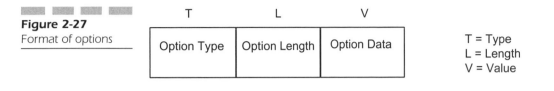

header extension is identified by the use of the value 60 in the Next Header field in the immediately preceding header.

Routing Extension   The Routing header extension includes a Routing Type field to enable various routing options to take place. For the moment, however, IPv6 only defines the routing type 0. The function of this routing type is to specify the nodes that should be visited on the path from the source of a packet to the ultimate destination. When used with routing type 0, the Routing header extension has the format shown in Figure 2-28.

As with all header extensions, the first two fields are the Next Header field and the Length field. The next field is the Routing Type field, which is an 8-bit field indicating a particular variant of the routing header (and which is 0 in the figure). The Segments Left field indicates the number of nodes that still need to be visited before the packet reaches its ultimate destination. Finally, a list of IP addresses that need to be visited along the way is included.

At first glance, one might think that this type of header would need to be analyzed by intermediate nodes between the source IP address and the destination IP address. After all, the purpose of the header is to specify a particular routing, that is which nodes to visit en route. Hence, one would think that this header is another exception to the rule that only destination nodes examine header extensions.

In fact, the way this header is used is quite clever. Let's assume that a packet is to be sent from address A to address Z and must visit addresses B, C, and D in that order along the way. In such a case, the source address in the IP header is address A and the destination address in the IP header is address B. Addresses C, D, and Z are listed in the Route header. Therefore, the node at address B is the destination of the IPv6 packet and as such is required to examine the route header. Upon examination of the header, it swaps address C into the destination address of the IP header, decrements the Segments Left field, and forwards the packet. At address C, the same process is repeated, but this time address D is placed in the destination address of the IP header. The process continues until the packet reaches

Figure 2-28
Type 0 routing
header

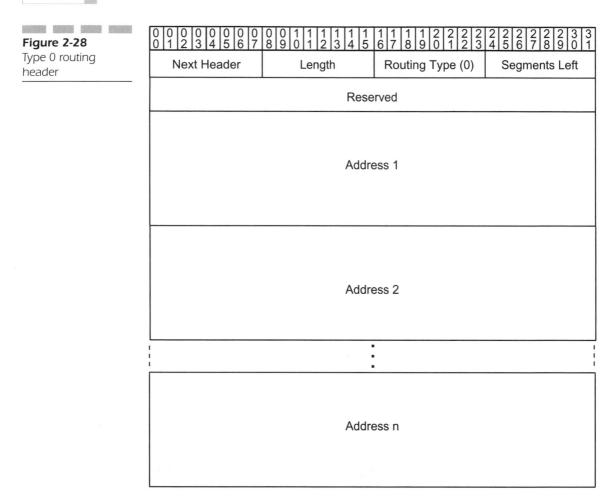

address Z, at which point the value in the Segments Left field is 0, indicating that the packet is at its final destination. Furthermore, the Route header now provides a list of nodes visited en route.

Interworking IPv4 and IPv6

IPv6 is the new version of IP, but IPv4 is used everywhere today, and it is not reasonable to expect that all IP networks will be changed from IPv4 to IPv6 at the same time. In fact, it is certain that IPv4 and IPv6 will coexist for a long time and it is likely that IPv6 networks will initially appear as islands in a sea of IPv4. Therefore, a clear requirement exists to support

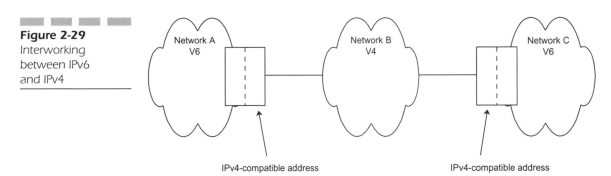

Figure 2-29
Interworking
between IPv6
and IPv4

seamless interworking between the two versions. In other words, IPv6 nodes must be able to send datagrams to IPv4 nodes, and IPv4 nodes must be able to send datagrams to IPv6 nodes. Moreover, IPv6 nodes must be able to send datagrams to each other via IPv4 networks, and IPv4 nodes must be able to send datagrams to each other via IPv6 networks.

These requirements can be met by the use of IPv4-compatible addresses. Through the use of IPv4-compatible nodes with IPv4-compatible addresses at the boundaries of IPv6 networks, interworking can be accomplished. Figure 2-29 provides an example of how this is done. By allocating an IPv4-compatible address to the interfaces on an IPv6 network towards the IPv4 network, the IPv6 network can appear as an IPv4 network.

Consider a scenario in which a packet has to go from one IPv6 network to another IPv6 network via an IPv4 network. At the edge of the source network, the packet can be wrapped in an IPv4 header. That header would contain an address for the remote IPv6 network, which would cause the packet to tunnel through the IPv4 network. At the destination network, the IPv4 header is removed, and the remaining IPv6 packet is routed locally within the destination network.

Speech-Coding Techniques

Chapter 1, "Introduction," mentions the fact that *Voice over IP* (VoIP) can take advantage of a variety of efficient speech-coding techniques and Chapter 2, "Transporting Voice by Using IP," delves into some detail regarding how digitally encoded speech is carried in an *Internet Protocol* (IP) network by using the *Real-Time Transport Protocol* (RTP). Chapter 2 also makes reference to various RTP payload types, which in turn refer to various speech-coding standards. If it were not already obvious, the discussions in these previous chapters should indicate that speech-coding techniques are of particular significance in VoIP networks. In fact, speech coding is a very important topic.

Though some analog systems still remain, they are rare. These days, voice is digitally encoded and shipped around the network (any network) as a stream of 1's and 0's. Speech/voice coding is simply the process by which a digital stream of 1's and 0's is made to represent an analog voice waveform. The process involves converting the incoming analog voice pattern into a digital stream and converting that digital stream back to an analog voice pattern at the ultimate destination. One of the reasons for implementing VoIP is the opportunity to take advantage of efficient voice coding, where fewer bits are used to represent the voice being transmitted, thereby reducing bandwidth requirements and reducing cost.

The initial conclusion could be that one should implement the most bandwidth-efficient coding scheme possible, thereby saving the most money. Unfortunately, however, there is no such thing as the proverbial "free lunch." As a rough guide, the lower the bandwidth, the lower the quality. So if we reduce the bandwidth significantly, then we run the risk of providing substandard voice quality—not a noble or smart goal and certainly not something that is consistent with a carrier-grade operation.

One should not assume, however, that anything like a linear relationship exists between bandwidth and quality. Another component also effects the relationship. That is the speech-coding scheme used, which relates to processing power.

For example, voice transmission at 64 Kbps sounds better than any voice transmission at 16 Kbps. Several coding schemes might exist, however, that result in a bit rate of 16 Kbps, and one of those techniques might provide better quality than another. You can be certain that the technique that provides the best quality uses a more complex algorithm and hence needs more processing power. Consequently, that technique might be more expensive to implement, emphasizing again that we get nothing for nothing. In the end, the choice of coding scheme is a balance between quality and cost.

In this chapter, we look at some of the issues surrounding speech coding in general and consider some of the coding standards that exist today. Note that this is not by any means an exhaustive analysis of coding standards.

It is meant only as an overview of some of the most common standards so that the reader may gain an understanding of some of the technologies involved. Given that this area of telecommunications has undergone great advances in recent years, further advances are likely in the near future.

Voice Quality

We would like to minimize bandwidth while maintaining sufficiently good voice quality. Bandwidth is easily quantified, but how do we quantify voice quality? Surely, this measurement is subjective rather than objective. Standardized methods exist for measuring voice quality, however.

The most common speech quality metric is called the *Mean Opinion Score* (MOS). This measurement scheme is described in *International Telecommunication Union Telecommunications Standardization Sector* (ITU-T) Recommendation P.800. Basically, MOS is a five-point scale as follows:

- Excellent—5
- Good—4
- Fair—3
- Poor—2
- Bad—1

The objective with any coding technique is to achieve as high a ranking on this scale as possible while keeping the bandwidth requirement relatively low. To determine how well a particular coding technique ranks on this scale, a number of people (a minimum of about 30 people and possibly many more) listen to a selection of voice samples or participate in conversations, with the speech being coded according to the technique to be evaluated. They rank each of the samples or conversations according the five-point scale and a mean score is calculated to give the MOS.

On the surface, this does not seem too scientific. Although the basis of the MOS test is a matter of people listening to voice samples, there is more to it than that. ITU-T P.800 makes a number of recommendations regarding the selection of participants, the test environment, explanations to listeners, the analysis of results, and so on. Consequently, different MOS tests performed on the same coding algorithm tend to give roughly similar results. However, because the tests are subjective, variations occur. One test may yield a score of 4.1 for a particular coding algorithm, whereas another test may yield a score of 3.9 for the same algorithm. Therefore, one should treat MOS values with care, and at the risk of sounding cynical, MOS scores should perhaps be treated with some skepticism depending on who is claming a particular MOS for particular *coder/decoder* (codec).

In general, and within design restrictions such as bandwidth, most coding schemes aim to achieve or approach toll quality. Although the definition of this term can vary, toll quality generally refers to an MOS of 4.0 or higher.

Although MOS is the most common voice quality measurement technique, a number of other methods have been developed. In fact, a great deal of effort has been expended on the development of subjective and objective quality-testing techniques. One of the most common techniques is *Perceptual Speech Quality Measurement* (PSQM) described in ITU-T Recommendation P.861. PSQM is effectively an algorithm that attempts to faithfully represent human judgement and perception processes. The result of PSQM testing is a PSQM score that can be converted to an MOS value.

PSQM testing involves the algorithmic comparison between the output signal of a communications system (such as a voice coder/decoder) and a known input signal. The testing includes an algorithm that determines the combined perceptual effects of a number of variables, such as the type of speaker (male, female, or child), the loudness of the input signal, the delay, the percentage of active/silent speech frames, clipping, and environmental noise. Tests are conducted with a variety of different speech samples that may have been produced artificially or by real speakers. If produced artificially, then the speech samples should be generated according to ITU-T Recommendation P.50.

A number of commercial systems exist for PSQM testing. These test systems are often used in the evaluation of a speech coder during its development and test. A VoIP carrier may also use the test systems in order to choose a coding scheme to use in a network.

A Little about Speech

Speech is generated when air is pushed from the lungs past the vocal cords and along the vocal tract. The basic vibrations occur at the vocal cords, but the sound is altered by the disposition of the vocal tract, that is, by the position of the tongue or the shape of the mouth. The vocal tract can be considered a filter and many codec technologies attempt to model the vocal tract as a filter. As the shape of the vocal tract changes relatively slowly, the transfer function of the filter needs to be changed relatively infrequently (every 20 milliseconds or so). If the vocal tract can be modeled as a filter, then the excitation of the vibrations at the vocal cords corresponds to the excitation signal applied to the filter.

Speech sounds can be classified into three main types as follows:

- Voiced sounds are produced when the vocal cords vibrate open and closed, thus interrupting the flow of air from the lungs to the vocal tract and producing quasi-periodic pulses of air. The rate of the opening and closing gives the pitch of the sound. This can be adjusted by varying the shape of, and the tension in, the vocal cords and the pressure of the air behind them. Voiced sounds show a high degree of periodicity at the pitch period, which is typically between 2 and 20 milliseconds. This long-term periodicity can be seen in Figure 3-1, which shows a segment of voiced speech sampled at 8 kHz. Here the pitch period is about 8 milliseconds or 64 samples. The power spectral density for this segment is shown in Figure 3-2.

- Unvoiced sounds result when the excitation is a noise-like turbulence produced by forcing air at high velocities through a constriction in the vocal tract while the glottis is held open. Such sounds show little

Figure 3-1
Typical segment of voiced speech

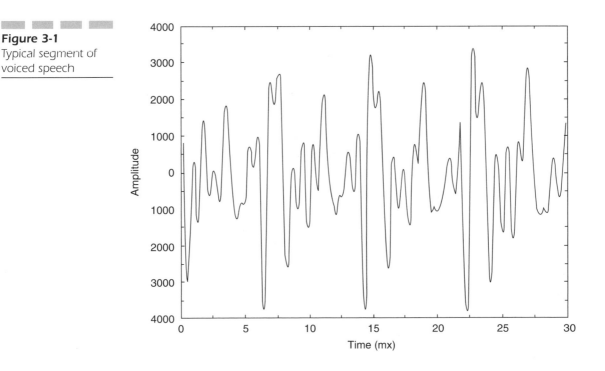

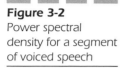

Figure 3-2
Power spectral
density for a segment
of voiced speech

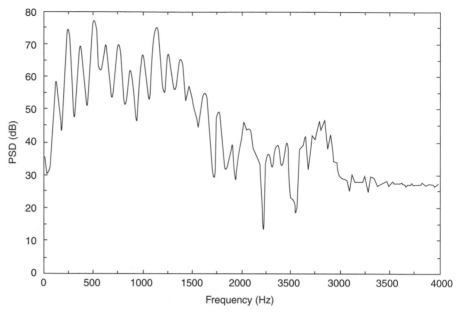

long-term periodicity, as can be seen in Figures 3-3 and 3-4, although short-term correlations due to the vocal tract are still present.

■ Plosive sounds result when a complete closure is made in the vocal tract, and air pressure is built up behind this closure and released suddenly.

Although the combined efforts of the vocal cords and the vocal tract can be made to create a vast array of sounds, the type of vibration at the vocal cords and the shape of the vocal tract change fairly slowly. We can see from the figures that the speech signal is relatively predictable over time. This predictability can be used in the design of voice-coding systems to minimize the amount of data to be transferred and hence reduce bandwidth. For example, if the systems at each end of a conversation can both make reasonably accurate predictions of how the speech signal will change over time, then it is not necessary to transfer the same amount of data as would be required if speech were totally unpredictable. With good voice-coding schemes, leveraging the predictability of speech can lead to a significant reduction in bandwidth, without a huge degradation in perceived quality.

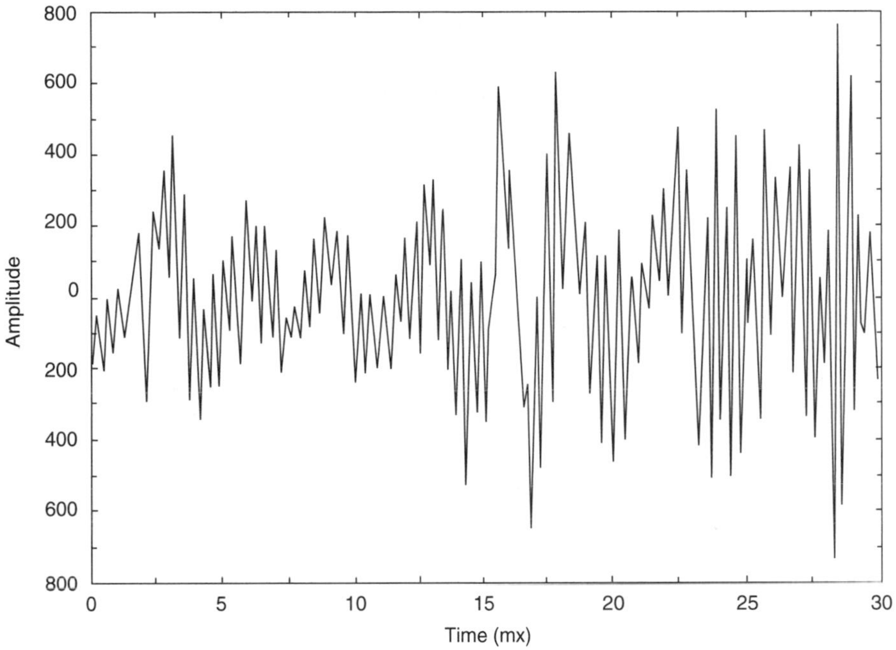

Figure 3-3
Typical segment of
unvoiced speech

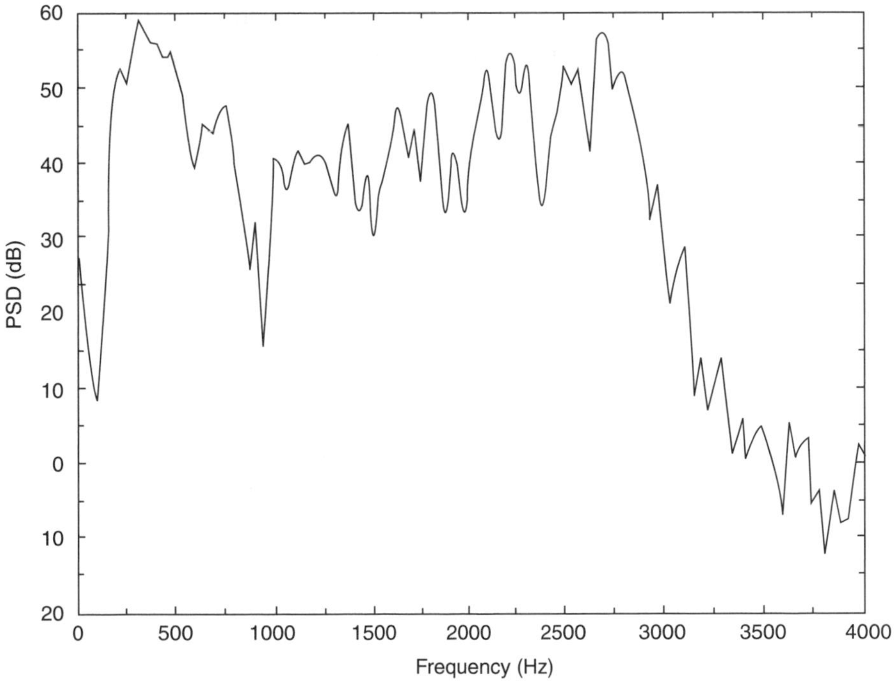

Figure 3-4
Power spectral
density for a segment
of unvoiced speech

Voice Sampling

In order to create a digital representation of an analog waveform (such as voice), it is first necessary to take a number of discreet samples of the waveform and then represent each sample by some number of bits.

Figure 3-5 gives an example of the sampling of a simple sine wave and what the samples might look like. Obviously, it would take an infinite number of samples in order for those samples to completely recreate the original signal in all respects. Of course, that is not possible, nor is it desirable. Instead, we would prefer to take a sufficient number of samples such that we can use the sample values and suitable mathematics to recreate the original signal (or at least come very close). The Nyquist sampling theorem can help in this regard. This theorem basically states that a signal can be reconstructed if it is sampled at a minimum of twice the maximum frequency of the signal. Thus, if the maximum frequency of a given signal is 4,000 Hz, then we need to take at least 8,000 samples per second.

In fact, this technique is used in all speech codecs. Human speech is generally in the frequency range of 300 and 3,800 Hz. We assume a maximum frequency of less than 4,000 Hz, apply a low-pass filter to take out any frequency components above 4,000 Hz, and then take 8,000 samples per second. Note that this procedure will not properly capture Luciano Pavarotti in full voice or even a less talented individual signing in the shower—just typical conversational speech. For this reason, someone singing on the telephone never sounds quite as good as he or she should, regardless of whether the singer has any talent.

Figure 3-5
Wave sampling

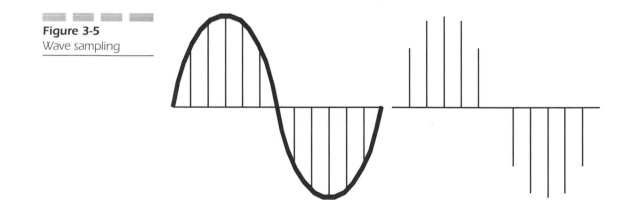

Quantization

So we sample at 8,000 Hz, but the next concern is how to represent each sample. In other words, how many bits are used to represent each sample? The answer is not as simple as one might think, largely because of something called *quantization noise*.

When we take a sample, we have a limited number of bits to represent the value of the sample. By using a limited number of bits to represent each sample, however, we quantize the signal. Figure 3-6 shows an example of quantization, where three bits represent various signal levels. Since we have only three bits, we can represent a maximum of eight different signal levels. If, at a given sampling instant in our example, the analog signal has a value of 5.3, then with our limited number of bits we can only apply the value 5 to the digitized representation. When we transmit the digitized representation to the far end, all that can be recovered is the value 5. The real signal level can never be recovered. The difference between the actual level of the input analog signal and the digitized representation is known as quantization noise.

The easiest way to minimize the effect of quantization noise is to use more bits, thereby providing better granularity. So, if we were to use 6 bits instead of 3, we could at least tell that the sampled value is closer to 5.5

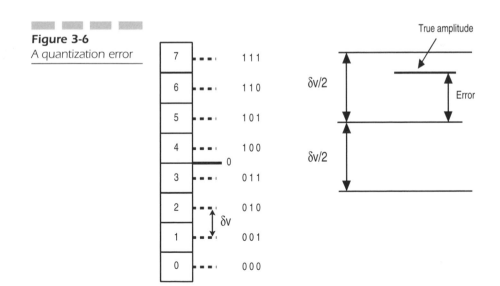

Figure 3-6
A quantization error

than to 5.0 and we could send the bit representation for 5.5. This method can easily approach the stage of diminishing returns. The more accurate we want the samples to be, the more bits we need to represent each sample and the more bandwidth we need. Alternatively, we can use relatively few bits, but we can use them in a more sophisticated manner.

If we look at the possible range of signal levels in speech and if we want to apply uniform quantization levels, then we have two effects. First, we would need many bits to represent each sample, and second, loud talkers would sound better than quiet talkers. The latter is because the effect of quantization noise is less at higher levels than at lower levels. For example, the detrimental effect of digitizing a sample of 11.2 as the value 11 (about 1.8 percent too low) is a lot less than the effect of digitizing a sample of 2.2 as the value 2 (about 9 percent too low). More formally, the signal-to-noise ratio is better for loud talkers than for quiet talkers.

Therefore, we use nonuniform quantization. This approach involves the usage of smaller quantization steps at smaller signal levels and larger quantization steps for larger signal levels. This process gives greater granularity at low signal levels and less granularity at high signal levels. The effect is to spread the signal-to-noise ratio more evenly across the range of different signals and to enable fewer bits to be used compared to uniform quantization. In a coder where the quantization levels are simply sent from one end to the other, this technique translates to a lower bandwidth requirement than for uniform quantization.

Types of Speech Coders

Before we get into descriptions of the various types of codecs available, it is first worth looking at the various categories of codecs. Three types exist: waveform codecs, source codecs (also known as vocoders), and hybrid codecs.

Waveform codecs basically sample and code the incoming analog signal without any thought as to how the signal was generated in the first place. They then transmit quantized values of the samples to the destination end, where the original signal is reconstructed, at least to a very good approximation. Generally, waveform codecs produce a high-quality output and are not very complex. The big disadvantage is that they consume large amounts of bandwidth compared to other codecs. When waveform coders are used at lower bandwidths, the speech quality degrades significantly.

Source codecs attempt to match the incoming signal to a mathematical model of how speech is produced. They usually use a linear predictive filter model of the vocal tract, with a voiced/unvoiced flag to represent the excitation applied to the filter. In other words, the filter represents the vocal tract and the voiced/unvoiced flag represents whether a voiced or unvoiced input is received from the vocal cords. The information that is sent to the far end is a set of model parameters rather than a representation of the signal itself. The far end, using the same modeling technique in reverse, takes the values received and reconstructs an analog signal.

Vocoders operate at low bit rates but tend to produce speech that sounds synthetic. Using higher bit rates does not offer much improvement due to the limitations in the underlying model. Although vocoders are used in private communications systems and particularly in military applications, they are generally not used in public networks.

Hybrid codecs attempt to provide the best of both worlds. Although they attempt to perform a degree of waveform matching, they also utilize knowledge of how people produce sounds in the first place. They tend to provide quite good quality at lower bit rates than waveform coders. Figure 3-7 shows a comparison of the three types of codec with respect to quality and bandwidth.

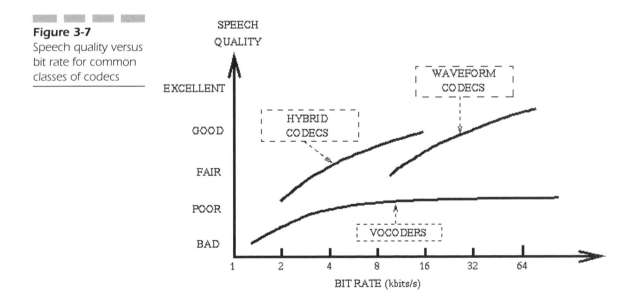

Figure 3-7
Speech quality versus bit rate for common classes of codecs

G.711

G.711 is the most commonplace coding technique used today. It is a wave-form codec and is the coding technique that is used in circuit-switched telephone networks all over the world. G.711 has a sampling rate of 8,000 Hz. If uniform quantization were to be used, then the signal levels commonly found in speech would be such that at least 12 bits per sample would be needed, leading to a bit rate of 96 Kbps. Nonuniform quantization is used, however, with 8 bits used to represent each sample. This quantization leads to the well-known 64 Kbps *digital signal 0* (DS0) rate that we have come to know and love. G.711 is often called *pulse code modulation* (PCM).

G.711 has two variants: A-law and μ-law. μ-law is used primarily in North America and A-law is used in most other countries. The difference between the two is the manner is which nonuniform quantization is performed. Both are symmetrical around zero. A-law is skewed to be a little friendlier to lower signal levels than μ-law in that it provides very small quantization intervals for a longer range of low-level signals at the expense of larger quantization levels for a longer range of higher-level signals.

Both A-law and μ-law provide very good quality and have an MOS of about 4.3. The main drawback with G.711, however, is the 64 Kbps bandwidth requirement. Achieving comparable speech quality at a lower bandwidth would be nice.

Adaptive Differential PCM (ADPCM)

PCM codecs such as G.711 send individual samples to the far end, where they are reconstructed to form something close to the original waveform. Since voice changes relatively slowly, however, it is possible to predict the value of a given sample based on the values of previous samples. In such a case, we need only transmit the difference between the predicted value and the actual value of the sample. Because the far end is performing the same predictions, it can determine the original sample value if told the difference between the prediction and the actual sample value. This technique is known as *differential PCM* (DPCM) and can significantly reduce the transmission bandwidth requirement without a drastic degradation in quality. In its simplest form, DPCM does not even make a prediction of the next sample; it simply transmits the difference between sample N and sample

N+1, providing all the information necessary for the far end to recreate sample N+1.

A slightly more advanced version of DPCM is *adaptive DPCM* (ADPCM). ADPCM typically predicts samples' values based on past samples while factoring in some knowledge of how speech varies over time. The error between the actual sample and the prediction is quantized, and this quantized error is transmitted to the far end. Assuming that predictions are fairly accurate, then fewer bits are required to describe the quantized error and hence the bandwidth requirement is lower. The ITU-T Recommendation G.721, which offers ADPCM-coded speech at 32 Kbps, is a good example of ADPCM.

G.721 has now been superseded by ITU-T recommendation G.726, which is a more advanced ADPCM codec. A G.726 coder takes A-law- or μ-law-coded speech and converts it to or from a 16, 24, 32, or 40 Kbps channel. Running at 32 Kbps, G.726 ADPCM has an MOS of about 4.0, which is pretty good.

Both PCM and ADPCM are waveform codecs. As such, they have effectively no algorithmic delay. In other words, the functioning of the algorithm itself does not require that quantized speech samples or quantized error values be held up for a brief time before being sent to the far end. Given that voice is a delay-sensitive form of communication, this absence of algorithm delay is a significant advantage. The disadvantage, of course, is the bandwidth that these codecs consume—especially at the higher end.

Analysis-by-Synthesis (AbS) Codecs

Hybrid codecs attempt to fill the gap between waveform and source codecs. As described previously, waveform coders are capable of providing good-quality speech at higher bit rates. They can even provide relatively acceptable quality at rates down to about 16 Kbps. They are of limited use at lower rates, however. Vocoders, on the other hand, can provide intelligible speech at 2.4 Kbps and lower, but they cannot provide natural-sounding speech at any bit rate. The objective with hybrid coders is to provide good-quality speech at a lower bit rate than is possible with waveform coders.

Although other forms of hybrid codecs exist, the most successful and most commonly used are time domain *Analysis-by-Synthesis* (AbS) codecs. Such codecs use the same linear prediction filter model of the vocal tract as found in *Linear Predictive Coding* (LPC) vocoders. Instead of applying a simple two-state, voiced/unvoiced model to find the necessary input to this filter, however, the excitation signal is chosen by attempting to match the reconstructed speech waveform as closely as possible to the original speech waveform. In other words, different excitation signals are attempted, and the one that gives a result closest to the original waveform is selected (hence the name AbS). AbS codecs were first introduced in 1982 with what was to become known as the *Multi-Pulse Excited* (MPE) codec. Later, the *Regular-Pulse Excited* (RPE) and the *Code-Excited Linear Predictive* (CELP) codecs were introduced. Variations of the latter are discussed briefly here.

G.728 Low-Delay CELP (LD-CELP)

CELP coders implement a filter, the characteristics of which change over time, and they contain a codebook of acoustic vectors. Each vector contains a set of elements, where the elements represent various characteristics of the excitation signal. This approach is obviously far more sophisticated than a simple voiced/unvoiced flag.

With CELP coders, all that is transmitted to the far end is a set of information indicating filter coefficients, gain, and a pointer to the excitation vector chosen. Because the far end contains the same codebook and filter capabilities, it can reconstruct the original signal to a good degree of accuracy.

ITU-T Recommendation G.728 specifies a *Low-Delay CELP* (LD-CELP coder), which is a backward-adaptive coder because it uses previous speech samples to determine the applicable filter coefficients. G.728 operates on five samples at a time. In other words, the coder takes 5 samples (sampled at 8,000 Hz) and determines a codebook vector and filter coefficients to best match the 5 samples. The choice of filter coefficients is based upon previous and current samples. Because the coder operates on 5 samples at a time, we have a delay of less than 1 millisecond.

Because the receiver also has access to the previous samples and can perform the same determination, all that is necessary to transmit is an indication of the excitation parameters. In fact, an index to the excitation vector is sent. Because there are 1,024 vectors in the codebook (and these are available at both the sender and receiver), the index is only 10 bits long. G.728 uses 5 samples at a time and these samples are taken at a rate of 8,000 per second. For each 5 samples, G.728 needs to transmit only 10 bits. A little arithmetic shows that G.728 results in a transmitted bit rate of 16 Kbps.

The following text is taken directly from the G.728 specification and provides some further information on the coding mechanism. The text and the content of Figures 3-8 and 3-9 are reproduced here with the prior authorization of the ITU as copyright holder.

After the conversion from A-law or μ-law PCM to uniform PCM, the input signal is partitioned into blocks of five-consecutive input signal samples. For each input block, the encoder passes each of 1024 candidate codebook vectors (stored in an excitation codebook) through a gain scaling unit and a synthesis filter. From the resulting 1024 candidate quantized signal vectors, the encoder identifies the one that minimizes a frequency-weighted mean-squared error measure with respect to the input signal vector. The 10-bit codebook index of the corresponding best codebook vector (or "codevector"), which gives rise to that best candidate quantized signal vector, is transmitted to the decoder.

The decoding operation is also performed on a block-by-block basis. Upon receiving each 10-bit index, the decoder performs a table look-up to extract the corresponding codevector from the excitation codebook. The extracted codevector is then passed through a gain scaling unit and a synthesis filter to produce the current decoded signal vector.

The decoded signal vector is then passed through an adaptive postfilter to enhance the perceptual quality. The postfilter coefficients are updated periodically using the information available at the decoder. The five samples of the postfilter signal vector are next converted to five A-law or μ-law PCM output samples.

The reason why the G.728 coder is termed LD-CELP is the very small algorithmic delay that it introduces. Due to the fact that it operates on 5 PCM samples at a time and given that these samples are taken at a rate of 8,000 Hz, the coder introduces a delay of 0.625 milliseconds. This delay is so small that it would not be noticeable even to the keenest ear.

The fact that only a 10-bit pointer value is transmitted for every 5 incoming PCM samples means that the actual bandwidth consumed between encoder and decoder is one-quarter of that used by regular PCM (that is, 16 Kbps). Given that the G.728 codec gives an MOS score of about 3.9, the quality reduction is not enormous compared to the bandwidth saved. Figure 3-8 provides a simplified block diagram of the LD-CELP encoder and decoder.

Although the G.728 coder provides high quality at a reasonably low bandwidth, it is quite rare to find the coder used in commercial VoIP products. The reason is the fact that the G.728 coder is quite processor intensive. Consequently, the *digital signal processors* (DSPs) used for G.728 are quite expensive. There are other coders that provide good quality at a low bandwidth and a lower cost than G.728.

Figure 3-8

A simplified block
diagram of the LD-
CELP coder

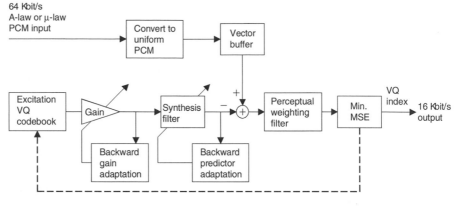

a) LD-CELP encoder

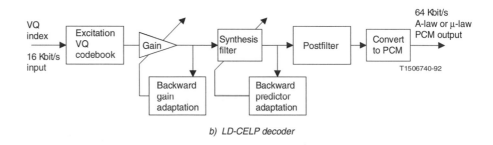

b) LD-CELP decoder

G.723.1 Algebraic Code-Excited Linear Prediction (ACELP)

ITU-T Recommendation G.723.1 specifies a speech coder that can operate at either 6.3 Kbps or 5.3 Kbps, with the higher bit rate providing higher speech quality. Both rates are mandatory parts of the codec and we can change from one mode to another during a conversation.

The coder takes a band-limited input speech signal that is sampled at 8,000 Hz and that undergoes uniform PCM quantization, resulting in a 16-bit PCM signal. The encoder then operates on blocks or frames of 240 samples at a time. Thus, each frame corresponds to 30 milliseconds of speech, which means that the coder automatically causes a delay of 30 milliseconds. The G.723.1 coder also utilizes a look-ahead of 7.5 milliseconds, resulting in a total algorithmic delay of 37.5 milliseconds. Of course, other small delays will take place within the coder itself as a result of the processing effort involved.

Each frame is passed through a high-pass filter to remove any DC component and then is divided into 4 subframes of 60 samples each. Various operations are performed on these subframes in order to determine the appropriate filter coefficients. *Algebraic Code-Excited Linear Prediction* (ACELP) is used in the case of the lower bit rate of 5.3 Kbps and *Multipulse Maximum Likelihood Quantization* (MP-MLQ) in the case of the higher rate of 6.3 Kbps.

The information transmitted to the far end includes linear prediction coefficients, gain parameters, and excitation codebook index values. The information transmitted comprises 24-octet frames in the case of transmission at 6.3 Kbps and 20-octet frames in the case of transmission at 5.3 Kbps.

Normal conversation involves significant periods of silence (or at least silence from one of the parties). During such periods of silence, it is desirable not to consume significant bandwidth by transmitting the silence at the same rate as speech is transmitted. For this reason, G.723.1 Annex A specifies a mechanism for silence suppression whereby *Silence Insertion Description* (SID) frames can also be used. These are only 4 octets in length, which means that transmission of silence occupies about 1 Kbps. This is significantly better than G.711 where silence is still transmitted at 64 Kbps.

Therefore, three different types of frame can be transmitted by using G.723.1: one for 6.3 Kbps, one for 5.3 Kbps, and an SID frame. Within each frame, the two least significant bits of the first octet indicate the frame size and the codec version in use (as shown in Table 3-1).

G.723.1 has an MOS of about 3.8, which is good considering the vastly reduced bandwidth that it uses. G.723.1 does, however, have the disadvantage of a minimum 37.5-millisecond delay at the encoder. Although this delay is well within the bounds of what is acceptable for good-quality

Table 3-1

G.723.1 frame size and codec version

Bits	Meaning	Octets/frame
00	High-rate speech (6.3 Kbps)	24
01	Low-rate Speech (5.3 Kbps)	20
10	SID frame	4
11		N/A

speech, we must remember that it is round-trip delay that is important, not just one-way delay. Moreover, there will be various other delays in the network, including processing delays and queuing delays such as at routers in a VoIP network, for example.

G.729

The basic ITU-T Recommendation G.729 specifies a speech coder that operates at 8 Kbps. This coder uses input frames of 10 milliseconds, corresponding to 80 samples at a sampling rate of 8,000 Hz. G.729 also includes a 5-millisecond look-ahead, resulting in an algorithmic delay of 15 milliseconds (significantly better than G.723.1). From each input frame, the coder determines linear prediction coefficients, excitation codebook indices, and gain parameters. These pieces of information are transmitted to the far end in 80-bit frames. Given that the input signal corresponds to 10 milliseconds of speech and results in a transmission of 80 bits, the transmitted bit rate is 8 Kbps. G.729 offers an MOS of about 4.0. Figure 3-9 shows a high-level block diagram of the G.729 encoder.

G.729 Annex A   G.729 is a complex codec. In order to reduce the complexity in the algorithm, a number of simplifications were introduced in Annex A to G.729. These include simplified codebook search routines and a simplification to the postfilter at the decoder among other things. G.729A uses exactly the same transmitted frame structure as G.729 and therefore uses the same bandwidth. In other words, the encoder may be operating according to G.729, while the decoder may operate using G.729A or vice versa. Note that G.729A can result in slightly lower quality than G.729. G.729A provides a MOS of about 3.7.

G.729 Annex B   Annex B to G.729 is a recommendation for *voice activity detection* (VAD), *discontinuous transmission* (DTX), and *comfort noise generation* (CNG). VAD is simply the decision as to whether voice or noise is present at the input. The decision is based on an analysis of several parameters of the input signal. Note that the determination is not done simply on the basis of one frame; rather, the determination is made on the basis of the current frame, plus the preceding two frames. This mechanism ensures that transmission occurs for at least two frames after a person has stopped speaking.

The next decision is whether to send nothing at all or to send a SID frame. The SID frame contains some information to enable the decoder to

Figure 3-9
Encoding principle of
the CS-ACELP model

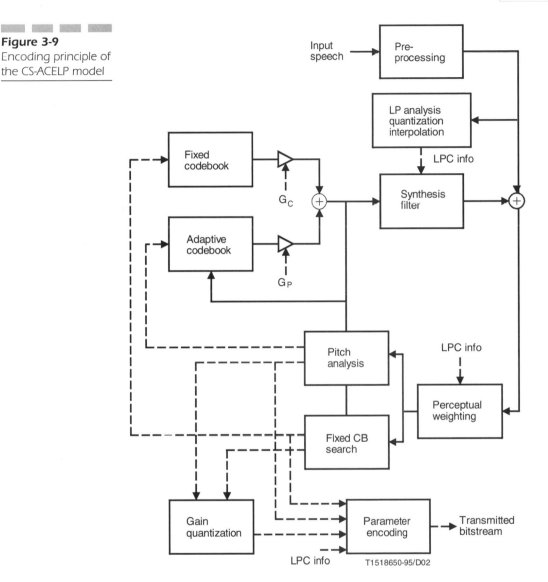

generate comfort noise that simulates the background noise at the transmission end. The G.729B SID frame is a mere 15 bits long, significantly shorter than the 80-bit speech frame.

Assuming that the silence continues for some time, the encoder keeps watch on the background noise. If no significant change occurs, then nothing is sent and the decoder continues to generate the same comfort noise. If, however, the encoder notices a significant change in the background noise energy, an updated SID frame is sent to update the decoder on the characteristics of the background noise. This avoids a comfort noise that is constant and that, if it persists for some time, might no longer be very comforting to the listener.

G.729 Annex D   G.729 Annex D is intended as a lower-rate extension to the basic G.729 algorithm. Like the basic G.729 algorithm, Annex D operates on 10-millisecond speech samples. Rather than sending 80 bits per frame, however, the Annex D algorithm uses 64 bits per frame, resulting in a bit rate of 6.4 Kbps.

G.729 Annex D provides a MOS that is similar to that of G.723.1 operating at 6.3 Kbps. This value is lower than the MOS of the basic G.729 algorithm. The slight reduction in quality can be considered the cost of achieving a lower bandwidth.

G.729 Annex E   G.729 Annex E offers a higher bit rate enhancement to the basic G.729 algorithm. The intention with this enhancement is to provide greater robustness in the presence of significant background noise (particularly music) at the input.

G.729 uses a tenth-order linear prediction filter, which means that the filter contains 10 coefficients. The G.729E coder uses 30 filter coefficients. Moreover, the codebook of G.729E is 44 bits, as opposed to 35 bits in G.729. The net effect of these changes is that G.729E transmits 118 bits for every 10 milliseconds of input signal, resulting in a bit rate of 11.8 Kbps.

Selecting Codecs

The codecs just discussed constitute a small selection of the coding techniques available. Many others add to the menu of choices when choosing a coding technology for use in a given network. Examples of other choices include the following:

- **CDMA QCELP, as defined in IS-733**   This is a variable-rate coder and is currently used in IS-95-based *Code Division Multiple Access* (CDMA) wireless systems. *Qualcom code-excited linear predictor* (QCELP) can operate at several rates, but the two most common are the high rate of 13,300 *bits per second* (bps) and the low rate of 6,200 bps. The codec supports silence suppression, however, such that the net bit rate transmitted is less in each case. The packetization of the QCELP codec for use with RTP is defined in RFC 2658.

■ *Global System for Mobile Communications* **(GSM)** *Enhanced Fullrate* **(EFR), as defined in GSM specification 06.60** This coder is an alternative to the original GSM *full-rate* (FR) coder. Although the GSM FR coder uses *Regular Pulse-Excited—Long Term Prediction* (RPE-LTP), the EFR coder uses an ACELP-based coding technique. The EFR coder operates at a rate of 12,200 bps, while the FR coder operates at a rate of 13,000 bps. Despite the lower bandwidth, the EFR coder provides better quality than the FR coder. Packetization of GSM EFR for use with RTP is documented in RFC 1890.

■ *Adaptive Multi-Rate* **(AMR) codec, as defined in GSM specification 06.90** AMR is a multimode codec that can operate in any of 8 different modes, with bit rates from 4.75 to 12.2 Kbps. When operating at 12.2 Kbps, it is effectively the GSM EFR coder. When operating at 7.4 Kbps, it is effectively the IS-641 coder used in *Time Division Multiple Access* (TDMA) cellular systems. The coder can change to any of its modes as often as every 20 milliseconds in response to channel conditions. AMR supports discontinuous transmission, thereby minimizing the bandwidth consumed during silent periods, which reduces the net bandwidth consumption. The AMR codec is the coding choice for many third-generation wireless systems. The packetization rules for use of this codec with RTP are in the form of an Internet draft as of this writing.

Clearly, many options exist from which to choose. Furthermore, new coding schemes are being developed at a rapid pace. The task of selecting the best codec or set of codecs to use for a given network implementation is a matter of balancing quality with bandwidth consumption. Given the range of codecs available, however, the choice might not be that easy.

When looking at MOS values, one should remember that the MOS value presented for a particular codec is normally the value that is achieved under laboratory situations. The likelihood is significant that a given codec will not provide the same quality in a particular network implementation. For example, G.711, while offering the highest quality under ideal circumstances, does not incorporate any logic to deal with lost packets. In contrast, G.729 does have the capability to accommodate a lost frame by interpolating from previous frames. On the other hand, this causes subsequent speech frames at the decoder to have errors because of the fact that the decoder bases its output on the parameters of previous frames and it can take some time to recover from a lost or corrupt frame.

An important consideration is the delay that a codec will introduce. While waveform codecs do not introduce any algorithmic delay, predictive coders introduce a delay at the coder and a somewhat smaller delay at the decoder.

Another consideration is the DSP power required, measured in *million instructions per second* (MIPS). Whereas G.728 or G.729 may need to run on a 40 MIPS processor, G.726 could run on a 10 MIPS processor. In general, the more MIPS, the greater the cost of the DSP. Fortunately, the price/performance of DSPs continues to improve such that this issue is becoming less of a factor.

At the end of the day, there is no hard and fast answer. The cost/quality tradeoff needs to be made for each individual circumstance.

Cascaded Codecs

In general, we should minimize the number of times that a given speech segment is coded and decoded. In the ideal case, the original analog signal should be coded only once and extracted only at the ultimate destination. It is certainly not a great idea to code a particular analog stream using one codec at, say, 16 Kbps, then convert that to G.711, and later encode again using G.729. In such cases, the voice quality might not even come close to the MOS of the lowest-quality codec in the chain. The simple reason is that each coder can only generate an approximation of the incoming signal that it receives. At each step along the chain, the approximation bears less and less resemblance to the original analog waveform.

In some cases, a VoIP implementation may be such that cascaded codecs are unavoidable. If that is the case, then a thorough trial of different configurations and codecs is well worth the effort to ensure that the ultimate quality does not degrade to an extent that the loss of quality would negate any benefits that VoIP might give.

Tones, Signals, and Dual-Tone Multifrequency (DTMF) Digits

The most sophisticated codecs available today achieve bandwidth efficiency without losing significant quality due to smart algorithms and powerful DSPs. Those smart algorithms, however, are based upon a detailed understanding of how voice is produced in the first place. In other words, the codecs are optimized for human speech. Unfortunately, human speech is not always the only information that needs to be transmitted.

Within the telephone network, in addition to voice, all sorts of tones and beeps are transmitted—not the kind of sounds produced naturally as part of human conversation. These tones include fax tones, various tones such as busy tones, and the *dual-tone multifrequency* (DTMF) digits used in two-stage dialing, voice mail retrieval, and other applications. G.711 can generally handle these tones and beeps quite well, because G.711 samples the incoming signal without considering the origin of that signal. Since G.711 is the coder used in traditional telephony, such tones present no problem in circuit-switched networks. The same tones cannot, however, be faithfully reproduced by lower-bandwidth codecs such as G.723.1 or G.729. If such tones are used as input, the output from those codecs can be unintelligible.

Many VoIP systems use gateways to provide an interface between the IP network and the circuit-switched network. Incoming speech, either in analog form or coded according to G.711, is converted to packetized speech at a lower bandwidth using an efficient coder. It is then passed though the IP network to the destination. The destination may be a native VoIP device or another gateway, in which case the packetized speech needs to be converted back to G.711. In either case, if DTMF digits or other tones arrive at the ingress gateway, they need to be intercepted by the gateway and communicated to the far end in a different manner. They should not be encoded in the same way as speech.

Assuming that a gateway has the capability to detect tones of different types, then it can convey those tones to the destination in one of two ways. The first is to use an external signaling system that is separate from the media stream itself. Although various signaling protocols exist (as discussed in Chapter 4, "H.323," Chapter 5, "The Session Initiation Protocol (SIP)," and Chapter 6, "Media Gateway Control and the Softswitch Architecture"), they are not optimized for transmitting tones under every circumstance. For example, it is one thing to collect DTMF digits at the start of a call and convey them in a signaling system as a called party number. That task is relatively straightforward if the VoIP network is expecting to receive digits at the start of a call and it understands that such digits represent a dialed number. DTMF digits in the middle of a call, however, such as those that are produced when someone is trying to retrieve voice mail, are less easy to handle. That is because the VoIP network itself is not involved in the interpretation of the real meaning of such digits.

The other alternative is to encode the tones differently from the speech but send them along the same media path. We can approach that task in two ways. First, if a gateway can recognize a given tone, then it can send one or more RTP packets that simply provide a name or other identifier for the tone, plus an indication of how long the tone is applied. This effectively means using an RTP packet to signal the occurrence of a tone rather than carrying the tone itself. Assuming that the far end understands the identifier, then it can reproduce the actual tone. The second approach is to use a dynamic RTP profile, where RTP packets carry information regarding the frequency, volume, and duration of the tone. That approach means that the ingress gateway does not need to recognize a particular tone; rather, it only needs to detect and measure the characteristics of a tone and then send a packet describing what was measured.

RFC 2198 describes an RTP payload format for redundant audio data, whereby speech quality can be maintained in the presence of packet loss by sending redundant codings for the same speech signal. Having two methods for communicating the same information regarding a tone means that we can use one of the methods as a backup for the other in accordance with RFC 2198. Thus, the entity (a gateway, for example) that detects a tone can send both types of RTP payload in a given RTP packet. The receiver can then select from the two representations of the tone.

RFC 2833 has been prepared to address the issue of carrying tones within the RTP stream. This specification addresses both of the two methods just described. To support the sending of RTP packets with identifiers for the individual tones, RFC 2833 specifies a large number of tones and events such as DTMF digits, busy tones, congestion tones, and ringing tones. It also provides a numeric identifier for each.

The payload format for these named events is shown in Figure 3-10. The event field contains a number indicating the event in question. The R bit is reserved and the E bit indicates the end of the tone. A sender may delay setting this bit until the tone has stopped. The volume field indicates the power level of the tone from 0 dBm0 to −63 dBm0, with the sign omitted. Therefore, a larger value in the volume field implies a lower volume. The volume field is relevant only for DTMF digits. The duration indicates, in timestamp units, how long the tone has lasted so far. Therefore, the tone event starts at the time indicated in the RTP timestamp, and the duration field indicates how long the tone has lasted. The size of this field can accommodate a duration up to 8 seconds, assuming a sampling rate of 8,000 Hz.

Note that a gateway that detects a tone should send an event packet as soon as the tone is recognized and every 50 milliseconds thereafter while the tone lasts. At least three event packets should be sent even if the tone is very brief, just to make sure that the far end has been informed of the event occurrence, even in the case of packet loss. When several packets are sent for the same tone, the RTP timestamp of each packet should be the

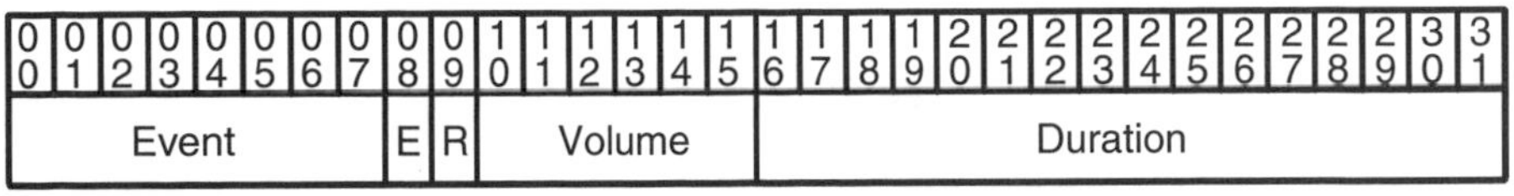

Figure 3-10
The RTP payload format for named events

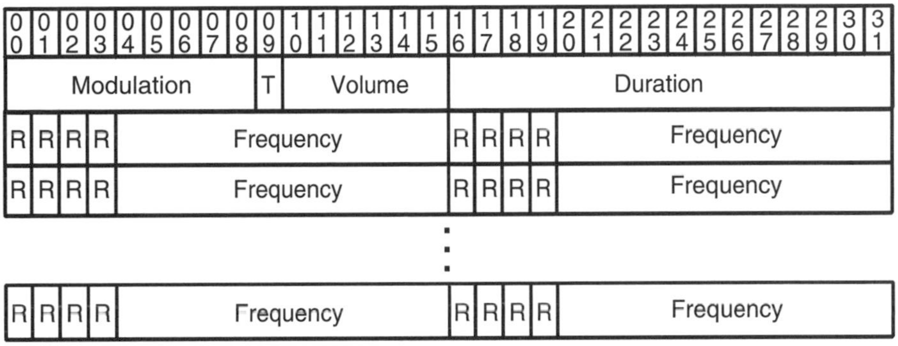

0 1 2 3 4 5 6 7	8 9	0 1 2 3 4 5	6 7 8 9 0 1 2 3 4 5 6 7 8 9 0 1
Event	E R	Volume	Duration

Figure 3-11
The RTP payload format for tones

0 1 2 3 4 5 6 7 8	9	0 1 2 3 4 5	6 7 8 9 0 1 2 3 4 5 6 7 8 9 0 1
Modulation	T	Volume	Duration
R R R R	Frequency	R R R R	Frequency
R R R R	Frequency	R R R R	Frequency

.
.
.

| R R R R | Frequency | R R R R | Frequency |

same, indicating the instant that the tone was detected, with the duration field increased in each packet.

Figure 3-11 shows the payload format to be used for the second method of representing a tone, that is, when sending a description of the tone characteristics. The modulation field indicates the modulation in Hz, up to a maximum of 511 Hz. If no modulation exists, this field is 0. The T bit indicates whether the modulation frequency is to be divided by 3. Some tones are in use where the modulation frequency is not a whole number of Hz. For example, cases occur where modulation at $16 \frac{2}{3}$ Hz is used.

The volume measures the volume of the tone from 0 dBm0 to -63 dBm0. The duration is in RTP timestamp units. The frequency indicates the frequency, in Hz, of one component of the tone. This field support values up to 4,095 Hz, which exceeds the frequency range telephone systems. Note that a given tone can have multiple frequency components. For example, MF (as used in R1-MF signaling) stands for multifrequency. The R field is reserved for future use.

CHAPTER 4

H.323

The foregoing chapters address the issues related to *Internet Protocol* (IP) networks in general and the mechanisms for transporting digitally encoded voice across those networks. Chapter 2, "Transport Voice by Using IP," in particular, describes how voice is carried in *Real-Time Transport Protocol* (RTP) packets between session participants. What is not addressed, however, is the setup and teardown of those voice sessions. Thus far, we have assumed that session participants know of each other's existence and that media sessions are somehow created such that they can exchange voice using RTP packets. So how are those sessions created and ended? How does one party indicate to another a desire to set up a call, and how does the second party indicate a willingness to accept the call? The answer is signaling.

In traditional telephony networks, specific signaling protocols are invoked before and during a call in order to communicate a desire to set up a call, to monitor call progress, and to gracefully bring a call to an end. Perhaps the best example is the *ISDN User Part* (ISUP), a component of the *Signaling System 7* (SS7) signaling suite. In *Voice over IP* (VoIP) systems, signaling protocols are necessary for exactly the same reasons.

The very first VoIP systems used proprietary signaling protocols. The immediate drawback was that two users could communicate only if they both used systems from the same vendor. This lack of interoperability between systems from different vendors was a major inconvenience and impeded the early adoption of VoIP. In response to this problem, the *International Telecommunications Union Telecommunications Standardization Sector* (ITU-T) recommendation H.323 was developed, which serves as a standardized signaling protocol for VoIP. H.323 is the most widely deployed standard in VoIP networks today and is the focus of this chapter.

The first version of H.323 was released in 1996 and bore the title "Visual Telephone Systems and Equipment for Local Area Networks Which Provide a Non-Guaranteed Quality of Service" (quite a mouthful). Its scope was multimedia communications over *local area networks* (LANs). As such, many found the first version of H.323 to be lacking the functions needed for supporting VoIP in a broader environment.

Consequently, the ITU-T produced version 2 of H.323, which was released in 1998. This bears the friendlier title, "Packet-based Multimedia Communications Systems," and has seen greater acceptance and deployment. Since the publication of H.323 version 2, additional enhancements have been specified, which have led to additional revisions of the specification. The most recent version is H.323 version 4. Wherever we make reference to H.323 in this chapter, we are implying version 4 (unless otherwise stated).

The H.323 Architecture

H.323 is one of those ITU recommendations that specifies an overall architecture and methodology, and that incorporates several other recommendations. H.323 should be read in conjunction with several other recommendations and, equally, those other recommendations should be read in conjunction with H.323. Among the other most important recommendations are H.225.0 and H.245, though many others exist.

The scope of H.323 is illustrated in Figure 4-1. The architecture involves H.323 terminals, gateways, gatekeepers, and *multipoint controller units* (MCUs). The overall objective of H.323 is to enable the exchange of media streams between H.323 endpoints, where an H.323 endpoint is an H.323 terminal, a gateway, or an MCU.

An H.323 terminal is an endpoint that offers real-time communications with other H.323 endpoints. Typically, a terminal is an end-user communications device that supports at least one audio *coder/decoder* (codec) and may optionally support other audio codecs and/or video codecs.

A gateway is an H.323 endpoint that provides translation services between the H.323 network and another type of network, such as an

Figure 4-1

The scope of H.323 and interoperability of H.323 terminals

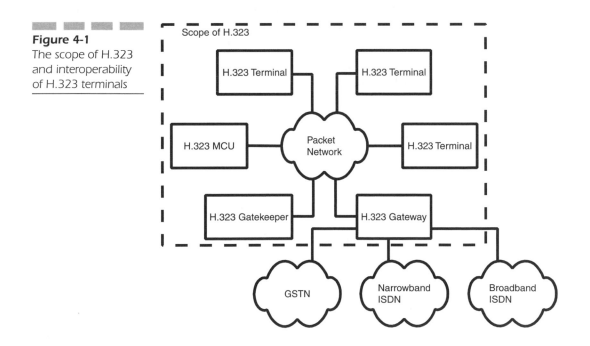

Integrated Services Digital Network (ISDN) or the regular phone network, which in ITU parlance is known as the *General Switched Telephone Network* (GSTN). One side of the gateway supports H.323 signaling and terminates packet media according to the requirements of H.323. The other side of the gateway interfaces with a circuit-switched network and supports the transmission characteristics and signaling protocols of that network. On the H.323 side, the gateway has the characteristics of an H.323 terminal. On the circuit-switched side, it has the characteristics of a node in the circuit-switched network. Translation between the signaling protocols and media formats of one side and those of the other side is performed internally within the gateway. This translation is transparent to other nodes in the circuit-switched network and in the H.323 network. Gateways can also serve as a conduit for communications between H.323 terminals that are not on the same network, where the communication between the terminals needs to pass via an external network such as the *Public Switched Telephone Network* (PSTN).

A gatekeeper is an optional entity within an H.323 network. When present, a gatekeeper controls a number of H.323 terminals, gateways, and *multipoint controllers* (MCs). By control, we mean that the gatekeeper authorizes network access from one or more endpoints and may choose to permit or deny any given call from an endpoint within its control. A gatekeeper may offer bandwidth control services, which can help to ensure high *quality of service* (QoS) if used in conjunction with bandwidth and/or resource management techniques. A gatekeeper also offers address translation services, enabling the use of aliases within the network.

The set of terminals, gateways, and MCs that a single gatekeeper controls is known as a *zone*. Figure 4-2 shows a representation of such a zone. A zone can span multiple networks or network segments, and it is not necessary that all entities within a zone be contiguous.

An MC is an H.323 endpoint that manages multipoint conferences between three or more terminals and/or gateways. For such conferences, the MC establishes the media that may be shared between entities by transmitting a capability set to the various participants. The MC can change the capability set in the event that other endpoints join the conference or if existing endpoints leave the conference. The MC itself does not, however, perform translation or mixing functions. Those functions are performed by a *multipoint processor* (MP), which the MC controls. An MC may reside within a separate MCU or may be incorporated within the same platform as a gateway, a gatekeeper, or an H.323 terminal.

For every MC, there is at least one MP that operates under the control of the MC. It is the MP that processes the actual media streams, creating a

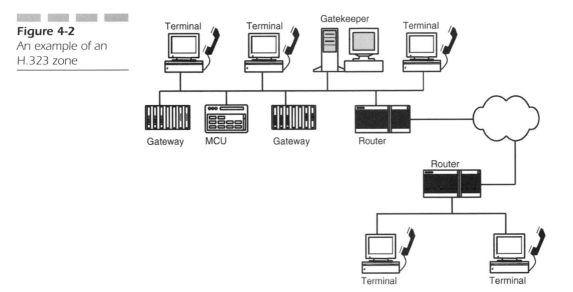

Figure 4-2
An example of an
H.323 zone

number of output media streams from a number of input streams. The MP performs such manipulation by switching, mixing, or a combination of the two. The control protocol between the MC and MP is not standardized.

MCs can support two main types of multipoint conferences: centralized and decentralized. These two arrangements are depicted in Figure 4-3. In a centralized configuration, every endpoint in the conference communicates with the MC in a sort of hub and spoke arrangement. In a decentralized configuration, each endpoint in the conference exchanges control signaling with the MC in a point-to-point manner, but they can share media with the other conference participants through multicast. The MC also enables a mixed conference (also shown in Figure 4-3), where some participants use multicast transmission while other participants use unicast.

Overview of H.323 Signaling

Figure 4-4 shows the H.323 protocol stack. Upon examination, we find a number of protocols already discussed, such as RTP, *RTP Control Protocol* (RTCP), *Transmission Control Protocol* (TCP), and *User Datagram Protocol* (UDP). It is clear from the figure that the exchange of media is performed using RTP over UDP and, of course, wherever we find RTP, we also find RTCP.

Figure 4-3
An example of H.323
conference
configurations

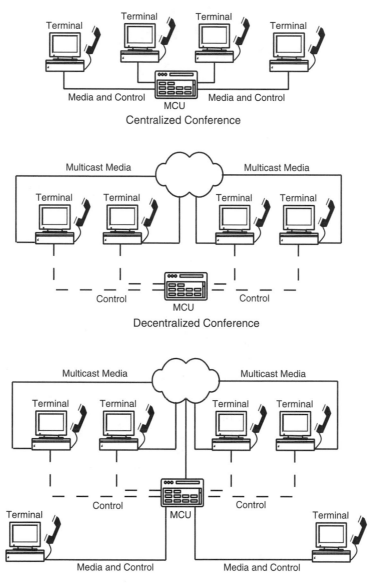

In Figure 4-4, we also find two protocols that have not yet been discussed: namely H.225.0 and H.245. These two protocols define the actual messages that are exchanged between H.323 endpoints. They are generic protocols in that they could be used in any of a number of network architectures. When it comes to the H.323 network architecture, the manner in

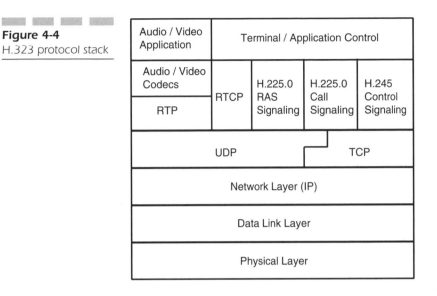

Figure 4-4

H.323 protocol stack

which the H.225.0 and H.245 protocols are applied is specified by recommendation H.323.

Overview of H.323 Protocols

As mentioned previously, the actual signaling messages exchanged between H.323 entities are specified in the ITU recommendations H.225.0 and H.245. H.225.0 is a two-part protocol. One part is effectively a variant of ITU-T recommendation Q.931, the ISDN layer 3 specification, and should be quite familiar to those with knowledge of ISDN. This signaling is used for the establishment and teardown of connections between H.323 endpoints. This type of signaling is known as call signaling or Q.931 signaling. The other part of H.225.0 is known as *Registration, Admission, and Status* (RAS) signaling. This signaling is used between endpoints and gatekeepers, and it enables a gatekeeper to manage the endpoints within its zone. For example, an endpoint uses RAS signaling to register with a gatekeeper, and a gatekeeper uses RAS signaling to allow or deny an endpoint access to network resources.

Note that RAS signaling is always carried over UDP, while call signaling may be carried over either UDP or TCP. H.323 version 2 mandates that call signaling use TCP. The establishment of a TCP connection takes a little time, however, which can lead to a delay in call setup. To speed things up, Annex E to H.323 version 4 specifies a mechanism whereby either TCP or UDP can be used for call signaling. In fact, both TCP and UDP can be used in parallel. The sending entity sends the first message using UDP and simultaneously establishes a TCP connection. If, by the time the TCP connection is established, no response has been received to the UDP-based message, then the TCP connection is used. If, however, a response has been received to the UDP-based message, then the original sender can assume that the far end supports H.323 version 4 and can continue to use UDP for call signaling. The sender then closes the TCP connection.

H.245 is a control protocol used between two or more endpoints. The main purpose of H.245 is to manage the media streams between H.323 session participants. To that end, H.245 includes functions such as ensuring that the media to be sent by one entity is limited to the set of media that can be received and understood by another. H.245 operates by the establishment of one or more logical channels between endpoints. These logical channels carry the media streams between the participants and have a number of properties such as media type, bit rate, and so on.

All three signaling protocols—RAS, call signaling, and H.245—can be used in the establishment, maintenance, and teardown of a call. The various messages may be interleaved. For example, consider an endpoint that wants to establish a call to another endpoint. First, the endpoint might use RAS signaling to obtain permission from a gatekeeper. The endpoint might then use call signaling to establish communication with the other endpoint to set up the call. Finally, the endpoint might use H.245 control signaling to negotiate media parameters with the other endpoint and set up the media transfer.

H.323 messages are sent on various types of channels, depending on the message (and, in some cases, the context). For example, RAS messages are sent over the RAS channel, call-signaling messages are sent over the call-signaling channel, H.245 control messages are sent over the H.245 control channel, and the actual media streams are sent on one or more logical channels. Although this approach might appear to involve a lot of different channels, it is worth noting that these channels are not necessarily related to particular physical interfaces or hardware. Instead, a channel, when used in an IP environment, is simply a reference to a socket address (that is, an IP address and port number). For example, if a given endpoint uses a particular IP address and port number to receive RAS messages, then any messages that arrive at that IP address and port are said to arrive on the endpoint's RAS channel.

Protocol Syntax   As we move forward in this chapter, we will describe the main points of each of the protocols. We will also provide a number of illustrations to explain how the different protocols fit together. We will not, however, describe the detailed syntax of the protocols—and for good reason.

As is the case for many ITU-T recommendations, H.323 signaling is specified using *Abstract Syntax Notation 1* (ASN.1). Although this syntax is eminently suitable for interpretation by software tools and is in fact a formalized language, it is certainly not overly friendly to the human reader. Explaining the syntax of the protocols in detail would require a great deal of ASN.1 syntax in this book and a tutorial on the interpretation of the syntax. Such material would most likely clutter the text and detract from the main objective, which is to explain the functionality offered and how the protocols support that functionality. Consequently, little ASN.1 syntax is included here; instead, we provide a textual description.

H.323 Addressing

Every entity in the H.323 network has a network address, which uniquely identifies that particular entity. In an IP environment, the network address is an IP address. If *Domain Name Service* (DNS) is available, this IP address may be specified in the form of a *Uniform Resource Locator* (URL), according to RFC 1738 and RFC 2396. For example, the URL ras://GK1 @somedomain would be a valid URL for a gatekeeper, since RAS is the protocol supported by a gatekeeper. Any URL can have a port number appended. It the case of RAS, a default port number (1719) exists and should be used in the case that a port number is not specified.

For convenience of identification, entities such as terminals, gateways, and MCUs should have a domain name that is common with their controlling gatekeeper. Note, however, that although applying URLs to endpoints and gatekeepers is a convenient means of identification, the actual IP address is passed in messages between H.323 entities.

For each network address, an H.323 entity can have one or more *Transport Service Access Point* (TSAP) identifiers. In generic terms, a TSAP identifier is an identifier for a particular logical channel at a given entity. In IP terms, it is the same as a socket address.

In general, the port numbers to be used for signaling transactions or media exchanges are assigned dynamically. Important exceptions are the gatekeeper UDP discovery port with value 1718, the gatekeeper UDP registration and status port with value 1719, and the call-signaling TCP or UDP port with value 1720. These port numbers are registered with the *Internet Assigned Numbers Authority* (IANA). The first port number is used when an endpoint wants to determine the gatekeeper with which it should be registered. The second port number is used for RAS signaling to a gatekeeper. The third port number is used when sending call-signaling (Q.931) messages.

In addition to network addresses and TSAP identifiers, terminals and gateways can have one or more aliases. Given that messages must be addressed with real IP addresses, the capability for nodes to have aliases implies that a means to translate between an alias and an IP address must be established. This translation is one of the functions of a gatekeeper and is supported by RAS signaling.

H.323 is flexible when it comes to the allocation of alias addresses. They can take any number of forms, and a given endpoint may have multiple aliases. The only real restriction is that a given alias must be unique within a zone, that is, within the set of nodes controlled by a given gatekeeper (which makes sense). One useful application of aliases is the use of E.164 numbers as aliases. For example, a given gateway that is connected to a regular *private branch exchange* (PBX) over a *primary rate interface* (PRI) could be allocated a set of E.164 numbers as aliases. These numbers could correspond to the telephone numbers that are reachable at the PBX. If another H.323 endpoint wants to call a number on the PBX, the use of the alias makes the task easier. Only the gatekeeper and the gateway need to know the relationship between the alias and the gateway address. The calling endpoint and other entities can avail themselves of the gatekeeper to take care of the necessary translation.

Codecs

In Figure 4-4 we see reference to audio and video codecs. Video support is optional, but when video is supported, an H.323 endpoint must, at a minimum, support video according to H.261 *Quarter Common Intermediate Format* (QCIF).

The support of audio is mandatory and H.323 mandates that the G.711 codec be supported (both A-law and μ-law). Given its relatively high bandwidth requirement, G.711 is certainly not considered the first choice for audio codecs. G.711 is specified as mandatory, however, so that all H.323 endpoints support at least one audio codec. In reality, most endpoints will support a range of more efficient codecs. After all, if a system implements only G.711, then the potential bandwidth efficiencies of VoIP are negated. After all, we would like to carry voice as efficiently as possible, provided that quality objectives can be met.

RAS Signaling

RAS signaling is used between a gatekeeper and the endpoints that it controls. RAS is the signaling protocol through which a gatekeeper controls the endpoints within its zone. Note that a gatekeeper is an optional entity in an H.323 network. Consequently, RAS signaling is also optional. If an endpoint wants to use the services of a gatekeeper, then it must implement RAS. If the endpoint chooses not to use the services of a gatekeeper, then the functions that a gatekeeper normally provides must be provided within the endpoint itself.

RAS signaling is defined in H.225.0 and supports the following functions:

- *Gatekeeper discovery* enables an endpoint to determine which gatekeeper is available to control it.

- *Registration* enables an endpoint to register with a particular gatekeeper and hence to join the zone of that gatekeeper.

- *Unregistration* enables an endpoint to leave the control of a gatekeeper or enables a gatekeeper to cancel the existing registration of an endpoint, thereby forcibly removing the endpoint from the zone.

- *Admission* is used by an endpoint to request access to the network for the purpose of participating in a session. A request for admission specifies the bandwidth to be used by the endpoint, and the gatekeeper may choose to accept or deny the request based on the bandwidth requested.

- *Bandwidth change* is used by an endpoint to request the gatekeeper to allocate extra bandwidth to the endpoint, or it is used by a gatekeeper to instruct an endpoint to reduce the amount of bandwidth consumed.

- *Endpoint location* is a function where the gatekeeper translates an alias to a network address. An endpoint will invoke this function when it wants to communicate with a particular endpoint for which it only has an alias identifier. The gatekeeper will respond with a network address to be used to contact the endpoint in question.

- *Disengage* is used by an endpoint to inform a gatekeeper that it is disconnecting from a particular call. Disengage can also be used from gatekeeper to endpoint to force the endpoint to disconnect from a call.

- *Status* is used between the gatekeeper and endpoint to inform the gatekeeper about the health of an endpoint or about certain call-related data, such as current bandwidth usage.

■ *Resource availability* is used between an endpoint and a gatekeeper to inform the gatekeeper of an endpoint's currently available capacity, such as the amount of bandwidth it has available. This procedure can also be used by an endpoint to inform a gatekeeper that the endpoint has run out of (or is about to run out of) capacity.

■ *Non-standard* is a mechanism by which proprietary information may be passed between an endpoint and a gatekeeper. Of course, the message contents and functions to be invoked are not defined in H.225.0.

Table 4-1 provides a list of the various RAS messages that are used to support these functions and gives a brief description of the purpose of each message. Subsequent sections of this chapter provide more detail regarding the most common messages and functions.

Table 4-1

RAS messages that support RAS signaling functions

Message	Function
GatekeeperRequest (GRQ)	Used by an endpoint when trying to discover its gatekeeper.
GatekeeperConfirm (GCF)	Used by a gatekeeper to indicate that it is available to be the gatekeeper for a given endpoint.
GatekeeperReject (GRJ)	Used by a gatekeeper to indicate that it will not be the gatekeeper for a given endpoint.
RegistrationRequest (RRQ)	Used by an endpoint to register with a gatekeeper.
RegistrationConfirm (RCF)	A positive response from gatekeeper to endpoint, indicating a successful registration.
RegistrationReject (RRJ)	A negative response to a RegistrationRequest.
UnregistrationRequest(URQ)	Used either by a gatekeeper or endpoint to cancel an existing registration.
UnregistrationConfirm (UCF)	Used by a gatekeeper or endpoint to confirm the cancellation of a registration.
UnregistrationReject (URJ)	A negative response to an UnregistrationRequest.
AdmissionRequest (ARQ)	Sent from an endpoint to a gatekeeper to request permission to participate in a call.
AdmissionConfirm (ACF)	Used by a gatekeeper to grant permission to an endpoint to participate in a call.
AdmissionReject (ARJ)	Used by a gatekeeper to deny an endpoint permission to participate in a call.
BandwidthRequest (BRQ)	Sent from an endpoint or a gatekeeper to request a change in allocated bandwidth.

Table 4-1 cont.

RAS messages that support RAS signaling functions

Message	Function
BandwidthConfirm (BCF)	Positive response to a BandwidthRequest.
BandwidthReject (BRJ)	Used by a gatekeeper or endpoint to deny a change in bandwidth. BRJ should be used by an endpoint only if the new bandwidth cannot be supported.
InfoRequest (IRQ)	Sent by a gatekeeper to an endpoint in order to request status information.
InfoRequestResponse (IRR)	Sent from an endpoint to a gatekeeper to provide status information. Can be sent on demand or autonomously.
InfoRequestAck (IACK)	Sent by a gatekeeper as confirmation of the receipt of an IRR.
InfoRequestNak (INAK)	Sent by a gatekeeper upon receiving an IRR in an error situation, such as from an unregistered endpoint.
DisengageRequest (DRQ)	Sent by an endpoint or gatekeeper to request disconnection of a call at an endpoint.
DisengageConfirm (DCF)	Positive response to a DRQ.
DisengageReject (DRJ)	Negative response to a DRQ, such as one from an unregistered endpoint.
LocationRequest (LRQ)	Sent to a gatekeeper to request translation from an alias to a network address.
LocationConfirm (LCF)	Response to an LRQ, including the required address.
LocationReject (LRJ)	Response to an LRQ when translation was not successful.
Non-Standard Message (NSM)	Vendor-specific.
UnknownMessage Response (XRS)	Sent in response to a message that is not recognized.
RequestInProgress (RIP)	Sent by an endpoint or gatekeeper as an interim response when a request is taking a long time to process.
ResourceAvailableIndicate (RAI)	Sent by a gateway to a gatekeeper to inform the gatekeeper of the gateway's current capacity.
ResourceAvailableConfirm (RAC)	Sent by a gatekeeper as acknowledgment to an RAI.
ServiceControlIndication (SCI)	Sent by a gatekeeper or endpoint in order to initiate a separate service control session related to advanced features and services. This message was introduced in H.323 version 4.
ServiceControlResponse (SCR)	Sent to acknowledge the receipt of an SCI message, but does not necessarily mean that the service client will initiate the session as given in SCI.

Gatekeeper Discovery

In a network that has one or more gatekeepers, an endpoint should register with one of the gatekeepers. In order to perform this registration, the endpoint must first find a suitably accommodating gatekeeper—one that is willing to take control of the endpoint. Of course, the endpoint may be configured in advance with the address of the gatekeeper that should be used. In that case, no discovery process takes place per se. Rather, the endpoint simply registers with the gatekeeper in question. Although such an approach might lead to a quicker registration, it lacks flexibility. In a given network, several gatekeepers may be in a load-sharing mode or a backup gatekeeper might need to assume the role of a failed gatekeeper. A static endpoint-gatekeeper relationship is not best suited for such scenarios. Therefore, an automatic gatekeeper discovery process is available.

The gatekeeper discovery process is available for an endpoint to determine its gatekeeper in the event that it does not already have that information. In order to discover which gatekeeper is to exercise control over a given endpoint, the endpoint sends a *gateway-request message* (GRQ). This message can be sent to a number of known addresses, or it can be sent to the gatekeeper discovery multicast address and port 224.0.1.41:1718.

The GRQ message contains a number of parameters. One of these is a gatekeeper identifier. If this parameter is empty, the GRQ indicates to those gatekeepers that receive the message that the endpoint is soliciting gatekeepers and is willing to accept any available gatekeeper. In other words, the endpoint is asking, "Will someone be my gatekeeper?" In the case that the parameter contains an identifier, the endpoint is asking a specific gatekeeper to be the endpoint's gatekeeper.

One or more of the gatekeepers may respond with a *gatekeeper-confirmation* (GCF) message, indicating that the gatekeeper is willing to control the endpoint. Alternatively, a gatekeeper may respond with a *gatekeeper-reject* (GRJ) message, indicating that the gatekeeper is not willing to assume control of that endpoint. In the case that a gatekeeper sends a GRJ message, it will include a reason for the rejection, such as lack of resources at the gatekeeper.

The process is illustrated in Figure 4-5. In this scenario, the terminal wanting to determine its gatekeeper sends just one GRQ message, but sends it to the gatekeeper discovery multicast address. Therefore, more than one gatekeeper can receive the message. In Figure 4-5, gatekeeper 1 sends a GRJ message, whereas gatekeeper 2 sends a GCF message. The terminal will now proceed to register with gatekeeper 2.

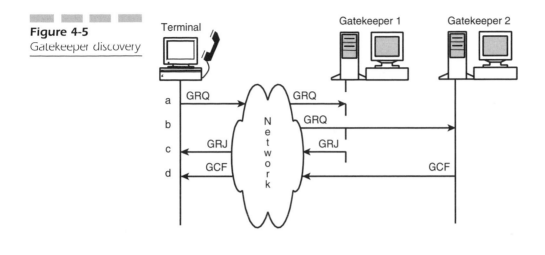

Figure 4-5
Gatekeeper discovery

Optionally, a given gatekeeper may send a GCF message indicating one or more other gatekeepers to try. This indication is included within the GCF message by the presence of the optional parameter AlternateGatekeeper. In such an event, the GCF message does *not* mean that the gatekeeper issuing the GCF message is willing to take control of the endpoint. Rather, the GCF message is somewhere between a positive response and a complete rejection and is equivalent to saying something like "I can't help you, but try the gatekeeper next door." This mechanism can be useful for load sharing or redundancy schemes.

Only one gatekeeper can control a given endpoint at a given time. If an endpoint receives multiple GCF messages from gatekeepers willing to accept control, then the choice of gatekeeper is left to the endpoint. If, on the other extreme, an endpoint receives no positive response to the GRQ message within a timeout period, then the endpoint can retry the GRQ, but should not issue a GRQ within 5 seconds of the previous request.

Endpoint Registration and Registration Cancellation

Gatekeeper discovery is simply a means for an endpoint to determine which gatekeeper is available to assume control of the endpoint. The fact that a given gatekeeper has responded to a GRQ with a GCF does not mean that the endpoint is now under control of that gatekeeper. The process by which an endpoint becomes controlled by a gatekeeper and joins the zone of that gatekeeper is known as *registration*.

Registration begins with the issuance of a *RegistrationRequest* (RRQ) message from an endpoint to a gatekeeper. The message is directed to a gatekeeper at an address configured within the endpoint or determined by the endpoint through the gatekeeper discovery procedure. The port number to be used for the message is 1719—the well-known RAS signaling port of a gatekeeper.

The endpoint includes in the RRQ an address to be used by the endpoint for RAS messages and an address to be used for call-signaling messages. The RRQ can include an alias, a name by which the endpoint wants to be called. The RRQ can also include one or more alternative addresses available for use as backup or redundant addresses. In such cases, these addresses can each have their own aliases.

The gatekeeper can choose to reject a registration request for any number of reasons and does so by responding with a *RegistrationReject* (RRJ) message. One reason, for example, would be if the endpoint wants to use an alias already in use within the zone. If all is well, however, and the gatekeeper is willing to accept the registration, then it responds with a *RegistrationConfirmation* (RCF) message. If the original RRQ does not specify an alias for the endpoint, then the gatekeeper can assign one or more aliases, in which case the assigned aliases are returned as a parameter in the RCF message.

Registrations can have a limited lifetime. In an RRQ, an endpoint can include a timeToLive parameter, indicating the requested duration for the registration in seconds to a maximum value of 4294967295 (hexadecimal FFFFFFFF), which translates to about 136 years. In the RCF, the gatekeeper may choose a lower timeToLive value than that received in the RRQ. If, after some time, the registration period is about to expire, then the endpoint can send another RRQ that includes the optional parameter keepAlive, which amounts to a request to reset the timeToLive timer and extend the registration. An RRQ message that contains the keepAlive parameter is known as a lightweight registration request, because it is designed only to extend an existing registration.

Once registered with a gatekeeper, the endpoint can choose to cancel its registration at some later time. The endpoint does so by sending an *UnregistrationRequest* (URQ) message. Normally, a gatekeeper will respond with a positive confirmation by sending an *UnregistrationConfirmation* (UCF) message. It might, however, send a rejection if the endpoint was not registered with that Gatekeeper in the first place or if the endpoint is attempting to cancel a registration while still involved in a call. The rejection takes the form of an *UnregistrationReject* (URJ) message.

The gatekeeper can also choose to cancel the registration of a particular endpoint, in which case the gatekeeper will send a URQ message to the endpoint. One reason for this would be when the registration timeToLive has expired. Upon receipt of a URQ from a gatekeeper, the endpoint should respond with a UCF.

Figure 4-6 depicts an example of registration and registration cancellation. Note that this is just an example and that a URQ message may be issued by an endpoint as well as by a gatekeeper.

Endpoint Location

Endpoint location is a service that enables an endpoint or gatekeeper to request a real address when only an alias is available. In other words, endpoint location is a translation service.

An endpoint that wants to obtain contact information for a given alias can send a *LocationRequest* (LRQ) message to a gatekeeper. The message may be sent to a particular gatekeeper or can be multicast to the gatekeeper discovery multicast address 224.0.1.41. The LRQ contains the aliases for which address information is required. In the multicast case, the procedure is equivalent to asking "Does anyone know who has alias XXX?"

Not only can an endpoint send the LRQ, but a gatekeeper can send an LRQ to another gatekeeper to determine the address of an endpoint. This would happen when the call signaling from an endpoint is to pass through the calling endpoint's gatekeeper.

Figure 4-6
An example of registration and registration cancellation

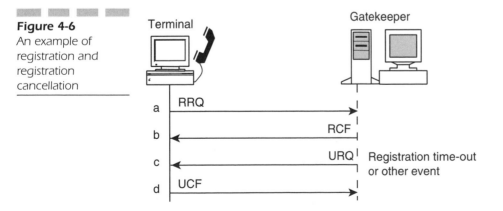

A *LocationConfirm* (LCF) message indicates a positive response to the LRQ. This message should be sent from the gatekeeper at which the endpoint is registered. Among some other optional parameters, the LCF contains a call-signaling address and an RAS signaling address for the located endpoint.

If a gatekeeper receives an LRQ message and the endpoint is not registered at that gatekeeper, then the gatekeeper should respond with a *LocationReject* (LRJ) if the original LRQ had been received by the gatekeeper via unicast on its RAS channel. If, however, the gatekeeper receives an LRQ by multicast on the gatekeeper discovery multicast address, then the gatekeeper should not send an LRJ.

Admission

Admission is the process by which an endpoint requests permission from a gatekeeper to participate in a call. The endpoint does so by sending an *AdmissionRequest* (ARQ) message to the gatekeeper. The endpoint indicates the type of call in question (two-party or multiparty), the endpoint's own identifier, a call identifier (a unique string), a call reference value (an integer value also used in Q.931 messages for the same call), and information regarding the other party or parties to participate in the call. The information regarding other parties to the call includes one or more aliases and/or a signaling address.

One of the most important mandatory parameters in the ARQ is the bandwidth parameter. This parameter specifies the amount of bandwidth required in units of 100 bits/second. Note that the endpoint should request the total media stream bandwidth needed, excluding overhead. Thus, if a two-party call takes place, with each party sending voice at 64 Kbps, then the bandwidth required is 128 Kbps and the value carried in the bandwidth parameter is 1,280. The purpose of the bandwidth parameter is to enable the gatekeeper to reserve resources for the call. Through the use of the optional parameter transportQOS, however, the endpoint may indicate that it is willing to undertake resource reservation rather than having the gatekeeper do it.

The gatekeeper indicates successful admission by responding to the endpoint with an *AdmissionConfirm* (ACF) message. This includes many of the same parameters that are included in the ARQ. The difference is that when a given parameter is used in the ARQ, it is simply a request from the endpoint, whereas a given parameter value in the ACF is a firm order from the gatekeeper. For example, the ACF includes the bandwidth parameter, which may be a lower value than that requested in the ARQ, in which case the endpoint must stay within the bandwidth limitations that the gatekeeper imposes. Equally, although the endpoint might have indicated that it will perform its own resource reservation, the gatekeeper might decide that it will take that responsibility.

Another parameter of interest in both the ARQ and the ACF is the callModel parameter, which is optional in the ARQ and mandatory in the ACF. In the ARQ, the parameter indicates whether the endpoint wants to send call signaling directly to the other party, or whether it prefers that call signaling be passed via the gatekeeper. In the ACF, the parameter represents the gatekeeper's decision as to whether call signaling is to pass via the gatekeeper. The first scenario is known as direct call signaling. In that case, the calling endpoint sends an ARQ to its gatekeeper. After the ACF is returned from the gatekeeper, the endpoint sends call signaling directly to the remote endpoint. In the case of gatekeeper-routed call signaling, the calling endpoint's gatekeeper returns a callModel parameter in the ACF, indicating that call signaling should be sent via the gatekeeper. The endpoint then sends the call signaling to the gatekeeper and the gatekeeper forwards the call signaling to the remote endpoint.

Figure 4-7 shows an example of direct call signaling, and Figure 4-8 shows an example of gatekeeper-routed call signaling. Both diagrams assume that the two endpoints are connected to the same gatekeeper, which

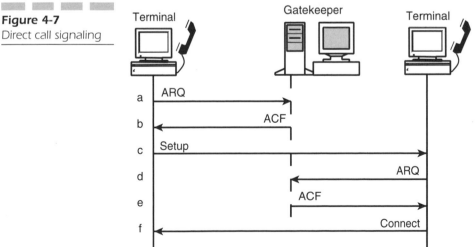

Figure 4-7
Direct call signaling

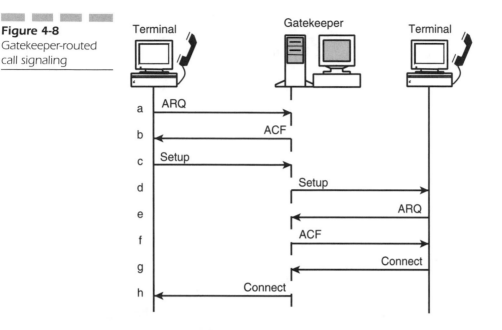

Figure 4-8
Gatekeeper-routed
call signaling

might or might not be the case. In the event that the endpoints are connected to different gatekeepers, then each gatekeeper determines independently whether it would like to be in the path of the call signaling. Therefore, in the case of two gatekeepers, the signaling may pass via none, one, or both gatekeepers.

Both Figure 4-7 and Figure 4-8 show Setup and Connect messages. These are part of call signaling (Q.931), as described later in this chapter. The Setup message is the first message in call establishment between endpoints, and the Connect message indicates that a call is accepted by the called endpoint. It is sent when the call is answered.

Of course, a gatekeeper may decide to deny a particular admission request. In such an event, it responds to the ARQ with an *AdmissionReject* (ARJ) message, which includes a reason for the denial. Possible reasons include a lack of available bandwidth, an incapability to translate a destination alias to a real address, or an endpoint not being registered.

Pregranted Admission In order to minimize call setup delay, a gatekeeper can provide an endpoint with admission in advance. When the endpoint registers with a gatekeeper, the RCF can include the parameter *preGrantedARQ*. This parameter can provide permission for the endpoint to make or receive a call without asking for explicit permission through the

ARQ/ACF mechanism. The parameter also indicates whether the endpoint should use direct call signaling or gatekeeper-routed call signaling.

Bandwidth Change

At any time during a call, an endpoint can request either an increase or a decrease in allocated bandwidth. An endpoint itself can change the bit rate of any given channel without approval from the gatekeeper, provided that the total bandwidth does not exceed the limit specified by the gatekeeper. Consequently, an endpoint is not required to request a reduction in bandwidth, though it should do so as a means of informing the gatekeeper of reduced bandwidth requirements, thereby allowing the gatekeeper to release certain resources for potential use in other calls. If a change is to occur in a call, however, and that change will lead to an aggregate bandwidth in excess of the limit specified by the gatekeeper, then the endpoint must first ask the gatekeeper for permission. The endpoint may increase the bandwidth only after confirmation from the gatekeeper.

A change in bandwidth is requested by the use of a *BandwidthRequest* (BRQ) message. The message indicates the new bandwidth requested and the call to which the bandwidth applies. The gatekeeper can either approve the new bandwidth with a *BandwidthConfirm* (BCF) message or deny the request with a *BandwithReject* (BRJ) message. If the request is denied, then the endpoint must live with whatever bandwidth has already been allocated, perhaps through the use of flow-control mechanisms.

Not only can an endpoint request a change in bandwidth, but the gatekeeper can also request that an endpoint change its bandwidth limit. For example, the gatekeeper might determine that network resource utilization is approaching capacity and it might order an endpoint to reduce its bandwidth usage. If the gatekeeper requests a reduction in bandwidth, then the endpoint must comply with the request and must respond with a BCF message.

Note that changes in bandwidth are closely tied to H.245 signaling. Recall that H.245 specifies the signaling that occurs between endpoints that are involved in a call and relates to the actual media streams that the endpoints share. Also recall that H.245 defines the use of logical channels that carry the media between the endpoints. If a change in bandwidth is to occur, then one or more logical channels will experience a direct impact. In fact, a reduction in bandwidth will require an existing logical channel to be closed and reopened with different parameter settings related to the new bandwidth limitation.

Figure 4-9 provides an example of the interaction between H.225.0 RAS signaling and H.245 control signaling for changes of bandwidth. In this example, an endpoint wants to increase the bandwidth used on a call. The endpoint first uses the BRQ message to request permission from its gatekeeper. Assuming that the request is acceptable, the gatekeeper responds with a BCF message. The endpoint closes the existing logical channel to the remote endpoint and opens a new logical channel, which has a new bit rate. The remote endpoint sees that a higher bit rate is required, but before confirming the new bit rate, it must ask its gatekeeper for permission. Therefore, the remote endpoint also sends a BRQ and receives a BCF. Only upon receipt of the BCF from its gatekeeper does the remote endpoint confirm the opening of the new logical channel. Note that Figure 4-9 shows that the two endpoints are controlled by the same gatekeeper. They could just as easily be controlled by different gatekeepers. H.245 signaling is described in greater detail later in this chapter.

Status

It is desirable for a gatekeeper to be informed of the status of an endpoint, simply to know whether the endpoint is still functioning and to obtain information regarding any call that the endpoint may currently have active.

Figure 4-9

Interaction between RAS bandwidth management and H.245 signaling

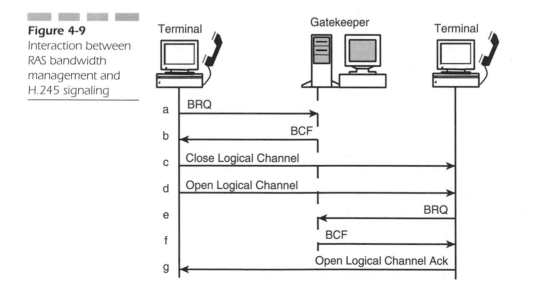

Information to be provided by an endpoint to a gatekeeper is carried in an *InformationRequestResponse* (IRR) message. This message carries information about the endpoint itself, plus information about one or more of the calls that the endpoint might currently have active. The per-call information includes such items as the call identifier, the call reference value, the call type (two-party or multiparty), the bandwidth in use, and RTP session information (the *canonical name* [CNAME], RTP address, RTCP address, and so on).

The gatekeeper can stimulate an endpoint to send an IRR in two ways. First, the gatekeeper can poll the endpoint by sending an *InformationRequest* (IRQ) message to the endpoint. The other way is for the gatekeeper to indicate within the ACF message that the endpoint should periodically send the information. This is done by including the optional parameter, irrFrequency, in an ACF message. This parameter indicates the frequency (in seconds) at which an endpoint in a call should send an IRR message to the gatekeeper. If the gatekeeper has pregranted admission in an RCF message, then the preGrantedACF information element will contain the irrFrequency parameter.

An IRR might or might not receive an acknowledgment. The gatekeeper and endpoint jointly determine whether an acknowledgement is to be sent. First, the gatekeeper indicates whether or not it is willing to send acknowledgments to IRR messages and does so by setting the BOOLEAN willRespondToIRR parameter in the ACF or RCF message. If that parameter has been set, then an endpoint can indicate that a particular IRR should receive a response, which it does by setting the BOOLEAN needResponse parameter in the IRR.

The response to an IRR can be positive or negative. A positive response takes the form of an *InfoRequestAck* (IACK) message and a negative response takes the form of an *InfoRequestNak* (INAK) message. The negative response should be returned only in cases when a gatekeeper has received an IRR message in error, such as from an unregistered endpoint.

Disengage

At the end of a call, the various parties involved stop transmitting media to each other and terminate the session. At this stage, each endpoint should send a *DisengageRequest* (DRQ) to its gatekeeper. The message contains, among other parameters, the call identifier, the call reference value, and a disengage reason. In the case that this is a normal call termination, the value of the reason will be normalDrop.

A gatekeeper that receives the DRQ request normally responds with a *DisengageConfirm* (DCF) message. The gatekeeper might respond with a *DisengageReject* (DRJ) message under some unusual circumstances. Such scenarios could include an endpoint requesting to disengage when the endpoint is not registered with that gatekeeper or if the endpoint wants to terminate a call that is handled by a different endpoint.

Situations might exist in which a gatekeeper, rather than an endpoint, decides to terminate a call. In such a case, the gatekeeper sends a DRQ message to the endpoint. Upon receiving such a request, the endpoint must stop transmitting media and must bring the session to a close using the appropriate H.245 control messages and Q.931 call-signaling messages. Once the session is terminated, the endpoint responds to the gatekeeper with a DCF message.

Resource Availability

H.225.0 defines two messages related to resource availability: the *ResourceAvailableIndicate* (RAI) message and the *ResourceAvailableConfirm* (RAC) message. The first message is sent from a gateway to a gatekeeper to inform the gatekeeper of the currently available call capacity and bandwidth for each protocol supported by the gateway. Of particular importance is the mandatory Boolean parameter, almostOutOfResources, that, when set, indicates that the gateway is almost out of resources. It is the manufacturer's choice as to how close to capacity the gateway needs to be before this parameter is set. Equally, any action taken as a result of the parameter being set is also at the discretion of the system manufacturer. The gatekeeper acknowledges the RAI message with the RAC message.

Service Control

H.323 version 4 includes the message *ServiceControlIndication* (SCI) and *ServiceControlResponse* (SCR). These two messages are used to initiate a session between H.323 entities for the purpose of enabling some advanced features. H.323 does not specify what those features might be; it simply provides the SCI and SCR as a mechanism to enable advanced features. A particular system vendor might exploit the SCI and SCR to implement a vendor-specific capability.

Request in Progress

Finally, H.225.0 defines the *RequestInProgress* (RIP) message. It is a means of indicating that a given request may take longer than expected. H.225.0 specifies recommended timeout periods for various messages. If an entity cannot respond to a request within the applicable timeout period, then it should send an RIP message indicating the expected delay before it expects to send a response to the original request. If the entity can subsequently respond before the delay period expires, then it should do so.

If an entity expecting a response to a RAS message receives an RIP message instead, then it should wait for a response or for the RIP delay time to expire before sending the original message again. Note that it is perfectly legal and sometimes necessary to retransmit a RAS message. After all, RAS messages are sent using unreliable transport, as indicated in Figure 4-4. Therefore, the possibility exists that RAS messages can become lost.

Call Signaling

We have already mentioned call signaling and the fact that it is the signaling used between endpoints to enable the establishment and teardown of calls. The messages used are Q.931 messages, as modified by recommendation H.225.0.

At first reading, such a statement can be somewhat confusing. After all, Q.931 is the layer 3 signaling protocol for an ISDN user-network interface, and the various messages are defined in that recommendation. The fact is that H.225.0 takes advantage of the protocol defined in Q.931 and simply reuses the messages, with some modifications necessary for use in the overall H.323 architecture. H.225.0 also uses one Q.932 message. This approach is a pretty smart way of doing things, since it has avoided a great deal of protocol development effort. H.225.0 does not use all the messages defined in Q.931, just those that are necessary for the support of call-signaling functions in the H.323 architecture. The messages used are specified in Table 4-2.

H.225.0 specifies the modifications that are applicable to Q.931 messages when used for call signaling in an H.323 network. In general, relatively few changes are made to the body of the Q.931 messages. For

Table 4-2

Messages used by H.225.0 for call-signaling functions in the H.323 architecture

Message	Function	Comment
Alerting	Sent by called endpoint to indicate that the called user is being alerted.	Must be supported.
Call-Proceeding	Optional interim response sent by called endpoint or gatekeeper prior to sending of connect message.	Should be sent if a called endpoint uses a gatekeeper.
Connect	Indication that the called user has accepted the call.	Must be supported.
Progress	Optional message sent by called endpoint prior to connect.	Can be used by a called gateway in the case of PSTN interworking.
Setup	Initial message used to begin call establishment.	Must be supported.
Setup Acknowledge	Optional response to setup message.	Can be forwarded from a gateway in the case of PSTN interworking.
Release Complete	Used to bring a call to an end.	Q.931 Release message is not used.
Information	An optional message that is used to send additional call-establishment information.	Can be used in overlap signaling or to enable proprietary features.
Notify	Optional message that may be used to provide information for presentation to the user.	Can be used by calling or called endpoint.
Status	Sent in response to status inquiry or in response to an unknown message.	An optional message that can be sent in either direction.
Status Inquiry	A message used to query the remote end as to the current call status from its perspective.	Can be sent in conjunction with RAS status procedures.
Facility (Q.932)	Used to redirect a call or to invoke a supplementary service.	Can be sent by either end at any time during a call and is useful for conveying information when no other message can be sent.

example, H.225.0 does not define any exotic new information elements to be included in particular messages. On the other hand, H.225.0 does not take each Q.931 message as is either. Instead, H.225.0 specifies a number of rules regarding the usage of information elements defined in Q.931. For example, H.225.0 does not permit the use of the Transit Network Selection information element, nor does it currently support the use of High-Layer Compatibility and Low-Layer Compatibility information elements. In fact, the changes specified by H.225.0 generally involve specifying that certain Q.931 mandatory information elements are either forbidden or optional when used in an H.323 network.

The big question, however, is how H.225.0's use of Q.931 messages enables the transfer of H.323-specific information. For example, a clear need exists to transport information regarding gatekeepers and H.245 addresses to be used for logical channels. These types of items are foreign to the usual ISDN environment of Q.931 signaling, which means that information must be added to the Q.931 messages. This task is performed through a rather clever use of the User-to-User information element. This information element in Q.931 ISDN messages is used to pass information transparently from one user to another using the D channel. The network does not interpret the information. In H.225.0, the User-to-User information element is used to convey all the extra information needed in an H.323 environment. Such information includes parameters such as a mandatory protocol identifier, endpoint aliases, H.245 addresses, and so on. The ASN.1 syntax specified in H.225.0 provides the exact syntax to be used in the User-to-User information element of the various messages.

The following sections of this chapter provide a brief overview of some of the most important call-signaling messages. For complete details, the user should refer to the Q.931 and H.225.0 recommendations.

Setup

The Setup message is the first call-signaling message sent from one endpoint to another to establish a call. If an endpoint uses the services of a gatekeeper, then this message will be sent after the endpoint has received admission from the gatekeeper providing permission for the endpoint to establish the call. The message must contain the Q.931 Protocol Discriminator, a Call ReferenceSetup, a Bearer Capability, and the User-to-User information element.

Although the Bearer Capability information element is mandatory, the concept of a bearer as used in the circuit-switched world does not map very well to an IP network. For example, no separate B-channel exists in IP, and the actual agreement between endpoints regarding the bandwidth requirements is done as part of H.245 signaling, where RTP information such as the payload type is exchanged. Consequently, many of the fields in the Bearer information element, as defined in Q.931, are not used in H.225.0. Of those fields that are used in H.225.0, many are used only when the call has originated from outside the H.323 network and has been received at a gateway, where the gateway performs a mapping from the signaling received to the appropriate H.225.0 messages.

A number of parameters are included within the mandatory User-to-User information element. These include the call identifier, the call type, a conference identifier, and information about the originating endpoint. Among the optional parameters, we might find a source alias, a destination alias, an H.245 address for subsequent H.245 messages, and a destination call-signaling address. Note that these parameters may supplement information already contained within the body of the Q.931 message. For example, the optional Calling Party Number information element in the body of the message will contain the E.164 number of the source, which the source might also use as an alias.

Call-Proceeding

Call-Proceeding is an optional message that can be sent by the recipient of a Setup message to indicate that the Setup message has been received and that call establishment procedures are under way.

As is the case for all call-signaling messages, the Protocol Discriminator, Call Reference, and Message Type elements are mandatory. The only other mandatory information element is User-to-User. Within that information element, the mandatory pieces of information, besides the Protocol Identifier, are the destination information (which indicates the type of endpoint) and the call identifier. Among the optional parameters is the H.245 address of the called end, which indicates where it would like H.245 signaling messages to be sent.

Alerting

The called endpoint sends this message to indicate that the called user is being alerted. Besides the Protocol Discriminator, Call Reference, and Message Type, only the User-to-User information element is mandatory. The optional Signal information element can be returned if the called endpoint wants to indicate a specific alerting tone to the calling party. The mandatory User-to-User information element in the Alerting message contains mostly the same parameters as those defined for the Call-Proceeding message.

Progress

The Progress message can be sent by a called gateway to indicate call progress, particularly in the case of interworking with a circuit-switched network. The Cause information element, though optional, is used to convey information to supplement any in-band tones or announcements that might be provided. The User-to-User information element contains the same set of parameters as defined for the Call-Proceeding message.

Connect

The Connect message is sent from the called entity to the calling entity to indicate that the called party has accepted the call. Although some of the messages from called party to calling party (such as Call-Proceeding and Alerting) are optional, the Connect message must be sent if the call is to be completed. The User-to-User information element contains the same set of mandatory parameters as defined for the Alerting message, with the addition of the Conference Identifier. This parameter is also used in a Setup message, and its use in the Connect message is to correlate this conference with that indicated in a Setup. Any H.245 address sent in a Connect message should match that sent in any earlier Call-Proceeding, Alerting, or Progress message.

Release Complete

The Release Complete message is used to conclude a call. Unlike ISDN, Release Complete is not sent as a response to a Release message. In fact, H.323 endpoints should never send a Release message. Instead, the Release Complete is all that is needed to bring the call to a close.

The Release Complete message contains the optional Cause information element. If the Cause element is not included, then the User-to-User information element will contain a Release reason. On the other hand, if no Release reason is specified in the User-to-User information element, then the Cause information element should be included. One way or the other, the party who releases the call should provide the other party with a reason for the release.

Facility

The Facility message is actually defined in ITU-T recommendation Q.932. In H.225.0, the message is used in situations where a call should be redirected. The Facility message can also be used to request certain supplementary services as described in recommendation H.450.1. Within the User-to-User information element, one important item is the mandatory reason parameter, which provides extra information to the recipient of the Facility message. An important use of the parameter is when a Setup message has been sent to an endpoint, and that endpoint's gatekeeper indicates to the endpoint that it wishes to be involved in the call signaling. In such a case, the called endpoint would return a Facility message, with the reason set to routeCallToGatekeeper. This would cause the recipient of the Facility message to release the call and attempt to set up the call again via the called endpoint's gatekeeper.

Interaction Between Call Signaling and H.245 Control Signaling

Before delving into some examples of call signaling, we should note a few points regarding the interplay between call signaling and H.245 control signaling. As mentioned previously, Q.931 call-signaling messages are used for the establishment and teardown of calls between endpoints, and H.245 messages are used for the negotiation and establishment of media streams between call participants. Obviously, the two are closely tied together.

Using Q.931, call establishment begins with a Setup message, and the call is considered established upon return of a Connect message from the called entity. In the interim, between the receipt of the Setup message and the sending of the Connect message, the called party might issue one or more progress-related messages (such as Call-Proceeding or Alerting). Given that H.245 messaging must also take place, at which point during call signaling should it occur? In other words, which of the call-signaling messages should be used as a trigger to begin the exchange of H.245 messages?

The answer to this issue is equipment dependent. H.323 requires that H.245 messages be exchanged, but it does not mandate that the exchange of these messages must occur at any specific point within the call-signaling process. Obviously, since the Setup message is the first call-signaling message to be sent, H.245 messages cannot be exchanged before the called endpoint has received a Setup message. H.245 messages can be exchanged at any point after that, however. The called endpoint can initiate the exchange of H.245 messages upon receipt of a Setup message or the calling endpoint can initiate the exchange upon receipt of a Call-Proceeding, Progress, or Alerting message. If an H.245 message exchange has not begun by the time a Connect message is sent, then it should begin immediately thereafter.

Call Scenarios

Now that we have described RAS signaling and call signaling, we will present a number of call scenarios to show how the different types of messages are used in the establishment and teardown of calls. Granted, H.245 messaging has not yet been described, and H.245 messaging is a prerequisite for media exchange between the endpoints. We can assume in the following scenarios, however, that a successful H.245 message exchange takes place immediately after a Connect message is sent.

Basic Call Without Gatekeepers

A gatekeeper is an optional entity in H.323. In the absence of gatekeepers, call signaling happens directly between endpoints. Figure 4-10 shows a typical call establishment and teardown in the absence of gatekeepers. The call is initiated with a Setup message, answered by a Connect message, and released with a Release Complete message. Either party can send the Release Complete message. Note that no acknowledgement is necessary for the Release Complete message.

A Basic Call with Gatekeepers and Direct Endpoint Call Signaling

In the event that the endpoints are registered with a gatekeeper, then prior to establishment of call signaling, an endpoint must first gain permission

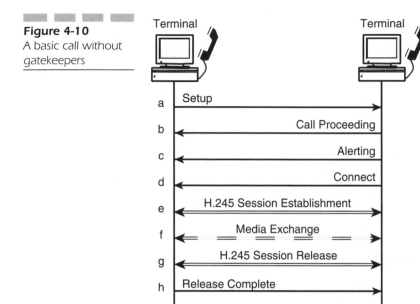

Figure 4-10

A basic call without gatekeepers

from its gatekeeper. The permission may be granted through the ARQ/ACF mechanism described earlier or may have been pregranted by the gatekeeper. Furthermore, at the end of the call, an endpoint must notify the gatekeeper that the call has been disconnected.

When gatekeepers are used in the network, a given endpoint may send call signaling directly to the other endpoint or it may send call signaling via its gatekeeper. Although an endpoint can indicate a preference, the gatekeeper chooses the path that the call signaling will take.

Figure 4-11 shows a scenario where two endpoints are controlled by different gatekeepers. Both gatekeepers have chosen that call signaling should be sent directly from endpoint to endpoint. We also assume that H.245 signaling goes from endpoint to endpoint. In the example of Figure 4-11, we assume that the gatekeeper has not pregranted access, which means that the originating endpoint must request access permission from its gatekeeper. It is only when permission is granted that the endpoint commences call signaling to the remote endpoint. Note that the remote endpoint immediately returns a Call-Proceeding message prior to the endpoint, asking its gatekeeper for admission to handle the call. Again, our example assumes that the gatekeeper has not pregranted admission as part of the registration of the endpoint with the gatekeeper. By sending back the Call-Proceeding message immediately, the endpoint signals to the call originator

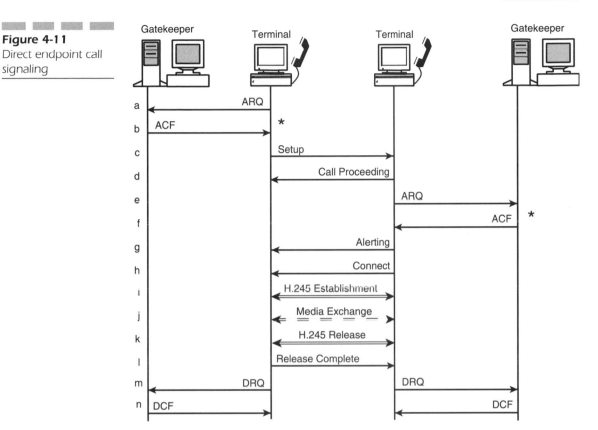

Figure 4-11
Direct endpoint call
signaling

* ARQ/ACF message exchange is required only if admission has not been pregranted.

that the Setup has been received successfully. Thus, the calling endpoint is
not left in the dark while the called endpoint requests permission from its
gatekeeper to handle the call.

In the example, it is assumed that H.245 message exchange begins
immediately after the Connect message. These messages could optionally
occur earlier.

A Basic Call with Gatekeeper/
Direct Routed Call Signaling

A gatekeeper might choose that call signaling be routed via the gatekeeper
rather than directly from endpoint to endpoint. In the case that the two

endpoints are connected to different gatekeepers, then it is very possible that one gatekeeper will choose to route the call signaling itself, while the other gatekeeper will not. Such a scenario is depicted in Figure 4-12, where the gatekeeper of the calling endpoint chooses to route the call signaling itself, while the gatekeeper of the called endpoint does not want to be in the path of call-signaling messages.

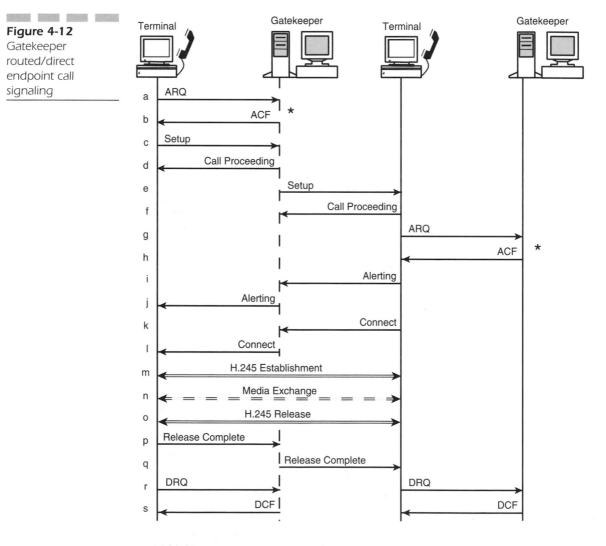

Figure 4-12
Gatekeeper routed/direct endpoint call signaling

★ ARQ/ACF message exchange is required only if admission has not been pregranted.

In the first ACF message, the gatekeeper of the calling endpoint indicates that call signaling should be sent to the gatekeeper. The gatekeeper includes a call-signaling address of itself in the ACF. Note that the endpoint might have received pregranted admission from the gatekeeper. In such a case, and if the gatekeeper wants to be in the path of call signaling, then the pregranted admission would specify that call signaling should be routed via the gatekeeper.

The endpoint sends a Setup message to the gatekeeper and the gatekeeper responds immediately with Call Proceeding. The gatekeeper then sends a Setup message to the destination endpoint using the call-signaling address of that endpoint.

If the called endpoint wants to handle the call, it first responds with a Call-Proceeding message and then sends an ARQ to its gatekeeper. In this case, the gatekeeper responds with an ACF indicating direct call signaling as opposed to gatekeeper-routed call signaling. Again, the ARQ/ACF procedure is not necessary if the endpoint has pregranted admission.

The called endpoint alerts the user and sends an Alert message to the gatekeeper of the calling endpoint. That gatekeeper forwards the Alerting message to the calling endpoint. Once the called party accepts the call, the called endpoint sends a Connect message, which follows the same path as the Alert message.

At this point in the example, H.245 messages are exchanged in order to set up the media streams between the endpoints. Note that an H.245 message exchange can occur earlier depending on the options implemented in the two endpoints.

In the example, the H.245 messages are routed directly from endpoint to endpoint. However, they could, like call-signaling messages, be sent via one or both gatekeepers, depending on whether each gatekeeper wants to be in the path of the H.245 messages. It is important to remember that the actual RTP streams will pass directly from endpoint to endpoint. This is regardless of whether a gatekeeper is in the path of call signaling and/or H.245 control signaling. It is not required that the media streams follow the same path as the H.245 control messages.

A Basic Call with Gatekeeper-Routed Call Signaling

The signaling process is slightly more complex when the gatekeeper controlling the called endpoint wants to be in the path of call-signaling messages. Such a scenario is shown in Figure 4-13, where both gatekeepers are

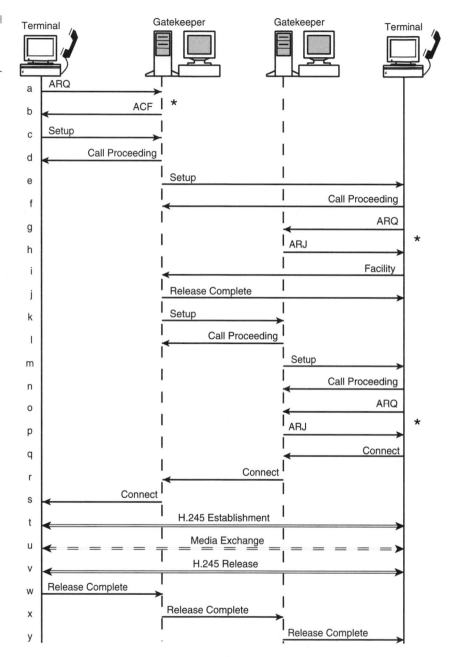

Figure 4-13
Gatekeeper-routed call signaling

* ARQ/ACF message exchange is required only if admission has not been pregranted.

involved in call signaling. Again, in the example, it is assumed that H.245 control messages are passed directly from endpoint to endpoint rather than via one or both gatekeepers.

Upon receipt of the Setup message from the calling endpoint's gatekeeper, the called endpoint requests permission from its own gatekeeper to handle the call (assuming that admission has not been pregranted). Since the gatekeeper wants to handle the call signaling, it returns an ARJ, with cause code of routeCallToGatekeeper. This action causes the called endpoint to send a Facility message to the gatekeeper of the calling endpoint. The Facility reason indicates that the call should be routed to the gatekeeper, and the call-signaling address of the gatekeeper is also included so that the calling gatekeeper knows where to send the call signaling. Note that a pregranted admission can include an indication that call signaling should be routed via a gatekeeper. If the called endpoint has pregranted admission with the requirement to route the call to the gatekeeper, then the endpoint will send the Facility message directly, without the need for the ARQ/ACF message exchange.

The gatekeeper on the originating side releases the call to the endpoint and attempts to establish the call to the gatekeeper on the terminating side instead. Upon receipt of the Setup message, the gatekeeper on the terminating side returns a Call-Proceeding message and forwards the Setup message to the called endpoint. The called endpoint immediately returns a Call-Proceeding message. If admission has not been pregranted, the called endpoint will again request permission to handle the call (ARQ).

Once the user is alerted, the Alerting message (not shown) is passed from the called endpoint, via the two gatekeepers to the calling endpoint. The same path is followed by the Connect message when the called user answers.

Once media have been exchanged, the Release Complete message is passed from one endpoint to the other via the two gatekeepers. Each endpoint then disengages using the DRQ/DCF message exchange with its gatekeeper (not shown).

Optional Called-Endpoint Signaling

The previous gatekeeper-routed example assumes that the gatekeeper of the calling endpoint knows the call-signaling address of the called endpoint.

As you can see in Figure 4-13, such a situation can lead to a rather cumbersome signaling exchange if the gatekeeper of the called endpoint wants to handle call signaling. Things do not need to be so complex, however.

Given that the two endpoints in our examples are controlled by different gatekeepers, it is very possible that the gatekeeper of the calling endpoint does not know the call-signaling address of the endpoint to be called. It may have only an alias for that endpoint, in which case it will have to issue an LRQ to the gatekeeper of the endpoint to be called, perhaps by multicast, in order to determine the call-signaling address of the endpoint.

If the gatekeeper of the terminating endpoint does not want to handle call signaling, then it will return the call-signaling address of the endpoint in the LCF. At that point, the call will proceed largely according to the example of Figure 4-12. On the other hand, if the gatekeeper wants to handle call signaling, it will return its own call-signaling address in the LRQ. In such an event, the sequence of events would be as depicted in Figure 4-14. Again, the Alerting message and the RAS Disengage sequence are not shown. As you can see, the call establishment from the beginning is in accordance with the desires of the remote gatekeeper. Therefore, the call establishment is much simpler and call redirection using the Facility message is not needed.

H.245 Control Signaling

We have mentioned the H.245 control protocol on several occasions in this chapter already. As described, H.245 is the protocol used between session participants to establish and control media streams. For a straightforward two-party voice call, this protocol ensures that participants agree on the media formats to be sent and received as well as the bandwidth requirements. For more complex multimedia calls, H.245 takes care of multiplexing multiple media streams for functions such as lip synchronization between audio and video.

You should note that H.245 is not responsible for carrying the actual media. For example, there is no such thing as an H.245 packet containing a sample of coded voice. That is the job of RTP. Instead, H.245 is a control protocol that manages the media sessions.

H.245 is not dedicated to use for VoIP. Rather it is a more generic protocol for the control of media streams and is designed to be used with a large number of applications. Consequently, the specification is a very large document. The intent in this book is not to describe H.245 in great

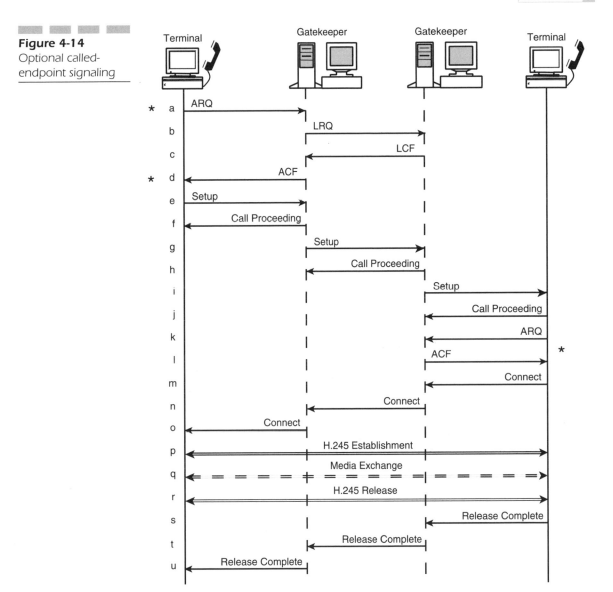

Figure 4-14
Optional called-endpoint signaling

* a ARQ
b LRQ
c LCF
* d ACF
e Setup
f Call Proceeding
g Setup
h Call Proceeding
i Setup
j Call Proceeding
k ARQ
l ACF *
m Connect
n Connect
o Connect
p H.245 Establishment
q Media Exchange
r H.245 Release
s Release Complete
t Release Complete
u Release Complete

* ARQ/ACF message exchange is required only if admission has not been pregranted.

detail, but rather to provide a general understanding of how it works in an H.323 environment and to provide descriptions of the most common messages and procedures.

H.245 Message Groupings

H.245 involves the sending of messages from one endpoint to another. These messages fall into four groups:[1]

- **Requests**   These are messages that require the recipient to perform some action and to send an immediate response.
- **Responses**   These messages are sent in reply to requests.
- **Commands**   These are messages that require the recipient to perform some action. No explicit response is required.
- **Indications**   These are messages that are of an informational nature only. The recipient of the message is not required to perform any specific action and no response is expected.

Only a selection of the most common H.245 messages is described in this chapter.

The Concept of Logical Channels

H.245 handles media streams through the use of logical channels. A logical channel is a unidirectional media path between two endpoints. Generally, in an IP environment, such a logical channel can be viewed as simply an IP address and port number supporting a particular type of media (such as G.729B-encoded voice). Each logical channel has a number that is specified by the sending entity.

Logical channels are unidirectional, from sender to receiver. Therefore, in a two-party conversation, two logical channels exist. This separation allows one terminal to potentially send voice in one format and receive voice in another format. Though logical channels are unidirectional, a perusal of the H.245 recommendation will find mention of a bidirectional channel. Such an entity is still two logical channels, though they are associated with each other.

To establish a media stream from one endpoint to another, the endpoint wanting to transmit information opens a logical channel. The endpoint does so by sending a message to the far end, indicating the logical channel num-

---

[1]The choices of names for these groups is a little unfortunate. At first glance, they have an uncanny resemblance to the service primitives request, indication, response, and confirm, as used in some SS7 applications. They are not the same thing.

ber and information about the media to be sent, such as RTP payload type. The message is called Open Logical Channel. If the far endpoint wants to receive the media, then it responds with a positive acknowledgement (Open Logical Channel Ack). Included in the acknowledgement will be an RTP port to which the media stream should be sent.

The H.245 messages are carried on the H.245 control channel. Each endpoint or gatekeeper shall establish one H.245 control channel for each call in which it is participating. The H.245 control channel is carried on a special logical channel, channel number 0. This is a special logical channel in that it is not opened and closed like other logical channels. Instead, logical channel 0 is considered permanently open as long as the endpoint is involved in a call.

H.245 Procedures

H.245 includes a number of procedures. For example, we have already briefly discussed the concept of opening logical channels. However, several other procedures are included in H.245, and some of them are briefly discussed here.

Before a logical channel can be opened, the sending endpoint should understand the capabilities of the receiving endpoint. There is no point in trying to send voice in a format that the receiving entity cannot handle. Consequently, H.245 provides a procedure known as Capabilities Exchange, which involves a set of messages that can be exchanged between endpoints to inform each other of their respective capabilities. Such capabilities include whether media can be sent or received (e.g. an endpoint might be able to receive video, but not send video), the media formats supported, and which formats can be supported simultaneously (particular combinations of voice and video).

In conference calls where two or more endpoints also have an MC, a potential problem can arise as to who is in control of the conference. Equally, imagine a situation where two endpoints have an urgent need to communicate with each other and both try to start a session with the other at the same time. To resolve these situations, H.245 has a procedure for master-slave determination.

Just because an endpoint can handle media in a particular format, it may have a preference for one format over another. Although capability exchange enables one endpoint to indicate to another which types of media it is able to handle, the Request Mode procedure enables an endpoint to rank formats in order of preference.

Capabilities Exchange Capabilities Exchange enables endpoints to share information regarding their receive and transmit capabilities. When a session is to be established between endpoints, the Capabilities Exchange procedure is started first. After all, there is no point in trying to open logical channels between two endpoints if they do not even speak the same language. Each endpoint should first determine what the other endpoint is able to support.

Receive capabilities indicate which media formats an endpoint is able to receive and process. Other endpoints should limit what they transmit to what the endpoint has indicated it can handle. Transmit capabilities indicate which media formats an endpoint is capable of sending. A remote endpoint can treat this information like a menu of possible media formats from which it can choose. The endpoint can subsequently indicate a preference for one format over another. If an endpoint does not indicate any transmit capabilities, then it must be assumed that the endpoint cannot and will not transmit any media.

An endpoint indicates its capabilities in a TerminalCapabilitySet message, which is a request message (and therefore requires a response). The message indicates a sequence number plus the types of audio and video formats that the endpoint can send and receive. The send and receive formats are indicated separately. Furthermore, the capabilities indicate what media can be handled simultaneously, such as a particular audio and video combination.

The TerminalCapabilitySet message is acknowledged by the TerminalCapabilitySetAck message. This message is empty except for a sequence number that matches the sequence number received in the original request. If the endpoint receiving the request finds a problem with the request, then it will respond with a TerminalCapabilitySetReject message indicating the reason for rejection. If no response is made within a timeout period, then the TerminalCapabilitySetRelease message is sent. This is an indication message and does not require a response.

As well as being able to unilaterally let another endpoint know about its capabilities, an endpoint can also request that another endpoint send information about its capabilities. This is done using the message SendTerminalCapabilitySet. Strictly speaking, this is a command message, and according to the definition of a command message, it does not require a specific response. However, the endpoint that receives the message should subsequently send a TerminalCapabilitySet message.

The SendTerminalCapabilitySet message contains two options. The sender can request that the far endpoint indicate all its capabilities. Alternatively, the sending endpoint can request confirmation about specific capabilities that the far endpoint has already described. This could be used, for example, in a situation where there has been a break in communication and one endpoint wants to make sure that it still has the latest information about the other.

Master-Slave Determination　　Situations can arise where two entities in a session want to initiate conflicting procedures. Consider, for example, a multiparty conference where several of the entities involved contain an MC and therefore could control the conference. For such situations, we need a method of deciding who is to be the master, which is done through the master-slave determination procedure.

The master-slave determination procedure involves two pieces of information at each entity. The first is a terminal type value, and the second is a random number between 1 and 16,777,215. The choice of the master is first made by comparing the values of terminal type. Whichever endpoint has the largest value of terminal type is automatically the master. H.323 specifies the values of terminal type that should be applied to different types of endpoints. For example, a terminal that does not have an MC has a terminal type value of 50. A gateway that does not contain an MC has a terminal type value of 60. An MCU that supports audio, video, and data conferencing has a terminal type value of 190 and an MCU that is currently managing a conference has the highest possible value, 240. If two entities have the same value of terminal type, then the random number is considered, and the terminal that chooses the highest random number value is the master.

In order to determine who is master, it is necessary that endpoints share their values of terminal type and random number. Such sharing is performed through the use of the Master-Slave Determination message and the Master-Slave Determination Ack message. Either entity may begin the master-slave determination procedure and does so by sending the Master-Slave Determination message containing a terminal type value and a random number. The receiving entity compares the value of terminal type with its own value and, if necessary, the received random number with another random number that the receiving entity chose itself. Based on these comparisons, the receiving entity makes the determination of who is master and returns a Master-Slave Determination Ack message.

The Master-Slave Determination Ack message indicates the sender's view of the other party's ranking. If it contains a master indication, it means that the receiver of the message is the master. If it contains a slave indication, it means that the receiver of the message is the slave.

Establishing and Releasing Media Streams   Media streams between participants are carried on one or more logical channels. These channels need to be established before media can be exchanged and need to be closed at the end of a call.

Opening Unidirectional Logical Channels   A logical channel is opened by sending an Open Logical Channel request message. This message contains a mandatory parameter called forwardLogicalChannelParameters, which relates to the media to be sent in the forward direction (from the endpoint issuing this request). The parameter contains information regarding the type of data to be sent (e.g., G.728 audio), an RTP session ID, an RTP payload type, an indication as to whether silence suppression is to be used, a description of redundancy encoding (if used), and so on.

If the recipient of the message wants to accept the media to be sent, then it will return an OpenLogicalChannelAck message containing at least the same logical channel number as received in the request and a transport address to which the media stream should be sent.

The process is shown in Figure 4-15, where a single logical channel is opened. The initiator of the logical channel has not indicated that it wants to receive media. We are not saying, however, that the initiator is unwilling to receive media. If the called endpoint also wants to send media, then after the first logical channel is opened, it can choose to open a logical channel in the reverse direction.

If an endpoint does not want to accept a request to open a channel, then it can reply with the OpenLogicalChannelReject message. This message contains the logical channel number received in the request and a reason for the rejection, such as an incapability to handle the proposed media format.

Opening Bidirectional Logical Channels   The foregoing discussion regarding the opening of unidirectional logical channels mentions the fact that two endpoints can send media to each other by opening a logical channel in each direction. This requires four messages: two requests and two responses. If, however, an endpoint wants to send media and is expecting to receive media in return, as would be the case for a typical voice conversation, then it can attempt to establish a bidirectional logical channel. This procedure is a means of establishing two logical channels, one in each direction, in a slightly more efficient manner.

The process is depicted in Figure 4-16 and it begins with the sending of an Open Logical Channel request message. In addition to forward logical channel parameters, however, the message also contains reverse logical

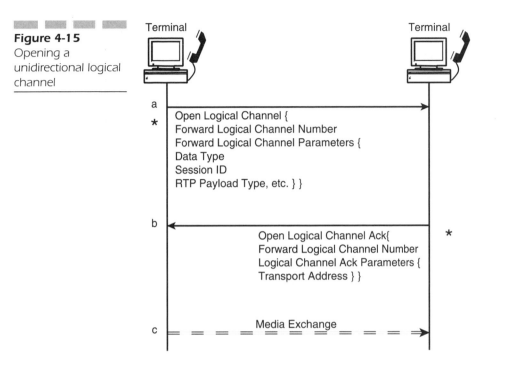

Figure 4-15
Opening a
unidirectional logical
channel

Terminal   Terminal

a
*
Open Logical Channel {
Forward Logical Channel Number
Forward Logical Channel Parameters {
Data Type
Session ID
RTP Payload Type, etc. } }

b
Open Logical Channel Ack{
Forward Logical Channel Number   *
Logical Channel Ack Parameters {
Transport Address } }

Media Exchange
c

* Not all parameters are shown.

channel parameters that describe the type of media that the endpoint is willing to receive and where that media should be sent.

Upon receiving the request, the far endpoint can send an Open Logical Channel Ack message containing the same logical channel number for the forward logical channel, a logical channel number for the reverse logical channel, and descriptions related to the media formats that it is willing to send. These media formats should be chosen from the options originally received in the request, thereby ensuring that the called end will only send media that the calling end supports.

Upon receipt of the Open Logical Channel Ack, the originating endpoint responds with an Open Logical Channel Confirm message to indicate that all is well. Media flow can now begin in both directions.

Closing Logical Channels and Ending a Session Closing a logical channel involves the sending of a CloseLogicalChannel message. In the case of a successful closure, the far end should send the response message Close LogicalChannelAck.

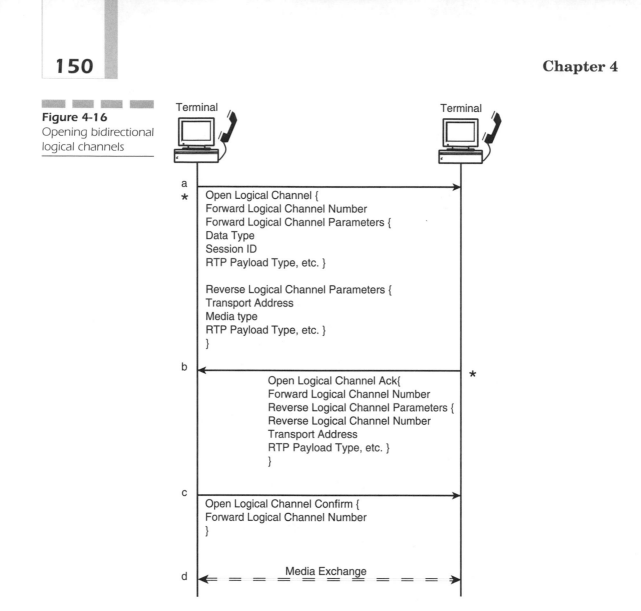

Figure 4-16
Opening bidirectional
logical channels

a
* Open Logical Channel {
   Forward Logical Channel Number
   Forward Logical Channel Parameters {
   Data Type
   Session ID
   RTP Payload Type, etc. }

   Reverse Logical Channel Parameters {
   Transport Address
   Media type
   RTP Payload Type, etc. }
   }

b
   Open Logical Channel Ack{   *
   Forward Logical Channel Number
   Reverse Logical Channel Parameters {
   Reverse Logical Channel Number
   Transport Address
   RTP Payload Type, etc. }
   }

c
   Open Logical Channel Confirm {
   Forward Logical Channel Number
   }

d   Media Exchange

* Not all parameters are shown.

In general, a logical channel can be closed only by the entity that created it in the first place. For example, in the case of a unidirectional channel, only the sending entity can close the channel. The receiving endpoint in a unidirectional channel can, however, request the sending endpoint to close the channel. It does so by sending the RequestChannelClose message, indicating the channel that the endpoint would like to have closed. If the sending entity is willing to grant the request, then it responds with a positive acknowledgment and then proceeds to close the channel.

One exception should be noted regarding the closure of a bidirectional channel. When an endpoint closes a forward logical channel, it also closes the reverse logical channel when the two are part of a bidirectional channel.

Once all logical channels in a session are closed, then the session itself is terminated when an endpoint sends an EndSession command message. The receiving endpoint responds with an EndSession command message. Once an entity has sent this message, it must not send any more H.245 messages related to the session.

Figure 4-17 provides an example of channel closure where one entity requests that the other close the channel, followed by the channel closure.

Fast Connect Procedure

Having reviewed RAS, Q.931, and H.245 signaling, it is possible to construct a call flow that incorporates more of the signaling required for call

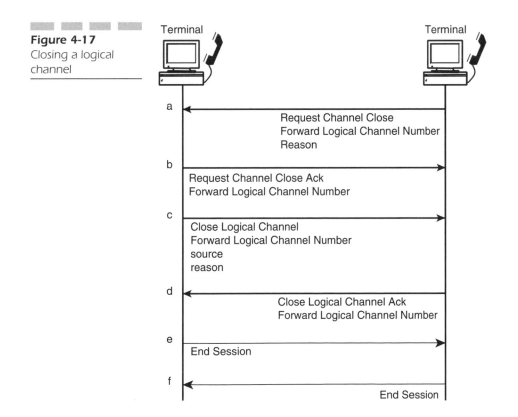

Figure 4-17
Closing a logical
channel

Terminal

Terminal

a — Request Channel Close
Forward Logical Channel Number
Reason

b — Request Channel Close Ack
Forward Logical Channel Number

c — Close Logical Channel
Forward Logical Channel Number
source
reason

d — Close Logical Channel Ack
Forward Logical Channel Number

e — End Session

f — End Session

establishment between two endpoints. Figure 4-18 shows an example of such a call establishment and can be considered a more complete version of the example depicted in Figure 4-11. One can see that many messages are required even for such a simple call (hence the term "slow start" in the caption). Furthermore, when we consider the fact that the figure does not show a Capabilities Exchange or possible master-slave determination, it becomes clear that even more messaging can be required.

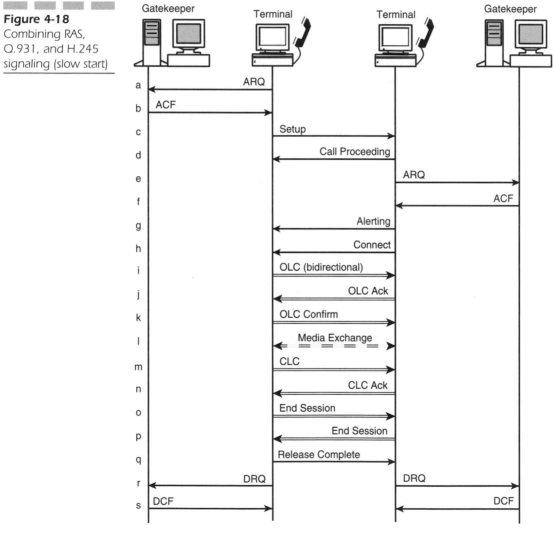

Figure 4-18
Combining RAS, Q.931, and H.245 signaling (slow start)

OLC = Open Logical Channel   CLC = Close Logical Channel

If one imagines a scenario with gatekeeper-routed call signaling and the need for the Facility message, as shown in Figure 4-13, plus the possibility of gatekeeper-routed H.245 control signaling, it is clear that the number of messages can be greater still. Call establishment becomes cumbersome and is certainly not conducive to fast call setup. Considering the fact that fast call setup is a requirement for carrier-grade operation, the situation is lamentable. Thankfully, the designers of H.323 recognized these issues and defined a procedure that reduces signaling overhead and speeds things up considerably. Appropriately, the procedure is known as the Fast-Connect procedure.

The Fast-Connect procedure involves setting up media streams as quickly as possible. To achieve this goal, the Setup message can contain a faststart element within the User-to-User information element. The faststart element is actually one or more OpenLogicalChannel request messages containing all the information that would normally be contained in such a request. The faststart element also includes reverse logical channel parameters if the calling endpoint expects to receive media from the called endpoint.

If the called endpoint also supports the procedure, then it can return a faststart element in one of the Call-Proceeding, Alerting, Progress, or Connect messages. That faststart element is basically another OpenLogicalChannel message, which appears like a request to open a bidirectional logical channel. The included choices of media formats to send and receive are chosen from those offered in the faststart element of the incoming Setup message. The calling endpoint has effectively offered the called endpoint a number of choices for forward and reverse logical channels, and the called endpoint has indicated those choices that it prefers. The logical channels are now considered open as if they had been opened according to the procedures of H.245. The use of faststart is depicted in Figure 4-19.

Note that the faststart element from the called party to the calling party may be sent in any message up to and including the Connect message. If the element has not been included in any of the messages, then the calling endpoint shall assume that the called endpoint either cannot or does not want to support Fast-Connect. In such a case, the standard H.245 methods must be used.

The use of the Fast-Connect procedure means that H.245 information is carried within the Q.931 messages and there is no separate H.245 control channel. Therefore, bringing a call to a conclusion is also faster. The call is released simply by the sending of the Q.931 Release Complete message. When used with the Fast-Connect procedure, this has the effect of closing

Figure 4-19
Fast connect
procedure

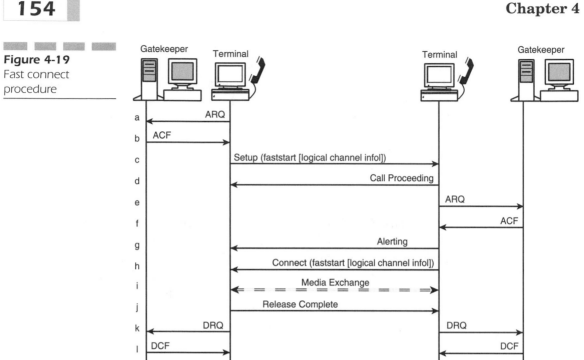

Figure 4-19
Fast connect
procedure

all the logical channels associated with the call and is equivalent to using the procedures of H.245 to close the logical channels.

H.245 Message Encapsulation

In addition to the Fast-Connect procedure, H.323 also enables H.245 messages to be encapsulated with Q.931 messages as octet strings. An endpoint that wants to encapsulate such messages with Q.931 messages sets the element h245Tunneling to true in the first Q.931 message that it sends and keeps the element set to true in every other message that it sends as long as it wants to support such encapsulation. For a calling endpoint, this means that the h245Tunneling element should be set to true in the Setup message. Note that a called endpoint can use encapsulation only if the calling endpoint indicates support for encapsulation in the Setup message. The encapsulated data is actually contained within the h245Control element, which, like all H.323 information, is included in the User-to-User information element of the Q.931 message.

Note that a conflict exists between encapsulation and faststart. After all, the faststart element itself is a type of encapsulated H.245 message. Consequently, a calling endpoint should not include a faststart element and an encapsulated H.245 message in a Q.931 Setup message. If, however, the endpoint wants to support encapsulation with later commands, then it should still set the h245Tunneling element to true such that the called endpoint knows that the calling endpoint can support encapsulation.

As is the case for the fast-connect procedure, the use of encapsulation means that there is no separate H.245 control channel, since H.245 messages are carried on the call-signaling channel. What happens if an endpoint needs to send an H.245 message at a time when it does not have a particular need to send a call-signaling message? This situation can be handled in one of two ways. First, an endpoint can send a Facility message with an encapsulated H.245 message. Alternatively, either endpoint can choose to change to a separate H.245 control channel.

An endpoint indicates a desire to switch to a separate H.245 control channel by populating the h245address element in a Q.931 message. In such a case, any existing logical channels established by Fast-Connect or through encapsulation shall be inherited by the H.245 control channel as if they had been established through normal H.245 procedures.

If the endpoint wants to change to a separate H.245 control channel and has no need to send a Q.931 message, then it has no means of indicating an H.245 address to the other endpoint. Again, the Facility message comes to the rescue, because that message can be sent at any time. All that is necessary is to send a Facility message with a FacilityReason of startH245 and with the h245address element populated with the endpoint's own H.245 address. This effectively establishes a separate H.245 control channel. Any subsequent Q.931 messages must have the h245Tunneling element set to false as anything else would be inconsistent with a separate H.245 control channel.

Conference Calls

The previous descriptions and examples have focused on the signaling necessary for the establishment and release of two-party calls. Situations will occur, however, when several parties are involved in a multipoint conference. In fact, H.323 defines the MC for the purpose of managing multipoint conferences. A conference call may be established in two ways. One is a prearranged conference, where participants call in to a separate MCU, which controls the conference. The other is when a two-party call needs to be expanded to include other parties. This is known as an ad hoc conference.

Pre-arranged Conference

In a pre-arranged conference, the individual participants are connected to the conference through the establishment of a call with the MCU. In other words, connections to the MCU are established through Q.931 call signaling. The sharing of media streams can be achieved in a centralized manner via the MCU or in a decentralized multicast manner. In the centralized scenario, both signaling and media pass via the MCU. In the decentralized scenario, only the signaling passes via the MCU, while the media streams are multicast from each endpoint to each other endpoint. The MCU determines the mode of the conference.

The MCU specifies the conference mode through the use of the H.245 command message Communication Mode command. This command specifies all the sessions in the conference (and for each session, various session data plus a unicast or multicast address). Note that the command only specifies the transmit requirements of each endpoint, not the receive requirements. The reason is because the receive requirements will be specified in separate Open Logical Channel commands that will be subsequently sent from the MC to the various endpoints.

Ad Hoc Conference

Quite often, the need will arise to expand an existing two-party call to a conference involving three or more participants. However, such an occurrence might not be foreseen at the beginning of the two-party call. Given that most PBXs and even residential phones support multiparty and the expansion of a call to include other parties, the same capability must be supported in a VoIP environment.

Many H.323 endpoints or gatekeepers include an MC function. It is not a callable entity like an MCU used for a call-in conference, but it can be utilized in the event that a two-party call needs to be expanded to three or more parties. In order for a two-party call to be expanded to an ad hoc conference, one of the endpoints must contain an MC or one of the endpoints' gatekeepers must contain an MC (in which case gatekeeper-routed call signaling must be used).

If these requirements are met, then the two-party call can be established as a conference call between two entities. This requires that the Setup message from endpoint 1 to endpoint 2 contain a unique *Conference ID* (CID).

If endpoint 2 wants to accept the call, then the Connect message that is returned should contain the same value of CID. H.245 capability exchange and master-slave determination procedures are performed, and the call is established as shown in Figure 4-20.

If, subsequently, the endpoint that is the master decides to invite another party (such as endpoint 3) to join the conference, then it sends a Setup message to endpoint 3, with the CID value of the conference and the conference goal parameter set to "invite." Assuming that endpoint 3 wants to join the conference, then it includes the same CID in the Connect message. Then capability exchange and master-slave determination is performed between endpoint 1 and endpoint 3. Finally, the MC at endpoint 1 sends the H.245 indication Multipoint Conference to each of the participants.

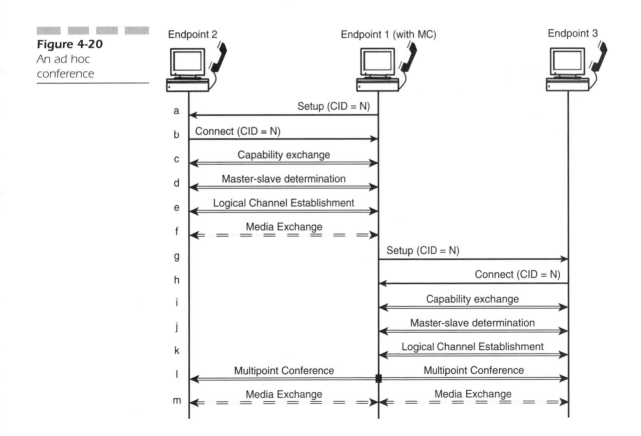

Figure 4-20
An ad hoc conference

The Decomposed Gateway

Previous descriptions have considered a gateway to be a single monolithic entity. In some cases, a gateway is a single entity. In fact, early versions of H.323 considered a gateway to be a single unit. As we shall see in Chapter 6, "Media Gateway Control and the Softswitch Architecture," however, a gateway can be comprised of several individual elements. An H.323 gateway can comprise a *media gateway* (MG) that handles media streams, a *media gateway controller* (MGC) that performs call control, and a *signaling gateway* (SG) that interfaces with external signaling networks such as the SS7 network. These components can reside on separate physical machines.

Depending on the specific application, an H.323 gateway might contain all the components in a single machine, it might contain two components in a single machine, with an interface to the third component on a separate machine, or the gateway might be completely disaggregated into three separate machines. For example, a gateway that interfaces towards PBXs will contain both a media gateway and signaling functionality (since PBXs usually employ channel-associated signaling that is carried on the same transmission facilities as the voice traffic), and it might interface with a separate MGC. A trunking gateway that interfaces with an SS7-based network might be decomposed into separate MGs, SGs, and MGC. Figure 4-21 depicts a situation where all three functions reside on separate machines.

Figure 4-21
A decomposed
gateway

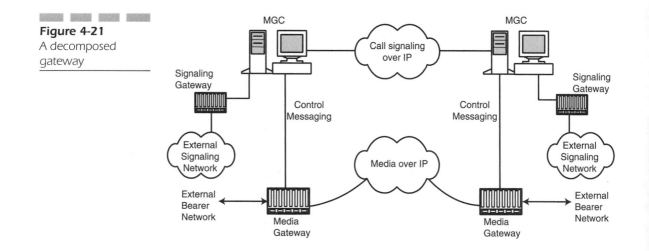

Note that the function of the MGC is separate from that of a gatekeeper. Although a gatekeeper is an optional entity, the functions of an MGC must be performed, whether on a separate machine or as part of a monolithic gateway machine. Standard protocols exist for the MGC-MG interface and for the MGC-SG interface. Those protocols are described in later chapters.

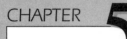

The Session Initiation Protocol (SIP)

Many consider the *Session Initiation Protocol* (SIP) a powerful alternative to H.323. They claim that SIP is a more flexible solution, simpler than H.323, easier to implement, better suited to the support of intelligent user devices, and better suited to the implementation of advanced features. These factors are of major importance to any equipment vendor or network operator. Simplicity means that products and advanced services can be developed faster and made available to subscribers sooner. The features themselves mean that operators are better able to attract and retain customers, and they also offer the potential for new revenue streams.

SIP is designed to be a part of the overall *Internet Engineering Task Force* (IETF) multimedia data and control architecture. As such, SIP is used in conjunction with several other IETF protocols such as the *Session Description Protocol* (SDP), the *Real-Time Streaming Protocol* (RTSP), and the *Session Announcement Protocol* (SAP).

The Popularity of SIP

Originally developed in the *Multiparty Multimedia Session Control* (MMUSIC) working group of the IETF, great interest in SIP led to the creation of a separate SIP working group. SIP was defined in *Request for Comments* (RFC) 2543, which has seen a number of revisions as the specification has proceeded from proposed standard status towards draft standard status. The number of proposed enhancements to SIP as it proceeds along the standards track is one indication of the attention it has received. Many developers are including SIP in their products without waiting for those enhancements, simply because SIP is seen as the way of the future for *Voice over IP* (VoIP) signaling. In fact, many believe that SIP, in conjunction with a *Media Gateway Control Protocol* (MGCP or MEGACO) as described in Chapter 6, "Media Gateway Control and the Softswitch Architecture," will be the dominant VoIP signaling architecture of the future.

A new SIP RFC has been produced with RFC number 3261. This specification includes a number of enhancements over the RFC 2543 specification. The descriptions provided in this chapter are based on the latest proposed enhancements to SIP as of May 2002 that are included within RFC 3261 or published as separate RFCs.

The development of SIP and its implementation by system developers has involved a number of events known as "bake-offs" or *SIP Interoperability Tests* (SIPit).[1] During these events, various vendors come together and test their products against each other to ensure that they have implemented the specification correctly and to ensure compatibility with other implementations. To give an indication of SIP's popularity, Figure 5-1 shows how many companies have been involved in the first eight events. Although attendance at the most recent events has diminished slightly, it is clear that significant industry support exists for SIP. In fact, although the number of attending companies might have peaked, the number of attending individuals continues to increase.

The SIP Architecture

SIP is a signaling protocol that handles the setup, modification, and teardown of multimedia sessions. SIP, in combination with other protocols, is used to describe the session characteristics to potential session participants. Although strictly speaking, SIP is written such that the media to be used in a given session could use any transport protocol, the media will

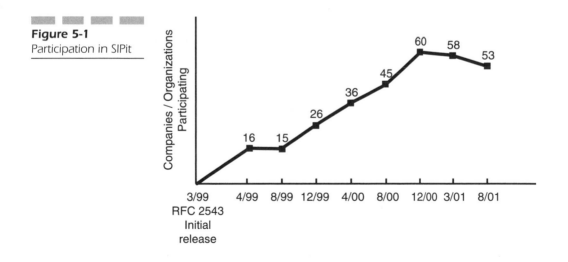

Figure 5-1
Participation in SIPit

---

[1]The first seven events were known as bake-offs. The Pillsbury company claimed, however, that the term "bake-off" is proprietary to that company. Therefore, the term was dropped for subsequent interoperability-testing events.

commonly be exchanged using the *Real-Time Transport Protocol* (RTP) as the transport protocol.

It is likely that SIP messages will pass through some of the same physical facilities as the media to be exchanged. SIP signaling should be considered separately from the media itself, however. Figure 5-2 shows the logical separation between signaling and session data. This separation is important, because the signaling may pass via one or more proxy or redirect servers while the media stream itself takes a more direct path. This approach can be considered somewhat analogous to the separation of signaling and media already described for H.323.

SIP Network Entities

SIP defines two basic classes of network entities: clients and servers. Strictly speaking, a client (also known as a user agent client) is an application program that sends SIP requests. A server is an entity that responds to those requests. Thus, SIP is a client-server protocol. VoIP calls using SIP are originated by a client and terminated at a server. A client may be found within a user's device, which could be a PC with a headset attachment or a SIP phone, for example. Clients may also be found within the same platform as a server. For example, SIP enables the use of proxies, which act as both clients and servers.

Four different types of servers exist: proxy server, redirect server, user-agent server, and registrar. A proxy server acts in a similar way to a proxy server used for web access from a corporate *local area network* (LAN). Clients send requests to the proxy and the proxy either handles those requests itself or forwards them on to other servers, perhaps after performing some translation. To those other servers, it appears as though the message is coming from the proxy rather than some entity hidden behind it.

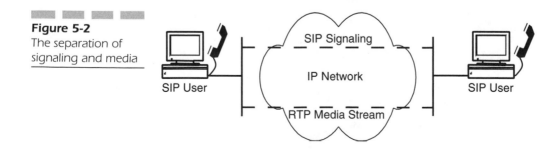

Figure 5-2
The separation of
signaling and media

SIP Signaling

IP Network

RTP Media Stream

SIP User

SIP User

Given that a proxy both receives requests and sends requests, it incorporates both server and client functionality.

Figure 5-3 shows an example of the operation of a proxy server. It does not take much imagination to realize how this type of functionality can be used for call forwarding, time-of-day routing, or follow-me services. For example, if the message from the caller to Collins in Figure 5-3 is an invitation to participate in a call, the net effect is that the call is forwarded to Collins at home. Of course, it is necessary that the proxy be aware that Collins happens to be at home instead of at work.

A redirect server is a server that accepts SIP requests, maps the destination address to zero or more new addresses, and returns the new address to the originator of the request. Thereafter, the originator of the request can send requests directly to the address(es) returned by the redirect server. A redirect server does not initiate any SIP requests of its own. Figure 5-4 shows an example of the operation of a redirect server. This can be another means of providing the call-forwarding/follow-me service that can be provided by a proxy server. The difference is that, in the case of a redirect server, the originating client does the actual forwarding of the call. The redirect server simply provides the information necessary to enable the originating client to correctly route the call, after which the redirect server is no longer involved.

A user-agent server accepts SIP requests and contacts the user. A response from the user to the user-agent server results in a SIP response on behalf of the user. In reality, a SIP device (such as a SIP-enabled telephone) will function as both a user-agent client and a user-agent server. Acting as a user-agent client, it is able to initiate SIP requests. Acting as a user-agent

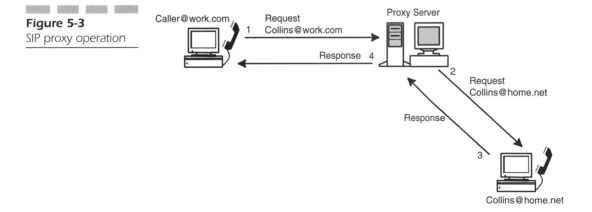

Figure 5-3
SIP proxy operation

Figure 5-4
SIP redirect operation

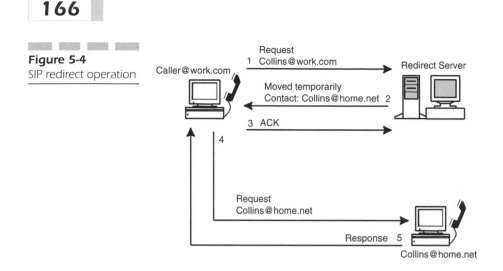

Figure 5-4
SIP redirect operation

server, it is able to receive and respond to SIP requests. In practical terms, this means that it is able to initiate and receive calls. This enables SIP (a client-server protocol) to be used for peer-to-peer communication.

A registrar is a server that accepts SIP REGISTER requests. SIP includes the concept of user registration, whereby a user indicates to the network that he or she is available at a particular address. The use of registration enables SIP to support personal mobility. For example, a user could have several SIP devices, one of which might be the user's office PC. When the user logs onto the office network, then the PC would issue a SIP REGISTER request to the appropriate registrar. Thereafter, calls can be routed to the user's office PC. When the user leaves the office, he or she might register at a different device, such as a home PC or SIP phone. A new registration would then be performed, enabling the user to be reached at the new device.

Typically, a registrar will be combined with a proxy or redirect server. Given that practical implementations will involve the combining of a user-agent client and a user-agent server as well as the combining of registrars with either proxy servers or redirection servers, a real network may well involve only user agents and redirection or proxy servers.

SIP Call Establishment

At a high level, a SIP call establishment is simple, as shown in Figure 5-5. The process starts with a SIP INVITE message, which is used from calling party to called party. The message invites the called party to participate in

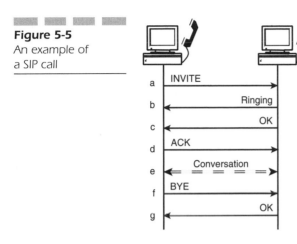

Figure 5-5
An example of
a SIP call

a session, that is, a call. A number of interim responses may be made to the INVITE prior to the called party accepting the call. For example, the caller might be informed that the call is queued and/or that the called party is being alerted (the phone is ringing). Subsequently, the called party answers the calls, which generates an OK response back to the caller. The calling client acknowledges that the called party has answered by issuing an ACK message. At this point, media are exchanged. This media will most often be regular speech, but could also be other media, such as video. Finally, one of the parties hangs up, which causes a BYE message to be sent. The party receiving the BYE message sends OK to confirm receipt of the message. At that point, the call is over.

SIP Advantages over Other Signaling Protocols

Given that the call establishment and release is pretty straightforward, one might wonder what is so special about SIP. After all, any call-signaling protocol must have a means for one party to call another, a means to indicate acceptance of the call, and a means to release the call. SIP performs these actions and does a little more, but unlike many traditional telephony protocols, SIP offers a great deal of flexibility. For example, SIP does not care what type of media is to be exchanged during a session or the type of transport to be used for the media. In fact, SIP itself can be carried over several different transport protocols. In other words, SIP provides more flexibility than is found in typical telecommunications protocols. That flexibility can be exploited to enable custom services and features.

SIP messages can include a number of optional fields, which can contain user-specified information. This approach enables users to share nonstandard user-specific information, which enables users and devices to make intelligent call-handling decisions. For example, a SIP INVITE can include a subject field. A person who receives an INVITE might decide to accept or reject the call depending on who is calling and what the subject happens to be. One can imagine a situation where an INVITE contains a text string such as "I know you are there. Please answer the phone." If the call is rejected, the response might contain a text string such as "Stop calling me!"

Imagine another scenario where a call is directed to a user who is currently not available. Obviously, the SIP response will indicate that the user is unavailable (no big surprise). The response could, however, include an indication that the user expects to be available again at 4 P.M. In such a case, the calling terminal could do two things. First, the terminal could tell the calling party that the called user expects to be available at 4 P.M. Second, at 4 P.M., it could ask the caller if he/she wants to make the call again and, if so, the terminal can automatically set up the call. Such a scenario is shown in Figure 5-6. This is a clever and simple mechanism in the IP world of providing a call-completion service, similar to the *Call Completion to Busy Subscriber* (CCBS) of the telephony world. It is also a lot more effi-

Figure 5-6

An example of a SIP-enabled service

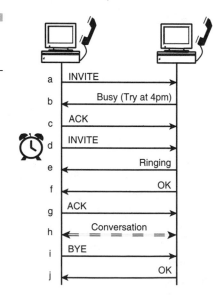

a	INVITE
b	Busy (Try at 4pm)
c	ACK
d	INVITE
e	Ringing
f	OK
g	ACK
h	Conversation
i	BYE
j	OK

cient than calling back every 15 minutes in the hope that the person will finally answer or paying the telephone company to keep trying to connect the call.

The foregoing are some simple examples of capabilities that SIP offers. Since SIP provides many pieces of information for inclusion in messages, and because additional, nonstandard information can also be included, the opportunity exists to offer numerous intelligent features to subscribers. Moreover, the control of those features is placed in the hands of the customer. No longer does a customer need to subscribe to a particular feature and have that feature actually controlled by the network operator.

The following sections of this chapter are devoted to describing the fundamentals of SIP. From this description, the reader should gain a good understanding of how SIP works, how it compares to other standards (such as H.323), and how it can be used to provide new and advanced features to users.

Overview of SIP Messaging Syntax

As mentioned, SIP is a signaling protocol. As such, it has a particular syntax. In the case of SIP, the syntax is text-based, using the *International Organization for Standardization* (ISO 10646) character set, and has a similar look and feel to the *Hypertext Transfer Protocol* (HTTP). One advantage of this approach is that programs designed for the parsing of HTTP can be adapted relatively easily for use with SIP. One obvious disadvantage, compared to a binary encoding, is that the messages themselves consume more bandwidth.

SIP messages are either requests from a client to a server or responses (which are also known as status messages) from a server to a client. Each message, whether a request or a response, contains a start-line, possibly followed by headers and a message body:

```
message = start-line
            *message-header
            CRLF
            [message-body]
```

Given that SIP defines only request and status messages, the start-line is

```
Start-line = request-line | status-line
```

The request-line specifies the type of request being issued, while the response line indicates the success or failure of a given request. In the case of failure, the status line indicates the type of failure or the reason for failure.

Message headers provide additional information regarding the request or response. Information that is obviously required includes the message originator and the intended message recipient. Message headers also offer the means for carrying additional information, however. For example, the Retry-after header indicates when a request should be attempted again. In the example of Figure 5-6, the Retry-after header would convey the fact that the called user expects to be available again at 4 P.M. Another example of a useful header is the Subject header, which enables a caller to indicate the reason for a call.

The message body normally describes the type of session to be established, including a description of the media to be exchanged. Thus, for a given call, the message body might indicate that the caller wants to communicate using voice, coded according to G.711 A-law. Note, however, that SIP does not define the structure or content of the message body. This structure and content is described using a different protocol. The most common structure for the message body is the *Session Description Protocol* (SDP) described later in this chapter. The message body could, however, contain information coded according to another standard. In fact, the message body could contain multiple parts, with each part coded according to a different format or structure.

This capability is used in some situations to carry an *ISDN User Part* (ISUP) message in binary format, thereby enabling SIP to carry ISUP information. Carrying an ISUP message within a SIP message body would be used, for example, in a scenario where the SIP network is interworking with the *Public Switched Telephone Network* (PSTN) using *Signaling System 7* (SS7) signaling. SIP does not greatly care about what the message body happens to say. It is concerned only with making sure that the message body is carried from one party to the other. It is at the two ends that the message body is examined. The message body can be considered to be within a sealed envelope. SIP carries it from one end to the other, but does not look inside the envelope.

SIP Requests

A SIP request begins with a request-line, which comprises a method token, a Request-URI, and an indication of the SIP version. The method token identifies the specific request being issued and the Request-URI is the address of the entity to which the request is being sent. The three components of the request-line are separated by spaces, and a CRLF character terminates the line itself. Thus, the syntax is as follows:

```
request-line = method SP Request-URI SP SIP-Version CRLF
```

RFC 3261 defines six different methods (and hence six different types of requests): INVITE, ACK, OPTIONS, BYE, CANCEL, and REGISTER. A number of extensions to SIP specify additional methods, such as INFO, REFER, and UPDATE.

The INVITE method is used to initiate a session. For a simple call between two parties, the INVITE is used to initiate the call, with the message including information regarding the calling and called parties as well as the type of media to be exchanged. For those who are familiar with ISUP, the INVITE can be considered akin to the *Initial Address Message* (IAM). As well as a means for initiating a simple two-party call, INVITE can also be used to initiate a multiparty conference call.

Once it has received a final response to an INVITE, the client that initiated the INVITE sends an ACK. The ACK method is used as a confirmation that the final response has been received. For example, if the response to an INVITE indicates that the called user is busy and the call cannot be completed at that time, then the calling client will send an ACK. On the other hand, if the response to the INVITE indicates that the called user is being alerted or that the call is being forwarded, then the client does not send ACK, because such responses are not considered final.

The BYE method terminates a session. The method can be issued by either the calling or called party and is used when the party in question hangs up.

The OPTIONS method queries a server as to its capabilities. This method could be used, for example, to determine whether a called user agent can support a particular type of media or how a called user agent would respond if sent an INVITE. In such a case, the response might indicate that the user can support certain types of media or perhaps that the user is currently unavailable.

The CANCEL method is used to terminate a pending request. For example, CANCEL could be used to terminate a session where an INVITE has been sent, but a final response has not yet been received. As discussed later in this chapter, it is possible to initiate a parallel search to multiple destinations (if a user is registered at multiple locations). In such a case, if a final response has been received from one of the destinations, then CANCEL can be used to terminate the pending requests at the other destinations.

The REGISTER method is used by a user-agent client to log in and register its address with a SIP server, thereby letting the registrar know the address at which the user can be reached. The user-agent client might register with a local SIP server at startup, with a known registrar server whose address is configured within the user agent, or by multicasting to the "all-SIP-servers" multicast address (224.0.1.175).

A client can register with multiple servers and a given client can have multiple registrations at a single server. Such a situation would occur in the event that a user has logged in at several terminals or devices. If a user has multiple active registrations, then calls to the user may be sent to all registered destinations. This capability can enable a "one-number" service where a user publishes just a single number, but when that number is called, the user's office phone, home office phone, and wireless phone all ring. Figure 5-7 shows an example of how this would work and also shows the usage of the CANCEL message in such a scenario.

Figure 5-7

Multiple registrations enabling a one-number service

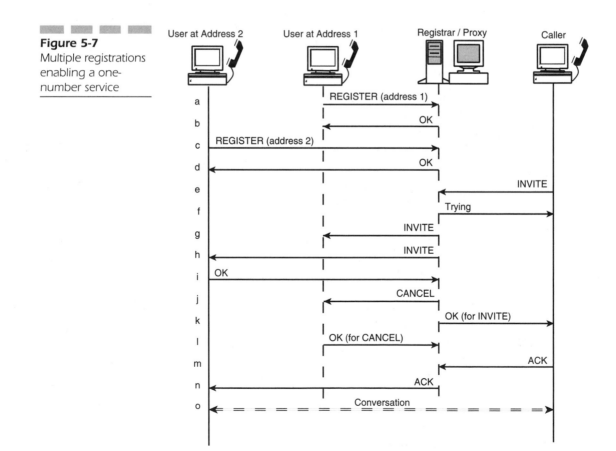

The SIP INFO method is specified in RFC 2976. This method is a means for transferring information during an ongoing session, that is, in the middle of a call. Examples of its use could include

- The transfer of *Dual-Tone Multifrequency* (DTMF) digits
- The transfer of account balance information
- The transfer of midcall signaling information generated in another network (such as the PSTN) and passed to the IP network via a gateway

Through the use of the INFO method, application-layer information could be transferred in the middle of a call, thereby affording user-agent servers, user-agent clients, or proxies the opportunity to act upon the information. For example, a prepaid subscriber could be informed in the middle of a call if the subscriber's prepaid account is nearing zero.

The SIP INFO method is a flexible tool and, as such, does not define the information that may or may not be sent using the INFO method. Typically, the INFO method would convey the user information within a message body, with the content and structure of the message body dependent upon the type of information to be conveyed. For example, an indication that the user's prepaid account balance is near zero could be conveyed in a message body in the form of text.

SIP Responses

The start line of a SIP response is a status line. This contains a status code, which is a three-digit number indicating the outcome of the request. The start line will also contain a reason phrase, which provides a textual description of the outcome. The reason code will be interpreted and acted upon by the client software, while the reason-phrase could be presented to the human user to aid in understanding the response. The syntax of the status line is

```
status-line = SIP version SP status code SP reason-phrase CRLF
```

Status codes defined in SIP have values between 100 and 699, with the first digit of the reason code indicating the class of response as follows. Thus, all status codes between 100 and 199 belong to the same class. The different classes are as follows:

- **1XX**   Provisional (for example, 181 indicates that the call is being forwarded).

- **2XX** Success (only code 200 is defined, which means that the request has been understood and has been performed. In the case of an INVITE, a response of 200 is used to indicate that the called party has accepted the call).

- **3XX** Redirection (such as 302, which indicates that the called party is not available at the addresses used in the request and that the request should be reissued to a new address included with the response).

- **4XX** Request Failure (such as 401, which indicates that the client is not authorized to make the request).

- **5XX** Server Failure (such as 505, which indicates that the server does not support the SIP version specified in the request).

- **6XX** Global Failure (such as 604, indicating that the called user does not exist anywhere).

All responses, except for 1XX responses, are considered final and should be acknowledged with an ACK message if the original message happened to be an INVITE. SIP specifies that 1XX responses are provisional and do not need to be acknowledged. Table 5-1 gives a complete listing of the status codes defined in the latest revision of SIP (as of this writing), along with proposed reason phrases.

Table 5-1

SIP status codes

Category	Status code	Reason phrase
Provisional	100	Trying
	180	Ringing
	181	Call is being forwarded
	182	Queued
	183	Session in progress
Success	200	OK
Redirection	300	Multiple choices
	301	Moved permanently
	302	Moved temporarily
	305	Use proxy
	380	Alternative service
Request Failure	400	Bad request
	401	Unauthorized
	402	Payment required
	403	Forbidden
	404	Not found
	405	Method not allowed
	406	Not acceptable

Table 5-1 cont.

SIP status codes

Category	Status code	Reason phrase
	407	Proxy authentication required
	408	Request timeout
	410	Gone
	413	Request entity too large
	414	Request-URI too long
	415	Unsupported media type
	416	Unsupported URI scheme
	420	Bad extension
	421	Extension required
	423	Interval too brief
	480	Temporarily not available
	481	Call leg/transaction does not exist
	482	Loop detected
	483	Too many hops
	484	Address incomplete
	485	Ambiguous
	486	Busy here
	487	Request terminated
	488	Not acceptable here
	491	Request pending
	493	Undecipherable
Server Failure	500	Server internal error
	501	Not implemented
	502	Bad gateway
	503	Service unavailable
	504	Server timeout
	505	SIP version not supported
	513	Message too large
Global-Failure	600	Busy everywhere
	603	Decline
	604	Does not exist anywhere
	606	Not acceptable

SIP Addressing

As with any signaling protocol, requests and responses are sent to particular addresses. In SIP, these addresses are known as SIP *Uniform Resource Indicators* (URIs). These addresses take the form of user@host, which is very similar to an e-mail address. In many cases, a given user's SIP address can, in fact, be guessed from the user's e-mail address. Although the two may look similar, however, they are different. Whereas an e-mail address uses a mailto *Uniform Resource Locator* (URL), such as mailto:collins @home.net, a SIP URI has the syntax sip:collins@home.net.

SIP deals with multimedia sessions, which could include voice. Furthermore, SIP networks need to interwork with traditional circuit-switched networks. For these reasons, SIP enables the user portion of the SIP address to be a telephone number. Thus, we could have a SIP address such as sip:3344556789@telco.net. In a given network, such a SIP address could be used to cause media to be routed to a gateway that interfaces with the traditional telephone company. A SIP URI may also be supplemented by a number of parameters that provide more information. For example, to clearly indicate that a call is to a telephone number, the URI could be supplemented by the term *user*=phone. In such a case, the URI would have the format sip:3344556789@telco.net;user=phone.

Message Headers

SIP includes a number of different message headers. These headers are items of information included in a request or response to provide further information about the message or to enable appropriate handling of the message. In that respect, message headers are akin to the message parameters or information elements included in any typical signaling protocol such as ISUP, Q.931, and so on. For example, the To: header in an INVITE message indicates the called party, and the From: header indicates the calling party. The intention in this chapter is to describe the usage of headers and to give an understanding of the most common and useful headers. For complete definitions, the user is referred to RFC 3261.

Some headers make sense only in certain requests or responses, and in some cases, the presence of a given header will depend on the context. Moreover, the presence of a given header in a response might be reasonable only if the response is issued to a specific request. Tables 5-2 and 5-3 list the various headers and specify the application of each header to each request or response. Table 5-2 provides a mapping between requests and headers. The information in Table 5-2 should be read as follows:

- M means mandatory.
- M* means that the header field should be present in the request, but a receiver should be prepared to process the request even if the header is absent.
- O means optional.
- T means that the header should be included in the request if a stream-based transport (such as TCP) is used.

- C means conditional; the presence of the header depends on the context of the message.
- N/A means not applicable; the header should not be sent in the request.

Table 5-2

A mapping between requests and headers

Header	ACK	BYE	CAN	INV	OPT	REG
Accept	N/A	O	N/A	O	M*	O
Accept-Encoding	N/A	O	N/A	O	O	O
Accept-Language	N/A	O	N/A	O	O	O
Alert-Info	N/A	N/A	N/A	O	N/A	N/A
Allow	N/A	O	N/A	O	O	O
Authentication-Info	N/A	N/A	N/A	N/A	N/A	N/A
Authorization	O	O	O	O	O	O
Call-ID	M	M	M	M	M	M
Call-Info	N/A	N/A	N/A	O	O	O
Contact	O	N/A	N/A	O	O	O
Content-Disposition	O	O	N/A	O	O	O
Content-Encoding	O	O	N/A	O	O	O
Content-Language	O	O	N/A	O	O	O
Content-Length	T	T	T	T	T	T
Content-Type*	C	C	C	C	C	C
CSeq	M	M	M	M	M	M
Date	O	O	O	O	O	O
Error-Info	N/A	N/A	N/A	N/A	N/A	N/A
Expires	N/A	N/A	N/A	O	N/A	O
From	M	M	M	M	M	M
In-Reply-to	N/A	N/A	N/A	O	N/A	N/A
Max-Forwards	M	M	M	M	M	M
Min-Expires	N/A	N/A	N/A	N/A	N/A	N/A
MIME-Version	O	O	N/A	O	O	O
Organization	N/A	N/A	N/A	O	O	O

Table 5-2 cont.

A mapping
between requests
and headers

Header	ACK	BYE	CAN	INV	OPT	REG
Priority	N/A	N/A	N/A	O	N/A	N/A
Proxy-Authenticate	N/A	N/A	N/A	N/A	N/A	N/A
Proxy-Authorization	O	O	N/A	O	O	O
Proxy-Require	N/A	O	N/A	O	O	O
Record-Route	O	O	O	O	O	O
Reply-to	N/A	N/A	N/A	O	N/A	N/A
Require	N/A	C	N/A	C	C	C
Retry-After	N/A	N/A	N/A	N/A	N/A	N/A
Route	C	C	C	C	C	C
Server	N/A	N/A	N/A	N/A	N/A	N/A
Subject	N/A	N/A	N/A	O	N/A	N/A
Supported	N/A	O	O	M*	O	O
Timestamp	O	O	O	O	O	O
To	M	M	M	M	M	M
Unsupported	N/A	N/A	N/A	N/A	N/A	N/A
User-Agent	O	O	O	O	O	O
Via	M	M	M	M	M	M
Warning	N/A	N/A	N/A	N/A	N/A	N/A
WWW-Authenticate	N/A	N/A	N/A	N/A	N/A	N/A

*The Content-Type header field must be included if the message contains a message body. Otherwise, the header can be omitted.

Table 5-3 provides a mapping between headers and responses. It should be noted that the inclusion of a particular header in a response is dependent upon both the status code of the response and the request that led to the response. The Status Code column indicates the status codes for which a given header may be included in the response. In some cases, a given header may be used only with certain status codes. In other cases, a given header may be used with all status codes. The six columns corresponding to the six request methods of RFC 3261 indicate whether a given header may be used in a response to that particular type of request. For example, the Allow header can be used with a 200 or a 405 status code; however, it can only be used with a 200 response if the response is sent as a result of an OPTIONS request.

The information in Table 5-3 should be read in the same way as the information in Table 5-2, with the addition that the indication (c) means that the value of a header is copied from the request to the response. A value of N/A in the Status Code column means that the header should not be included in any response.

Table 5-3

A mapping between headers and responses

Header	Status Code	ACK	BYE	CAN	INV	OPT	REG
Accept	2XX	N/A	N/A	N/A	O	M*	O
Accept	415	N/A	C	N/A	C	C	C
Accept-Encoding	2XX	N/A	N/A	N/A	O	M*	O
Accept-Encoding	415	N/A	C	N/A	C	C	C
Accept-Language	2XX	N/A	N/A	N/A	O	M*	O
Accept-Language	415	N/A	C	N/A	C	C	C
Alert-Info	180	N/A	N/A	N/A	O	N/A	N/A
Allow	2XX	N/A	O	N/A	M*	M*	O
Allow	405	N/A	M	N/A	M	M	M
Allow	All except 2XX, 415	N/A	O	N/A	O	O	O
Authentication-Info	2XX	N/A	O	N/A	O	O	O
Authorization	N/A	N/A	N/A	N/A	N/A	N/A	N/A
Call-ID	All (c)	M	M	M	M	M	M
Call-Info	All	N/A	N/A	N/A	O	O	O
Contact	1XX	N/A	N/A	N/A	O	N/A	N/A
Contact	2XX	N/A	N/A	N/A	M	O	O
Contact	3XX, 485	N/A	O	N/A	O	O	O
Content-Disposition	All	O	O	N/A	O	O	O
Content-Encoding	All	O	O	N/A	O	O	O
Content-Language	All	O	O	N/A	O	O	O
Content-Length	All	T	T	T	T	T	T
Content-Type*	All	C	C	N/A	C	C	C
CSeq	All (c)	M	M	M	M	M	M

Table 5-3 cont.

A mapping
between headers
and responses

Header	Status Code	ACK	BYE	CAN	INV	OPT	REG
Date	All	O	O	O	O	O	O
Error-Info	300–699	N/A	O	O	O	O	O
Expires	All	N/A	N/A	N/A	O	N/A	O
From	All (c)	M	M	M	M	M	M
In-Reply-to	N/A	N/A	N/A	N/A	O	N/A	N/A
Max-Forwards	N/A	N/A	N/A	N/A	N/A	N/A	N/A
Min-Expires	423	N/A	N/A	N/A	N/A	N/A	M
MIME-Version	All	O	O	N/A	O	O	O
Organization	All	N/A	N/A	N/A	O	O	O
Priority	N/A	N/A	N/A	N/A	N/A	N/A	N/A
Proxy-Authenticate	401	N/A	O	O	O	O	O
Proxy-Authenticate	407	N/A	M	N/A	M	M	M
Proxy-Authorization	N/A	N/A	N/A	N/A	N/A	N/A	N/A
Proxy-Require	N/A	N/A	N/A	N/A	N/A	N/A	N/A
Record-Route	2XX, 18X	N/A	O	O	O	O	N/A
Require	All	N/A	C	N/A	C	C	C
Retry-After	404, 413, 480, 486, 500, 503, 600, 603	N/A	O	O	O	O	O
Route	N/A	N/A	N/A	N/A	N/A	N/A	N/A
Server	All	N/A	O	O	O	O	O
Subject	N/A	N/A	N/A	N/A	N/A	N/A	N/A
Supported	2XX	N/A	O	O	M*	M*	O
Timestamp	All	O	O	O	O	O	O
To	All (c)	M	M	M	M	M	M
Unsupported	420	N/A	M	N/A	M	M	M
User-Agent	All	O	O	O	O	O	O
Via	All (c)	M	M	M	M	M	M
Warning	All	N/A	O	O	O	O	O
WWW-Authenticate	401	N/A	M	N/A	M	M	M
WWW-Authenticate	407	N/A	O	N/A	O	O	O

*The Content-Type header field must be included if the response contains a message body. Otherwise, the header can be omitted.

General Headers Some headers can be used in both requests and responses, and they are known as *general headers*. Such headers contain basic information needed for the handling of requests and responses. Examples include the To: header field, which indicates the recipient of the request, the From: header field, which indicates the originator of the request and the Call-ID: header field, which uniquely identifies a specific invitation to a session.

One of the most useful general headers is the Contact: header, which provides a URI for use in future communication regarding a particular session. In a SIP INVITE, the Contact: header might be different than the From: header. This means, for example, that the initiator of a SIP session need not be a participant in the session. An example of such usage would be a case where a multiparty session is organized and initiated by an administrator who does not participate in the session itself. Used in responses, the Contact: header is useful for directing further requests (such as ACK) directly to the called user when the original request passed through one or more proxies. This header can also be used to indicate a more appropriate address if an INVITE issued to a given URI failed to reach the user. An example of such usage would be with a 302 response (moved temporarily) where the Contact: header in the 302 response gives the current URI of the user.

Request Headers Request headers apply only to SIP requests. They are used to provide additional information to the server regarding the request itself or regarding the client. Examples include the Subject: header field, which can be used to provide a textual description of the topic of the session, and the Priority: header field, which is used to indicate the urgency of the request (emergency, urgent, normal, or nonurgent).

Response Headers Response header fields apply only to response (status) messages. These header fields are used to provide further information about the response that cannot be included in the status line. Examples of response header fields include the Unsupported: header field used to identify those features not supported by the server and the Retry-After header field, which can indicate when a called user will be available if the user is currently busy or unavailable.

Entity Headers In SIP, the message body is used to contain information about the session or information to be presented to the user. In the case of information regarding a session, the session description is most often specified according to SDP, in order to indicate information such as the RTP payload type as well as an address and port to which media should be sent. The purpose of the entity headers is to indicate the type and format of information included in the message body, so that the appropriate application can be called upon to act on the information within the message body. The entity header fields are Content-Length, Content-Type, Content-Encoding, Content-Disposition, and Content-Language.

The Content-Length header field specifies the length of the message body in octets. The Content-Type header field indicates the media type of the message body. For VoIP, this header will usually indicate SDP, in which case the header field will appear as Content-Type: application/sdp.

The Content-Encoding header field is used to indicate any additional codings that have been applied to the message body and hence which decoding actions need to be taken by the recipient in order to obtain the media type indicated by the Content-Type header field. For example, it is perfectly legal to compress the content of the message body. In such a case, the Content-Encoding header field would be used to pass information related to the compression scheme used.

The Content-Disposition header field describes how the message body or, for multipart messages, a message body part should be interpreted. If the message body, or part of a message body, describes the characteristics of a session, then the associated Content-Disposition field would have the value *session*. Non-session data can be carried in a message body, however. For example, a message body might contain an image of the caller as part of an advanced Caller-ID service. In such a case, the Content-Disposition would have the value *icon*. The value *render* is used to indicate that the message body part should be displayed to the user. This value would apply, for example, if the sender included some text that he or she wanted to present to the other party. The value *alert* indicates that the message body part contains information such as an audio clip that would alert the user to the receipt of a request. For example, a given SIP user could choose his or her own personal ring tone that is played to a called party, so someone could tell who is calling based on their personal ring tone.

Examples of SIP Message Sequences

Building upon the information provided already in this chapter, a number of examples are provided to demonstrate how SIP messages might appear in various situations. We do not explain every possible option in these examples, just some of the most common occurrences. For details of all options available, the reader is referred to RFC 3261.

Registration

The very first request issued by a client is likely to be REGISTER, because this is the request that provides the server with an address at which the user can be reached for SIP sessions. The REGISTER request is somewhat similar to the Registration Request between a terminal and a gatekeeper in H.323.

Figure 5-8 shows an example registration scenario. In this scenario, Collins has logged in to host station1.work.com. This causes a REGISTER request to be sent to the local registrar. The Via: header field contains the path taken by the request so far, which requires that the originating client insert its own address in this field. Note the format of the Via: header and in particular the fact that it specifies the transport being used. The default is the *User Datagram Protocol* (UDP). Note also that the Via: header

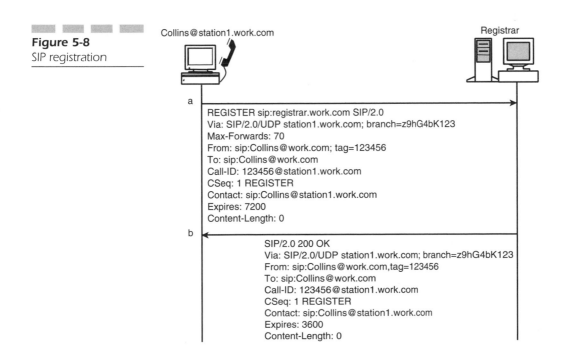

Figure 5-8
SIP registration

Collins@station1.work.com

Registrar

a
```
REGISTER sip:registrar.work.com SIP/2.0
Via: SIP/2.0/UDP station1.work.com; branch=z9hG4bK123
Max-Forwards: 70
From: sip:Collins@work.com; tag=123456
To: sip:Collins@work.com
Call-ID: 123456@station1.work.com
CSeq: 1 REGISTER
Contact: sip:Collins@station1.work.com
Expires: 7200
Content-Length: 0
```

b
```
SIP/2.0 200 OK
Via: SIP/2.0/UDP station1.work.com; branch=z9hG4bK123
From: sip:Collins@work.com,tag=123456
To: sip:Collins@work.com
Call-ID: 123456@station1.work.com
CSeq: 1 REGISTER
Contact: sip:Collins@station1.work.com
Expires: 3600
Content-Length: 0
```

contains a *branch parameter*. The purpose of the branch parameter is discussed later when we describe the operation of proxy servers.

The From: header field indicates the address of the individual who has initiated the registration. The To: header field indicates the "address of record" of the user being registered and is the address that the registrar will store for that user. In other words, this is the address at which the user wants to be contactable. The From: and To: fields will be identical if a user is registering himself. The two fields need not be identical, however, which means that one individual can perform a registration on behalf of another. Note that the To: header field is not used to contain the address of the registrar. That address is indicated in the first line of the request. Note that the From: field in our example contains a tag parameter, which is a pseudorandom value generated at the sender of the request and is used for identification purposes.

The Call-ID: header field is set by the originating client. All REGISTER requests for an individual client should use the same value of Call-ID. In order to avoid the possibility that different clients might choose the same Call-ID, the recommended syntax for the Call-ID is local-id@host, thereby making the Call-ID host specific.

In Figure 5-8, Collins has inserted a Contact: header field that indicates that future SIP messages should be routed to sip:collins@station1 .work.com. In other words, messages that are addressed to sip:Collins @work.com should be forwarded to sip:Collins@station1.work.com.

The REGISTER request does not contain a message body, since the message is not used to describe a session of any kind. Therefore, the Content-Length: field is set to 0.

The response to the REGISTER request is positive, as indicted by the 200 (OK) in the response line. Note that the content of the Via: header field is copied from request to response. The content of the CSeq: (or command sequence) header field is also copied from request to response. The CSeq: header field indicates the method used in the request and an integer number. Consecutive requests for the same Call-ID must use contiguous increasing integer numbers. Although not as critical for REGISTER requests, the use of the CSeq is very important in other requests where different requests can be issued for the same Call-ID, in which case the CSeq is used to avoid ambiguity. A response to any request must use the same value of CSeq as used in the request itself.

Through the use of the Expires: header, Collins has requested that the registration be effective for 2 hours. The registrar has chosen to override this aspect of the request and has limited the duration of the request to 1 hour. A registrar can change the length of time for which a given registration is effective, but if a registrar chooses to do so, it will normally set a lower registration interval than that requested. It will not set a higher interval. The Expires: header may be specified as a duration in seconds or as a specific data and time. If it is specified as a number of seconds, the maximum value is such that a registration may be active for up to approximately 136 years, which should be long enough for anyone. If a REGISTER request contains an Expires: header with a value that is too short (the registration is to be active for too short a time), then the REGISTER request will be rejected with the status code 423 (Interval too brief). The response will include the Min-Expires: header field, which specifies the minimum registration interval that the registrar will accept.

Having registered as shown in Figure 5-8, Collins could subsequently register at another terminal. In such a case, both registrations would become active and subsequent invitations destined for Collins would be routed to both terminals. Of course, before registering at the second location, Collins could cancel the existing registration. This would be done by sending another REGISTER request for the same address of record and Contact and specifying a registration interval of 0. In this case, the REGISTER request would be identical to the first REGISTER request, with the exceptions that the CSeq integer is incremented and the Expires header field contains the value 0. If Collins wanted to cancel all existing registrations, then he would send a REGISTER message with the Expires header field set to 0 and the Contact header field populated with the wildcard character *. The Expires header value of 0 indicates the cancellation of a Registration and the Contact: header value of * indicates that the request is applicable to all contact information for Collins.

Invitation

The INVITE request is the most fundamental and important SIP request, as it is the request used to initiate a session (i.e., establish a call). Figure 5-9 shows a two-party call, where Collins initiates the session. The INVITE request is issued to manager@station2.work.com, as can be seen from the Request-URI. The reason why the address is very specific to the host where the callee is located is because this example's signaling flow does not traverse any proxies. Under normal circumstances, one might expect that the message would be addressed to sip:manager@work.com and would be forwarded by a proxy to sip:manager@station2.work.com. Since our example does not include a proxy to forward the INVITE to the specific host, the INVITE itself is addressed to the specific host. In our example, the To:

Figure 5-9
SIP establishment of a
two-party call

Daniel<sip:Collins@work.com>

Boss<sip:Manager@station2.work.com>

a
```
INVITE sip:manager@station2.work.com SIP/2.0
Via: SIP/2.0/UDP station1.work.com; branch=z9hG4bK123
Max-Forwards: 70
From: Daniel<sip:Collins@work.com>; tag=44551
Contact: sip:Collins@station1.work.com
To: Boss<sip:Manager@station2.work.com>
Call-ID: 123456@station1.work.com
CSeq: 1 INVITE
Subject: Vacation
Content-Length: xxx
Content-Type: application/sdp
Content-Disposition: session
(message body)
```

b
```
SIP/2.0 180 Ringing
Via: SIP/2.0/UDP station1.work.com; branch=z9hG4bK123
From: Daniel<sip:Collins@work.com>;tag=44551
To: Boss<sip:Manager@station2.work.com>; tag=11222
Contact: sip:manager@station2.work.com
Call-ID: 123456@station1.work.com
CSeq: 1 INVITE
Content-Length: 0
```

c
```
SIP/2.0 200 OK
Via: SIP/2.0/UDP station1.work.com; branch=z9hG4bK123
From: Daniel<sip:Collins@work.com>;tag=44551
To: Boss<sip:Manager@station2.work.com>;tag=11222
Contact: sip:manager@station2.work.com
Call-ID: 123456@station1.work.com
CSeq: 1 INVITE
Subject: Vacation
Content-Length: xxx
Content-Type: application/sdp
Content-Disposition: session
(message body)
```

d
```
ACK sip:manager@station2.work.com SIP/2.0
Via: SIP/2.0/UDP station1.work.com; branch=z9hG4bK123
Max-Forwards: 70
From: Daniel<sip:Collins@work.com>;tag=44551
To: Boss<sip:Manager@station2.work.com>;tag=11222
Call-ID: 123456@station1.work.com
CSeq: 1 ACK
Content-Length: 0
```

e
Conversation

header field has the same value as the Request-URI, because we are not traversing any proxies. (As described later in this chapter, the Request-URI can be changed when a message passes through a proxy.)

The From: header field indicates that the call is from collins@work.com. As can be seen from Figure 5-9, SIP enables a display name to be used with the SIP URI. Thus, the called user's terminal could display the name Daniel when alerting the called user rather than displaying the SIP URI. The optional header field Subject: is used to indicate the nature of the call, and the type of media that Daniel wishes to use is described within the message body. The Content-Type: entity header field indicates that this particular message body is described according to SDP. The Content-Disposition header field indicates that the message body is to be treated as session information (as opposed to being displayed to the user, for example). In the example, the length of xxx is used just to indicate some nonzero value. The actual length will depend upon the content of the message body.

In Figure 5-9, the first response received is an indication that the user is being alerted. This is conveyed through the use of the 180 status code. Note that most of the header fields are copied from the request to the response, with the exception of the Content-Length and Content-Type, since this response does not contain a message body.

Note that the To: header field from the called party contains a tag parameter, whereas the To: field in the original invite did not contain a tag parameter. In fact, the tag parameter should be inserted only by the party that "owns" the address in question. In combination, the content of the tag in the From: field, the tag in the To: field, and the Call-ID constitute a dialog ID, which identifies a peer-to-peer relationship between two user agents. Once both parties have inserted their tag values into the To: and From: fields according to which of the fields they "own," then a dialog has been established. From that point onwards, each party should include the appropriate tag value in the To: and From: fields of every subsequent request and response until the dialog is terminated.

An important aspect of the dialog is the dialog state. If the response to an INVITE is a 1XX (provisional) response and it contains a tag parameter in the To: header field, then a session has not yet been established, but a dialog has. Such a dialog is said to be in *early* state. If a 2XX response is received with a tag in the To: header field, then the dialog is *confirmed*.

Subsequently in our example, the called user answers and a 200 (OK) response is returned. This response to an INVITE includes a number of header fields that have been copied from the original request. In addition, it contains a message body describing the media that the called party wants to use.

Finally, the caller sends an ACK to confirm receipt of the response. Note that the content of the CSeq: header field has changed to reflect the new request. Note also that, although most requests are answered with a 200 (OK) response if the request has been received and handled correctly, this is not the case for an ACK request. Once the ACK has been sent, the parties can exchange media.

Note that the requests shown in Figure 5-9 all include the Max-Forwards header field. This header should be included in all requests to ensure that a given request does not traverse too many proxies. The value of Max-Forwards is decremented at each proxy in the chain between requester and responder. Seventy is the recommended initial value for Max-Forwards. We discuss proxy servers later in this chapter.

Termination of a Call

A user agent that wants to terminate a call does so through the issuance of a BYE request. Any party to a call can issue this request. Upon receipt of a BYE request, the recipient of the request should immediately stop transmitting all media directed at the party who has issued the BYE request.

Figure 5-10 shows a call termination corresponding to the call origination shown in Figure 5-9. As can be seen, the BYE request from Daniel has many of the same header values as in the original INVITE request, with the important exception that the CSeq has been changed to reflect the BYE request, as opposed to the INVITE request. As with any successfully completed request (besides ACK), a 200 response is returned.

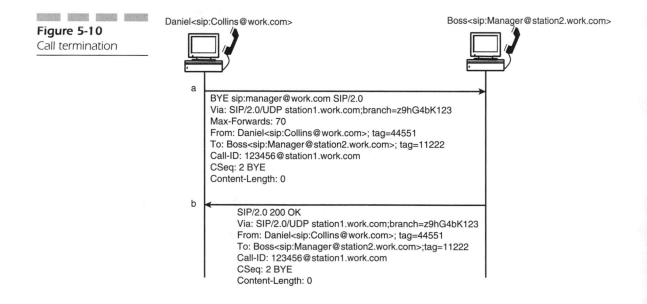

Figure 5-10
Call termination

Daniel<sip:Collins@work.com>   Boss<sip:Manager@station2.work.com>

a

BYE sip:manager@work.com SIP/2.0
Via: SIP/2.0/UDP station1.work.com;branch=z9hG4bK123
Max-Forwards: 70
From: Daniel<sip:Collins@work.com>; tag=44551
To: Boss<sip:Manager@station2.work.com>; tag=11222
Call-ID: 123456@station1.work.com
CSeq: 2 BYE
Content-Length: 0

b

SIP/2.0 200 OK
Via: SIP/2.0/UDP station1.work.com;branch=z9hG4bK123
From: Daniel<sip:Collins@work.com>; tag=44551
To: Boss<sip:Manager@station2.work.com>;tag=11222
Call-ID: 123456@station1.work.com
CSeq: 2 BYE
Content-Length: 0

Redirect and Proxy Servers

The examples shown so far have involved one user agent communicating directly with another. Although that is certainly possible, a more likely scenario will involve the use of one or more proxy or redirect servers.

Redirect Servers

A redirect server normally responds to a request with an alternative address to which the request should be directed. The exceptions are cases where the server has received a request that cannot be supported, in which case it will return a 4XX, 5XX, or 6XX response, or when the server receives a CANCEL request, to which it should respond with a 200 response.

Consider again the example call from Daniel to Boss as depicted in Figure 5-9. Instead of the call-completion scenario depicted in Figure 5-9, let us assume that Boss is out of the office and has registered at a different location. Using a redirect server, the call scenario would be as shown in Figure 5-11.

The initial INVITE is sent to Boss at the redirect server. In this case, the Request-URI and the To: address do not include the specific host where Boss has logged in. That is likely to be the most common situation. After all, a caller is unlikely to know exactly where a given user has logged in.

The server responds with the status code 302 (moved temporarily) and within the response includes the Contact: header field. The content of this field is the address that the caller should try as an alternative.

The client then creates a new INVITE. This new INVITE uses the address received from the redirect server as the Request-URI while the To: header remains the same. The call-ID also remains the same, but the CSeq number is incremented. This number is always incremented when a second request of the same type is made for a given call. In this case, this is the second INVITE for the call. The fact that the To: header is the same allows Boss (at home) to recognize that the call was originally made to his or her address at work. Based on this information (plus the identity of the caller and the subject of the call), Boss might or might not decide to accept the call. This is another example of how SIP provides information to enable a user to make intelligent call-handling decisions.

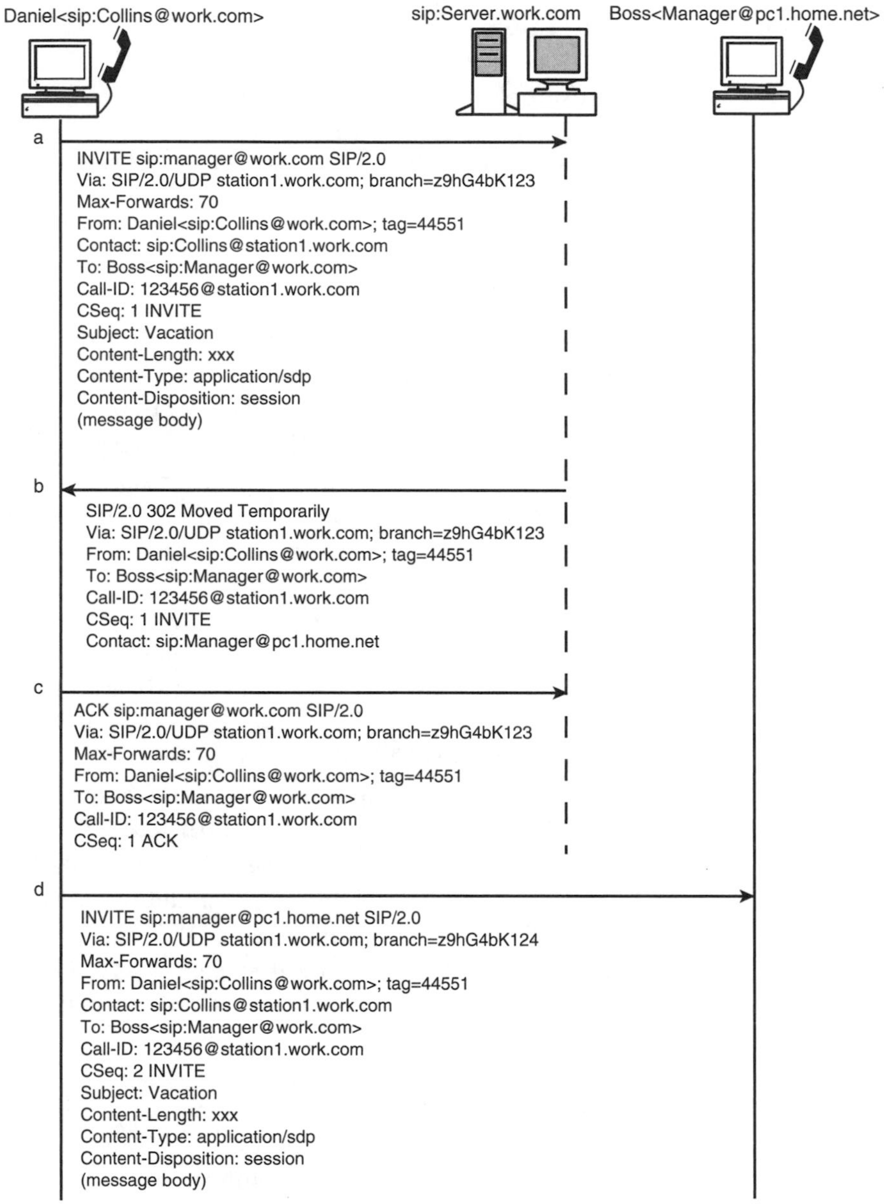

Figure 5-11
Application of a redirect server

Daniel<sip:Collins@work.com>   sip:Server.work.com   Boss<Manager@pc1.home.net>

a
```
INVITE sip:manager@work.com SIP/2.0
Via: SIP/2.0/UDP station1.work.com; branch=z9hG4bK123
Max-Forwards: 70
From: Daniel<sip:Collins@work.com>; tag=44551
Contact: sip:Collins@station1.work.com
To: Boss<sip:Manager@work.com>
Call-ID: 123456@station1.work.com
CSeq: 1 INVITE
Subject: Vacation
Content-Length: xxx
Content-Type: application/sdp
Content-Disposition: session
(message body)
```

b
```
SIP/2.0 302 Moved Temporarily
Via: SIP/2.0/UDP station1.work.com; branch=z9hG4bK123
From: Daniel<sip:Collins@work.com>; tag=44551
To: Boss<sip:Manager@work.com>
Call-ID: 123456@station1.work.com
CSeq: 1 INVITE
Contact: sip:Manager@pc1.home.net
```

c
```
ACK sip:manager@work.com SIP/2.0
Via: SIP/2.0/UDP station1.work.com; branch=z9hG4bK123
Max-Forwards: 70
From: Daniel<sip:Collins@work.com>; tag=44551
To: Boss<sip:Manager@work.com>
Call-ID: 123456@station1.work.com
CSeq: 1 ACK
```

d
```
INVITE sip:manager@pc1.home.net SIP/2.0
Via: SIP/2.0/UDP station1.work.com; branch=z9hG4bK124
Max-Forwards: 70
From: Daniel<sip:Collins@work.com>; tag=44551
Contact: sip:Collins@station1.work.com
To: Boss<sip:Manager@work.com>
Call-ID: 123456@station1.work.com
CSeq: 2 INVITE
Subject: Vacation
Content-Length: xxx
Content-Type: application/sdp
Content-Disposition: session
(message body)
```

Proxy Servers

A proxy server sits between a user-agent client and the far-end user-agent server. A proxy accepts requests from the client and forwards them onwards, perhaps after some translation. Numerous proxies can reside in a chain between the caller and callee. The most common scenario will have at least two proxies: one at the caller end and one at the callee end.

Let's assume that Boss (at home) calls Collins (at work). If we assume that the call will pass via a proxy, as might be expected, then the sequence of messages could appear as in Figure 5-12. To avoid clutter, the Subject and Message Body headers have been omitted. Those headers would normally be included.

The proxy receives the INVITE and forwards it onwards. In doing so, the proxy can change the Request-URI. The proxy will do so if it knows that the Request-URI should be mapped to a different address. For example, if a proxy is also a registrar and is aware of a Contact address for a given address of record, then the proxy will change the Request-URI of an INVITE to the appropriate contact address. In the example of Figure 5-12, the proxy is aware that the address sip:Collins@work.com should be mapped to sip:Collins@station1.work.com, and it uses sip:Collins@station1.work.com as the Request-URI when it forwards the request. In many cases, however, there might be several proxies in the chain from caller to callee. It is likely that only the last proxy in the chain is aware of where a user has logged in, and it is only that proxy that would map the Request-URI to a different value. The other proxies in the chain would simply use the domain part of the received Request-URI as input to a location function (such as a *Domain Name System* [DNS]) to determine the next hop in the path from caller to callee, but they would not change the value of the Request-URI.

In Figure 5-12, the proxy returns a provisional response of 100 (trying) to the initial INVITE, and the proxy does this in parallel with forwarding the INVITE towards the called party. The requirement to issue the provisional 100 response is mandatory in some cases and optional in others. In general, it is a good idea to provide the provisional response in all cases.

Of major importance for proxy servers is the Via: header field. The Via: header is used to indicate the path taken by a request so far. When a request is generated, the originating client inserts its own address in a Via: header field. Each proxy along the way also inserts its address in a new Via: header field, placed in front of any existing Via: headers, as can be seen in Figure 5-12. Therefore, the collection of Via: headers provides a map of the path taken through the network by a given request.

When a proxy receives a request, it checks to see if its own address is already included in one of the Via: header fields. If so, then the request has already passed through the proxy and a loop condition might exist. A loop condition will exist if those header fields that affect the routing of the request (most notably the Request-URI) are the same as a request that has

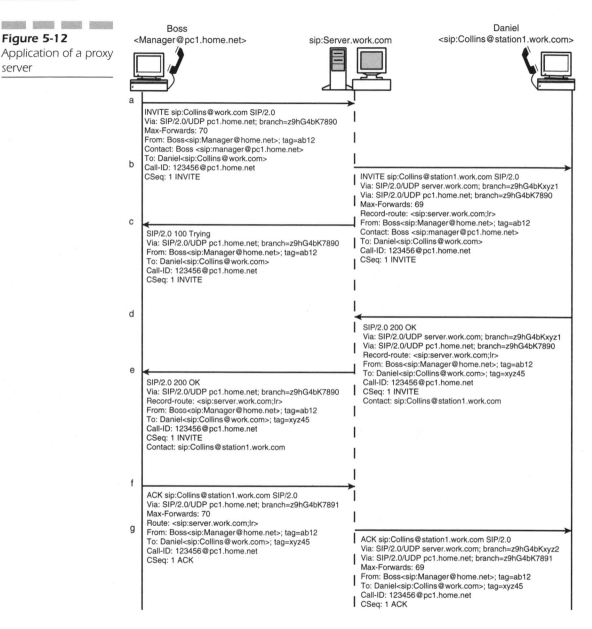

already been processed at the proxy. In other words, the fact that a proxy has received a request with a Via: header field indicating the proxy itself does not necessarily mean that a loop condition does exist, just that it *might* exist. Examination of the REQUEST-URI and the other header fields is needed to be certain.

For example, assume a request had been processed at a proxy and that request led to a certain routing that ultimately led back to the proxy. A loop condition would exist here only if the proxy would forward the request the same way both the first time and the second time. On the other hand, if the request had been modified such that the proxy would route the request differently the second time it arrives, then we have a spiral condition and not a loop condition. A spiral condition is not an error. In the event that a loop exists, the proxy can respond to the request with status code 482 (loop detected). Otherwise, the proxy will pass the request onwards after inserting its own address in a new Via: header field.

Responses also include Via: header fields, which are used to send a response back through the network along the same path that the request used (in reverse, of course). When a server responds to a request, the response contains the list of Via: headers exactly as received in the request. When a proxy receives a response, the first Via: header should refer to itself. If it does not, a problem has occurred and the message is discarded. Assuming that the first Via: header field indicates the proxy itself, then the proxy removes the header and checks to see if a second Via: header exists. If not, then the message is destined for the proxy itself. If there is a second Via: header field, then the proxy passes the response to the address in that field. In this way, the response finds its way back to the originator of the request along the path that was originally taken by the request.

Note that the Via: header in the example of Figure 5-12 contains a branch parameter. In the original RFC 2543, the primary objective of the branch parameter was to distinguish between multiple responses to the same request. As we shall describe shortly in a discussion regarding *forking*, a proxy might forward a single request to multiple destinations. In such an event, multiple responses might be received and those need to be distinguished.

In the latest SIP specification (as of this writing), greater significance is applied to the branch parameter. Specifically, the branch parameter is used to uniquely identify a transaction created at a given entity (such as a proxy) and assists in the recognition of loop conditions. Thus, the branch parameter must be unique. Moreover, the latest SIP specification requires that the branch parameter begin with the characters z9hG4bK, which is known as the *magic cookie*. This strange set of characters at the beginning of the branch parameter indicates that a client or server complies with the latest SIP specification, rather than the base RFC 2543. In the base RFC 2543 specification, the branch parameter is optional and does not uniquely identify a transaction. Consequently, if a proxy receives a request with a Via: header and the branch parameter, it will only assume that the branch parameter is globally unique if it begins with the magic cookie. The chances of a system that complies only with RFC 2543 selecting a branch parameter that begins with the magic cookie are miniscule.

The branch parameter is unique for every request issued by a client, with the exception of CANCEL and ACK for non-2XX responses. The branch parameter for a CANCEL or ACK for a non-2XX response should be the same as the branch parameters for the original request (the request being cancelled, or the request for which the non-2XX response was received).

Proxy State   A proxy can be either stateless or stateful. If stateless, then the proxy takes an incoming request, performs whatever translation is necessary, forwards the corresponding outgoing request, and forgets that anything ever happened. If a message is retransmitted through a stateless proxy, then it must be forwarded in exactly the same manner as the original request. This action requires that the choice of Via: header (specifically the tag value) be computed based on header values in a request that do not change on retransmission.

If a proxy is stateful, then the proxy remembers incoming requests and corresponding outgoing requests, and it is able to act more intelligently on subsequent requests and responses related to the same session.

Two important headers in Figure 5-12 are the Record-Route: and the Route: headers. In the example, all messages and responses pass through the same proxy. This need not necessarily be the case, however. Although the response to a given request must return along the same route as used by the request, two consecutive requests need not follow the same path. Given that an initial INVITE will contain a Contact: header and responses can also contain a Contact: header, the two ends of the call might have the addressing information necessary to communicate directly. Therefore, after the initial INVITE request and response, subsequent requests and responses might be sent end to end. A proxy might require that it remain in the signaling path for all subsequent requests, however. For example, a proxy might need to remain in the signaling path in order to provide some advanced service. A proxy can ensure that it remains in the signaling path through the use of the Record-Route: header.

Each proxy that wants to remain in the signaling path for future requests inserts its address in the Record-Route: header field of the INVITE, such that the Record-Route: header field ultimately contains a list of proxy addresses, each of the form server.domain. Each proxy places its address at the head of the list such that, at the end of the signaling path, the first entry in the list is that of the last proxy used.

A 200 response to the INVITE will include the Record-Route: header as it was received by the ultimate destination of the INVITE. Hence, this response contains an ordered list of all the proxies used in the INVITE. The content of the header is propagated back through the network along with the 200 response. When the 200 response is received at the client that originated the INVITE, the information contained in the Record-Route: header is used in subsequent requests (such as ACK or BYE) related to the same call. The client takes the information in the Record-Route: header and places it in a Route: header in reverse order so that the Route: header contains a list of proxies in the direction from calling client to called server. The client uses the first entry in the Route: header as the destination to send the next request (such as ACK or BYE). The proxy that receives that request removes the first entry in the Route: header field (which will indicate the proxy itself) and forwards the message to the next proxy in the chain, or to the user-agent server if the proxy in question is the last proxy in the chain. The result is that each request follows the same path as the original INVITE, thereby allowing each proxy in the chain to be aware of each request and response.

As can be seen from Figure 5-12, the Record-Route: and Route: header fields contain the *loose routing* (lr) parameter. Systems compliant with the latest SIP specification should insert this parameter in the Record-Route: header. The initial SIP specification did not require the presence of the lr parameter, and its absence indicates strict routing. The difference between the loose and strict routing is in how the Route: header is populated and in the value that is placed in the Request-URI.

In Figure 5-12, lr is depicted. A client or proxy that supports loose routing uses the content of the Route: header field as the address to which a message is forwarded, but uses the ultimate destination (as specified in a Contact: header) as the Request-URI. In Figure 5-12, we can see that Boss sends an ACK message to the proxy, but uses Collins@station1.work.com as the Request-URI (not the SIP address of the proxy). Moreover, the ACK from Boss contains the Route: header value of sip:server.work.com. If strict routing were used, then the ACK from Boss would have a Request-URI of sip:server.work.com, and the Route: header would not include the proxy server's SIP address; rather, it would contain Daniel's Contact: header information (sip:Collins@station1.work.com).

Forking Proxy A proxy can "fork" requests. This would happen if a particular user is registered at several locations. An incoming INVITE for the user would be sent from the proxy to each of the registered locations. If one of the locations answered, then the proxy could issue a CANCEL to the other locations so that they do not continue "ringing" when it is known that the called user has answered elsewhere. In order to handle such forking, a proxy must be stateful.

Figure 5-13 gives an example of a forked request. For the sake of clarity, the Subject: and message body header fields have been omitted. In this scenario, Boss (at home) calls Collins (at work). However, Collins happens to be logged in to two different work stations: pc1.work.com and pc2.work.com. When the proxy receives the INVITE, it immediately returns a 100 (trying) response. The proxy then forwards the INVITE request to the two locations in parallel. Not only does the proxy add a Via: header to each outgoing INVITE, but it ensures that the branch parameter is different in each of the outgoing INVITE requests, which enables the proxy to distinguish between the responses to those requests.

Collins answers the incoming call at pc2.work.com. Therefore, a 200 OK response is returned from pc2.work.com. Upon receiving the response, the proxy immediately issues a CANCEL to pc1.work.com so that any resources allocated by that machine may be released. Note that the CANCEL contains only a single Via: header, that of the server. Since a CANCEL request is always sent hop by hop, a Via: header identifying the caller is not included in the CANCEL request.

The response from pc2.work.com is then relayed back to the originator of the INVITE request. The remainder of the call (ACK, BYE, and so on) will proceed as normal. Meanwhile, however, pc1.work.com still owes the proxy server two responses: one for the initial INVITE and one for the CANCEL. Upon receipt of the CANCEL request, pc1.work.com responds to the INVITE with the status code 487 (request cancelled). It will also respond to the CANCEL request with 200 OK. These two responses are not shown in Figure 5-13.

The Session Description Protocol (SDP)

The examples so far have focused on various SIP requests, responses, and header fields. Although some mention has been made of the message body and entity header fields, the message body has not been described in any detail. Although a SIP message body can carry many different types of information (such as an image or text to display), the most common mes-

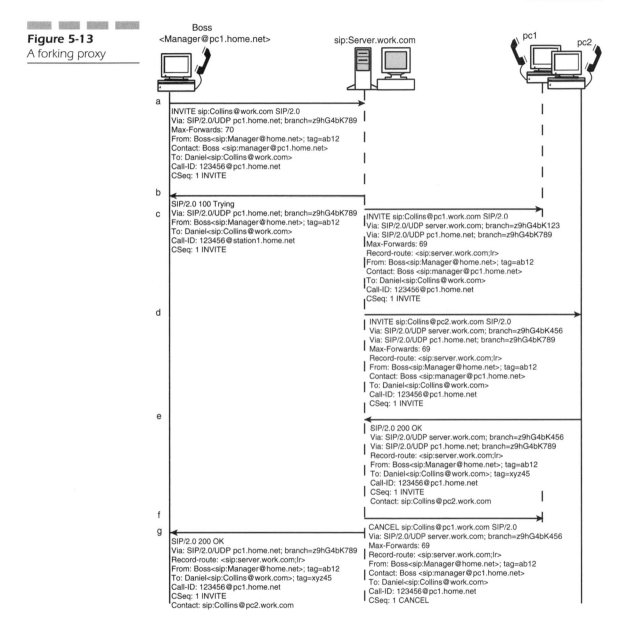

Figure 5-13
A forking proxy

sage body will be session information describing the media to be exchanged between the parties. The session description will include media information such as the RTP payload type, addresses, and ports. The format of that description will normally be according to SDP.

SIP uses SDP in an answer/offer mode. A caller sends an INVITE with an SDP description that describes the set of media formats, addresses, and ports that the caller is willing to use. This set of media formats comprises an offer by the caller. The called party responds with an SDP description that aligns with the offered SDP description, but which includes an acceptance or rejection of each of the offered media formats. The result of this exchange is an agreement between the two parties as to the types of media they are willing to share.

The Structure of SDP

SDP is specified in RFC 2327. Since the initial publication of RFC 2327, however, a number of modifications to the protocol have been suggested and a draft update to the specification has been prepared within the IETF. In addition, RFC 3264 entitled "An Offer/Answer Model with SDP" has been prepared to describe how SDP and SIP should be used together.

SDP simply provides a format for describing session information to potential session participants. Basically, a session is comprised of a number of media streams. Therefore, the description of a session involves the specification of a number of parameters related to each of the media streams. Information also exists that is common to the session as a whole, however. Therefore, we have session-level parameters and media-level parameters. Session-level parameters include information such as the name of the session, the originator of the session, and the time(s) that the session is to be active. Media-level information includes the media type, port number, transport protocol, and media format. Figure 5-14 illustrates this general structure.

Because SDP simply provides session descriptions and does not provide a means for transporting or advertising the sessions to potential participants, it must be used in conjunction with other protocols (such as SIP). For example, SIP carries SDP information within the SIP message body.

Similar to SIP, SDP is a text-based protocol, utilizing the ISO 10646 character set in *Unicode Standard Transmission Format* (UTF)-8 encoding (RFC 2044). This coding enables the usage of multiple languages and includes US-ASCII as a subset. Although SDP field names use only US-ASCII, textual information may be passed in any language. For example, the field name that indicates the name of a session must be in US-ASCII, but the actual string of characters that comprises the session name may be in any language.

Figure 5-14
*SDP session
description structure*

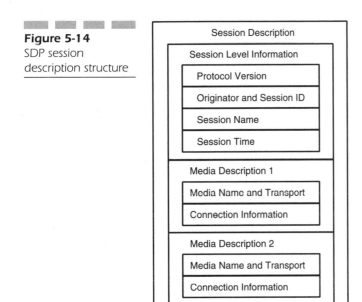

Although the use of ASCII coding in SDP, as opposed to binary coding, is a little bandwidth greedy, SDP is written in a compact form to counteract bandwidth inefficiency. Thus, where English words might have been used for field names, single characters are used instead. For example, *v* is version, *s* is session name, and *b* is bandwidth information.

SDP Syntax

SDP conveys session information using a number of lines of text. Each line uses the format field=value, where field is exactly one character (case-significant) and value is dependent on the field in question. In some cases, value may comprise a number of separate pieces of information, separated by spaces. No spaces are allowed between field and the = sign or between the = sign and value.

Session-level fields must be included first, followed by media-level fields. The boundary between session-level data and media-level data occurs at the first occurrence of the media description field (m=). Each subsequent occurrence of the media description field marks the beginning of data related to another media stream in the session.

Mandatory Fields SDP includes mandatory fields, which must be included in any session description, and optional fields, which may be omitted. The following fields are mandatory:

- **v=** Meaning the protocol version, this field marks the start of a session description and the end of any previous session description. When SDP is used as part of a SIP message body, only one SDP description should be present, which means that the v= field should appear only once in an SDP-based SIP message body.

- **o=** The session origin or creator and session identifier.

- **s=** Standing for the session name, this is a text string that could be displayed to potential session participants. The use of the session name is quite appropriate for a multicast session, but not for a unicast session such as a simple call between two parties. Therefore, when SDP is used in a SIP message body, the session name should be just a single space. SDP does not permit the s= line to be absent, nor does it permit the session name to be empty.

- **t=** This provides the start and stop times for the session. Unlike a pre-arranged multicast conference, however, sessions established by SIP are typically established in real time and can be of any duration. Consequently, the notion of predefined start and stop times does not apply to a SIP-established session. For that reason, an SDP description in a SIP message body should have a t= line of the form t=0 0. Strictly speaking, SDP interprets such time values to mean that the session is permanent. Such an interpretation does not matter in a SIP-established session, however, since SIP signaling will be responsible for the setup and teardown of the session, regardless of the time values indicated in the SDP description.

- **m=** This field indicates the media type, the transport port to which the data should be sent, the transport protocol (such RTP), and the media format (typically an RTP payload format). The appearance of this field also marks the boundary between session-level and media-level information or between different media descriptions.

Optional Fields Of the optional fields in SDP, some apply only at the session level or the media level, while some optional fields can be applied at both. In such cases, the value applied at the media level overrides the session-level value for that media instance. Thus, if a given field has value X for the session and value Y for media type 1, then value X applies to all media except for media type 1, where value Y applies. The following are the optional fields:

- **i=** Representing the session information, this is a text description of the session topic, meant to provide greater detail than the session name. This field may be specified both at the session and media levels. When SDP is used in a SIP message body, the i= field could be included, but would be somewhat superfluous, since SIP already supports the Subject: header field to convey session topic/subject information.

- **u=** This is a URI (such as a web address) where further session information may be obtained. For example, a session agenda might be posted on a web page, in which case the URI of the web page would be specified. Only one URI may be specified per session. Unlike pre-arranged conferences, SIP sessions are typically established on demand. Therefore, one is unlikely to find the u= line in an SDP description within a SIP message body.

- **e=** This is the e-mail address of the individual responsible for the session. Several instances of this field may exist and would be used if several individuals could be contacted for more information. This field applies only at the session level.

- **p=** This is a phone number for the individual responsible for the session. As for the e-mail address field, several instances of this field may be included. This field applies only at the session level. Similar to several of the other optional fields, this field is unlikely to be found in an SDP-based SIP message body.

- **c=** This field provides connection data, including the connection type, network type, and connection address. This field can be applied either at the session or media level. Although we are listing the c= field as optional here, it is more correct to say the field is conditional. Because this field defines the network address to which media should be sent, it must be included somewhere in the session description either at the session level or at the media level, or at both. If the field appears at the session level, then it is optional at the media level and vice versa.

- **b=** This optional field specifies the bandwidth required in kilobits per second. It may be specified at either the session or media level, or both. b= indicates the bandwidth required either by the whole conference or by a given media type. When used at the conference level, it can be considered analogous to a traditional dial-in conference where a maximum number of lines are available.

- **r=** In the case of a regularly scheduled session, this field specifies how often and how many times a session is to be repeated. For a SIP-initiated session, this field should not be included.

- **z=** Indicating the timezone adjustments, this field is used with regularly scheduled sessions. In such cases, the schedule of sessions may span a change from standard time to daylight savings time or vice versa. Since such changes occur at different times in different parts of the world, confusion might exist as to the actual start time of a specific session, particularly if it is to occur in the spring or fall. This field enables the session creator to specify up front when such timing changes will occur. As with other time-related fields, this field should not be included when SDP is used as a SIP message body.

- **k=** This field provides an encryption key or specifies a mechanism by which a key may be obtained for the purposes of encrypting and decrypting the media. The field may be applied either at the session or media level or both.

- **a=** This field is used to describe additional attributes to be applied to the session or to individual media.

Ordering of Fields Because certain fields can appear both at the session and media levels, the ordering of fields is important to avoid ambiguity. SDP demands that the following order be used.

Session Level
- Protocol version (v)
- Origin (o)
- Session name (s)
- Session information (i) (optional)
- URI (u) (optional)
- E-mail address (e) (optional)
- Phone number (p) (optional)
- Connection information (c) (conditional, depending on whether specified at the media level)
- Bandwidth information (b) (optional)
- Time description (t) (should be set to "t=0 0" when SDP is used with SIP)
- Repeat information (r) (optional) (should be omitted when SDP is used with SIP)
- Time zone adjustments (z) (optional) (should be omitted when SDP is used with SIP)

- Encryption key (k) (optional)
- Attributes (a) (optional)

Media Level
- Media description (m)
- Media information (i) (optional)
- Connection information (c) (optional if specified at session level)
- Bandwidth information (b) (optional)
- Encryption key (k) (optional)
- Attributes (a) (optional)

Subfields Of the various fields specified in SDP, a number are structured in several subfields. In such cases, the value of the field is a combination of several values, separated by spaces. The SDP specification's structure is as follows:

```
Field=<value of subfield 1> <value of subfield 2> <value of
subfield 3>.
```

A number of the most important fields are described here. The reader is referred to the SDP specification for a complete definition of all fields.

Origin Origin (o) has six subfields—username, session ID, version, network type, address type, and address.

The username is the originator's log-in identity at the host machine or is "-" if no log-in identity applies.

The session ID is a unique ID for the session, most likely created by the host machine of the originator. In order to ensure that the ID is unique, the SDP specification recommends that the ID makes use of a *Network Time Protocol* (NTP) timestamp.

The version is a version number for this particular session. When a session is advertised, the version number is important for the purpose of distinguishing a revised version of the session from an earlier version.

The network type is a text string indicating the type of network. The string IN refers to Internet.

The address type is the type of address in the network. SDP defines IP4 and IP6 to denote IP version 4 and IP version 6 respectively.

The address is the network address of the machine where the session was created. This address can be a fully qualified domain name or the actual IP address. The IP address can be the actual IP address only if the

IP address is unique as far as other session participants are concerned, which might not be the case if it is a local address, such as an IP address on a LAN behind a proxy.

Connection Data Connection Data (c) has three subfields: network type, address type, and connection address. Although some of these have the same names as the subfields defined for the origin, they have different meanings. They refer to the network and address at which media data are to be received, as opposed to the network and address at which the session was created.

Network type indicates the type of network in question. Currently, only the value IN is defined, meaning Internet.

Address type will have either the value IP4 or IP6 referring to IP version 4 or IP version 6.

The Connection address is the address where data should be sent. Although it may be a dotted decimal IP address, a fully qualified domain name is also possible and can be a better choice as it enables more flexibility and less ambiguity.

Media Information Media Information (m) has four subfields: media type, port, transport, and format. The media type may be audio, video, application, data, or control. For voice, the media type will be audio.

The port subfield indicates the port number to which media should be sent. The port number depends on the type of connection in question and the transport protocol. However, for VoIP, the media will normally be carried using RTP over UDP. Thus, the port number will be an even value between 1,024 and 65,535.

The format subfield lists the various types of media format that can be supported. If a given user can support voice coded in one of several ways, each supported format is listed with the preferred format listed first. Normally, the format will be an RTP payload format, with a corresponding payload type. In this case, it is necessary only to specify that the media is according to an RTP audio/video profile and specify the payload type.

If we have a system that wants to receive voice on port number 45678 and can only handle speech coded according to G.711 μ-law, then RTP payload type 0 applies and the media information is as follows:

```
m=audio 45678 RTP/AVP 0
```

If we have a system that wants to receive voice on port 45678 and can handle speech coded according to G.728 (payload type 15), *Global System for Mobile Communications* (GSM) (payload type 3), or G.711 μ-law (payload type 0) and would prefer to use G.728, then the media information is as follows:

```
m=audio 45678 RTP/AVP 15 3 0
```

Attributes SDP includes a field called attributes (a), which enables additional information to be included. Attributes may be specified at the session or media level or both. Furthermore, multiple attributes may be specified for the session as a whole and for a given media type. Therefore, numerous attribute fields may be present in a single session description, with the meaning and significance of the field dependent upon its position within the session description. If an attribute is listed prior to the first media information field, then it is a session-level attribute. If it is listed after a given media information field (m), then it applies to that media type.

Attributes can have two forms. The first form is a property attribute, where the attribute specifies that a session or media type has a specific characteristic. The second form is a value attribute, which is used to specify that a session or media type has a particular characteristic of a particular value. The SDP specification describes a number of suggested attributes.

Examples of property attributes in SDP are "sendonly" and "recvonly." The first of these specifies that the sender of the session description wants to send data but does not want to receive data. If RTP is being used, the entity in question will not receive an RTP stream, but should still receive RTCP messages. In the case of recvonly, the entity in question will not send any RTP streams, but should still send RTCP data.

An example of a value attribute would be "orient," used in a shared whiteboard session and indicating whether the whiteboard has a portrait or landscape orientation.

These examples would be included in SDP using the following syntax:

```
a=sendonly
a=recvonly
a=orient:landscape.
```

One important use of attributes in a VoIP application is provided by the *rtpmap* attribute. This attribute can be applied to a media stream and is particularly useful when the media format is not a static RTP payload type. rtpmap has the format

```
a=rtpmap:<payload type> <encoding name>/<clock rate>[/<encoding
parameters>].
```

The payload type corresponds to the RTP payload type. The encoding name is a text string. RTP profiles that specify the use of dynamic payload types must define the set of valid encoding names and/or a means to register encoding names if that profile is to be used with SDP. The clock rate is the rate in Hz. The optional encoding parameters may be used to indicate the number of audio channels.

Strictly speaking, the rtpmap attribute is only necessary for dynamic payload types. For example, standard G.711 speech is a static RTP payload type and would be fully described by a media field such as

```
m=audio 45678 RTP/AVP 0
```

A dynamic RTP payload type needs greater information for the remote end to fully understand the media coding, however. An example of such a dynamic payload type is 16-bit linear encoded stereo (2 channels) audio sampled at 16 kHz. If we want to use dynamic RTP/AVP payload type value 98 for such a stream, then the media description must be followed by an rtpmap attribute as follows:

```
m=video 45678 RTP/AVP 98
a=rtpmap 98 L16/16000/2
```

If multiple payload types are indicated by the media description, then a separate rtpmap attribute may be specified for each.

Although the rtpmap attribute is not necessary for static RTP payload types, SIP recommends that the attribute always be included when SDP is used for a SIP message body, even in the case of static payload types.

Usage of SDP with SIP

SIP and SDP make a wonderful partnership for the transmission of session information. SIP provides the messaging mechanisms for the establishment of multimedia sessions and SDP provides a structured language for describing those sessions. The message body in SIP, identified by the entity headers, provides a neat slot where SDP can be used. It is this usage that this section will discuss.

SIP uses SDP in an offer/answer model. The initiator of a session offers a selection of media formats to be used in the session. The receiver of the offer either rejects the offer completely or selects those media formats that have been offered and which the responder is willing to accept. If the responder agrees to participate in the session, then the response to the initial offer (a SIP INVITE) will include a message body of the same form as the received message body, with an indication as to the acceptance or rejection of each of the offered media types.

If we take the example outlined in Figure 5-9 and assume that Collins can support voice coded according to G.726, G.723.1, or G.728 and we assume that Boss can only support voice coded according to G.728, then the messages would appear as shown in Figure 5-15. For avoidance of clutter, a number of header fields (such as Via:) have been omitted, as has the 180 (Ringing) response. These would be included according to Figure 5-9.

In Figure 5-15, the message body of the INVITE lists the individual media formats separately, port numbers at which Daniel wants to receive each of the formats, and an rtpmap attribute exists for each media format. The message body of the 200 (OK) response also lists the media formats individually in the same order as received in the INVITE. In this example, Boss can support only G.728 and is rejecting the other two media formats. This rejection is indicated by specifying port number zero for each rejected media format. The fact that Boss is willing to accept G.728 is indicated by the fact that the associated port number (6666 in our example) is nonzero. The 6666 port number is the port at which Boss is willing to receive the G.728 RTP media stream. Since both parties can support the same media format, and each knows the port number at which the other party wishes to receive the media, media exchange (conversation) can take place.

In Figure 5-15, each of the offered media formats was listed in a separate line with a separate port number associated with each format. If Daniel wants to receive only one of the offered media formats and to reserve just a single number (port number 4444, for example), then the media part of the offer could be as follows:

```
m=audio 4444 RTP/AVP 2 4 15
a=rtpmap 2 G726-32/8000
a=rtpmap 4 G723/8000
a=rtpmap 15 G728/8000
```

This syntax indicates that Daniel wants to use G.726 (2), G.723 (4), or G.728 (15) in that order of preference, and he wants to receive the media stream at port number 4444.

If Boss is to respond to such an offer, then the response would have the following format (again assuming that Boss chooses only G.728):

```
m=audio 6666 RTP/AVP 15
a=rtpmap 15 G728/8000
```

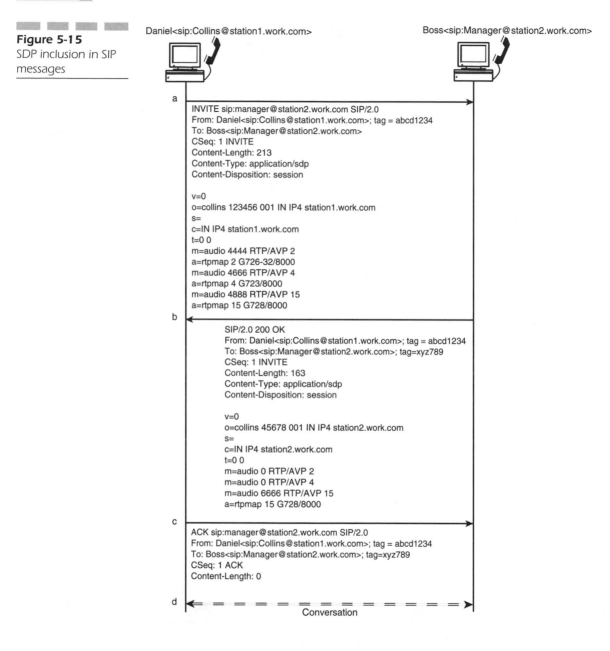

Figure 5-15
SDP inclusion in SIP messages

Negotiation of Media

Situations can arise where one party offers a number of media types and the responding party can support several of the offered media types. In the example of Figure 5-15, we assume that Boss supports only G.728. If Boss were to indicate that both G.723 and G.728 can be supported, then we have reduced the selection of *coder/decoders* (codecs) from three to two, but we have still not finalized exactly which codec will be used. Ideally, Boss's device should select just a single codec from the offered set. Such a selection would eliminate any confusion as to the codec to be used.

If, however, Boss were to reply with more than one codec, then the initiator of the session should choose which of the two to use and should generate a new offer that includes just a single codec. For SIP, this reoffer would mean the issuance of a new INVITE request with a message body that includes just the single selected codec. This new INVITE would exist within an already established dialog and, as such, it would use the same dialog identifier (From and To headers, including tag values), Call-ID, and Request-URI as would be used by any other request (such as INFO or BYE) within an established dialog. This scenario would appear as shown in Figure 5-16.

In Figure 5-16, we note that the initial INVITE specifies the offered media formats, but it uses the inactive attribute to specify that the media formats are set such that the media in question will neither be sent nor received (at least not for the present). As in Figure 5-15, Daniel offers the three coding schemes G.726, G.723, and G.728. In this case, Boss replies with G.723 and G.728. Collins acknowledges the response and issues a re-INVITE with G.728 as the only coding choice, and this time the inactive attribute is not present. The o= line of the re-INVITE indicates the same sender and session number, but the session version is increased by 1.

Although an ACK can carry a message body, we cannot use the ACK request in Figure 5-16 to carry an updated version of the offer that was in the initial INVITE. In SIP, two mechanisms are used for the answer/offer exchange. In the first case, an offer is made within an INVITE and is answered with a 2XX response. In the second case, an offer is made within a 2XX response and the ACK carries the answer. The second case would apply if the initial INVITE does not contain an offer (a valid situation). If the first INVITE does contain an offer in the form of an SDP description, however, the ACK cannot be used to modify that offer.

Not only can there be overlap between the supported media formats of the two parties in a session, but it is quite possible that the receiver of an offer (the receiver of an INVITE that contains an SDP description) cannot support any of the media formats proposed in the offer. If, for example, in Figure 5-15, Boss could only support voice coded according to G.711 μ-law (RTP payload type 0), then there would be a mismatch between the capabilities of the called party and those of the calling party. In such a situation,

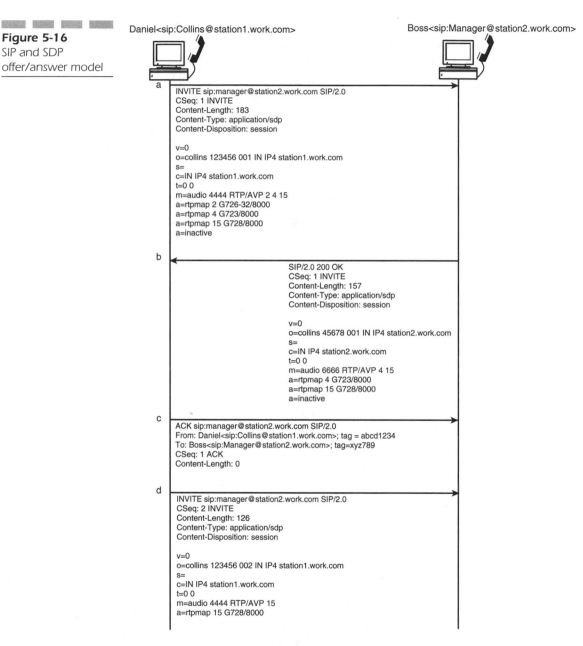

Daniel<sip:Collins@station1.work.com>   Boss<sip:Manager@station2.work.com>

a
INVITE sip:manager@station2.work.com SIP/2.0
CSeq: 1 INVITE
Content-Length: 183
Content-Type: application/sdp
Content-Disposition: session

v=0
o=collins 123456 001 IN IP4 station1.work.com
s=
c=IN IP4 station1.work.com
t=0 0
m=audio 4444 RTP/AVP 2 4 15
a=rtpmap 2 G726-32/8000
a=rtpmap 4 G723/8000
a=rtpmap 15 G728/8000
a=inactive

b
SIP/2.0 200 OK
CSeq: 1 INVITE
Content-Length: 157
Content-Type: application/sdp
Content-Disposition: session

v=0
o=collins 45678 001 IN IP4 station2.work.com
s=
c=IN IP4 station2.work.com
t=0 0
m=audio 6666 RTP/AVP 4 15
a=rtpmap 4 G723/8000
a=rtpmap 15 G728/8000
a=inactive

c
ACK sip:manager@station2.work.com SIP/2.0
From: Daniel<sip:Collins@station1.work.com>; tag = abcd1234
To: Boss<sip:Manager@station2.work.com>; tag=xyz789
CSeq: 1 ACK
Content-Length: 0

d
INVITE sip:manager@station2.work.com SIP/2.0
CSeq: 2 INVITE
Content-Length: 126
Content-Type: application/sdp
Content-Disposition: session

v=0
o=collins 123456 002 IN IP4 station1.work.com
s=
c=IN IP4 station1.work.com
t=0 0
m=audio 4444 RTP/AVP 15
a=rtpmap 15 G728/8000

the response to the INVITE should contain a 488 (Not Acceptable) or a 606 (Not Acceptable) status code. In addition, the response should contain a Warning: header field, with the warning code of 304 (media type not available) or 305 (incompatible media type). In that case, the caller could choose to issue a new INVITE request.

The fact that VoIP enables the use of a range of voice-coding techniques means that there can be a mismatch between the abilities of different SIP users. When some doubt exists about a party's ability to handle a particular media type, then the OPTIONS method provides a useful mechanism for finding out in advance. This avoids the partial establishment of a session that is doomed to fail.

OPTIONS Method   A potential caller can use the OPTIONS method to determine the abilities of a potential called party. The recipient of the OPTIONS request should respond with the abilities supported.

Figure 5-17 shows an example of the OPTIONS request. In this case, Collins queries the abilities of Boss, who responds with an indication that he can accept speech coded according to G.723 and G.728 (RTP payload formats 4 and 15 respectively).

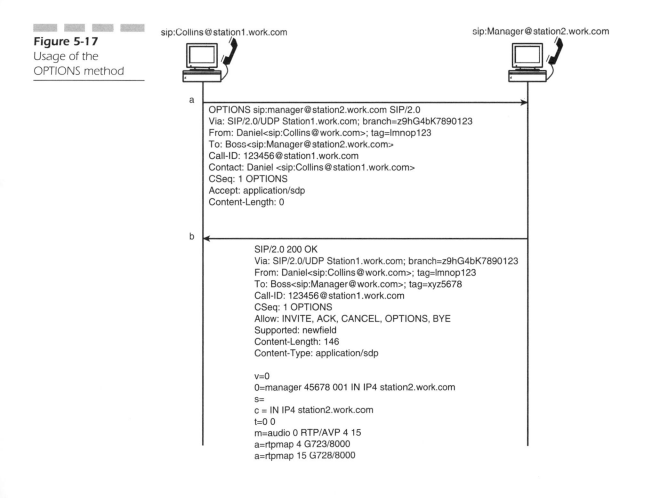

Figure 5-17
Usage of the
OPTIONS method

sip:Collins@station1.work.com

sip:Manager@station2.work.com

a

```
OPTIONS sip:manager@station2.work.com SIP/2.0
Via: SIP/2.0/UDP Station1.work.com; branch=z9hG4bK7890123
From: Daniel<sip:Collins@work.com>; tag=lmnop123
To: Boss<sip:Manager@station2.work.com>
Call-ID: 123456@station1.work.com
Contact: Daniel <sip:Collins@station1.work.com>
CSeq: 1 OPTIONS
Accept: application/sdp
Content-Length: 0
```

b

```
SIP/2.0 200 OK
Via: SIP/2.0/UDP Station1.work.com; branch=z9hG4bK7890123
From: Daniel<sip:Collins@work.com>; tag=lmnop123
To: Boss<sip:Manager@work.com>; tag=xyz5678
Call-ID: 123456@station1.work.com
CSeq: 1 OPTIONS
Allow: INVITE, ACK, CANCEL, OPTIONS, BYE
Supported: newfield
Content-Length: 146
Content-Type: application/sdp

v=0
0=manager 45678 001 IN IP4 station2.work.com
s=
c = IN IP4 station2.work.com
t=0 0
m=audio 0 RTP/AVP 4 15
a=rtpmap 4 G723/8000
a=rtpmap 15 G728/8000
```

We see the Accept: header field in the OPTIONS request. This field indicates the type of information that the sender is hoping to receive in the request. By specifying the value "application/sdp" in the Accept: header field, Collins is requesting that Boss respond with an SDP description for the type of SDP sessions that Boss can accept. If the OPTIONS request does not include the Accept: header, the application/sdp is used as a default.

In the response, we see the use of the Allow: header field. This field indicates those SIP methods that Boss can handle. We also see the Supported: header field, which indicates any extensions to SIP that Boss is able to support. For example purposes, Figure 5-17 shows that Boss can support an extension called "newfield." The Supported: header field provides a means for different parties to indicate those SIP extensions that they support. If both parties support the same extensions, then they can use them to provide additional services or features. SIP extensions are discussed later in this chapter.

SIP Extensions and Enhancements

Since RFC 2543 was released in March 1999 as an Internet standards track protocol, SIP has attracted enormous interest. Many companies, including those from the traditional telecommunications environment and others such as cable TV providers and *Internet Service Providers* (ISPs), have recognized that SIP offers great potential. In the process, however, several individuals and companies have found that SIP would be better if some enhancements or extensions were added. A large number of extensions to SIP have been proposed and are being handled in the SIP working group of the IETF.

As of this writing, numerous revisions have been made to the SIP specification as part of its progress along the standards track from proposed standard to standard. In addition, a number of Internet drafts have been prepared that we can expect to be included in the SIP specification in the future or that may become RFCs in their own right. The following descriptions address some of the most significant extensions to SIP.

One example of an extension that has been included within the revised SIP specification is the 183 (session progress) response code, which is typically used to open a one-way audio path from called end to calling end. The 183 response is often used when interworking with the SS7 network, where the Address Complete Message is used to open such a one-way path so that ring-back tone can be relayed from the called end to the calling end.

Another addition to the latest SIP specification (RFC 3261) is the addition of the Supported: header field. The base RFC 2543 specification includes the Require: header field, which enables a client to specify those extensions or options that a server must support in order to process a request. RFC 2543 does not, however, enable a server to independently indicate what extensions or options that it supports. Thus, if a client wants to know whether a server supports a given extension, it specifies that extension within a Require: header of a request (such as INVITE) and waits to see if an acceptable response is returned. If a 420 (bad extension) response is returned with an Unsupported: header field, then the client will know that the server does not understand the extension or option in question. Clearly, this is a cumbersome way of determining what extensions a server does or does not support. The Supported: header removes this shortcoming and provides a more elegant means for a client to determine what extensions a server supports. If a client issues an OPTIONS request to a server, then the server should include a Supported: header to indicate what extensions it has available.

The Supported: header can also be used in requests and should be used in requests from a client that supports SIP extensions. Associated with the Supported: header is the 421 (extension required) response. This response is returned by a server when it requires a particular extension in order to process a request and the request itself did not indicate that the client supports the extension.

The SIP INFO Method

The SIP INFO method is specified in RFC 2976. This method is a means for transferring information during an ongoing session, such as in the middle of a call. Examples of its use could include

- Transferring DTMF digits
- Transferring account balance information
- Transferring midcall signaling information generated in another network (such as the PSTN) and passed to the IP network via a gateway

Through the use of the INFO method, application-layer information could be transferred in the middle of a call, thereby affording user-agent servers, user-agent clients, or proxies the opportunity to act upon the information. For example, a prepaid subscriber could be informed in the middle of a call if the subscriber's prepaid account is nearing zero.

The SIP INFO method is a flexible tool and, as such, does not define the information that may or may not be sent using the INFO method. Typically, the INFO method would convey the user information within a message body, with the content and structure of the message body dependent upon the type of information to be conveyed. For example, an indication that the user's prepaid account balance is near zero could be conveyed in a message body in the form of text.

SIP Event Notification

Several SIP-based applications have been devised based on the concept of a user being informed of some event or events. For example, significant work has been performed related to use of SIP for instant messaging, and an important part of instant messaging is presence, that is, knowing when a member of a buddy list is available. The concept of event notification, however, can be applied in numerous applications. For example, a user might want to be informed of any schedule changes for a flight or of a change in a stock price.

To address these issues, RFC 3265 has been prepared to address the issue of event notification. That RFC specifies a general mechanism for event notification, but it does not specify the specific events. Any application that would want to avail itself of event notification would require a separate specification of the details of the events in question.

SIP-based event notification is achieved through the use of two new SIP methods: SUBSCRIBE and NOTIFY. The SUBSCRIBE method enables a user to subscribe to certain events, which means that the user should be informed when such events occur. The NOTIFY method is used to inform the user of the occurrence of a subscribed event. Figure 5-18 shows a basic example of the use of these two SIP methods.

The user who wants to subscribe to certain events sends a SUBSCRIBE method to a server that is aware of the events in question. The SUB-SCRIBE request must include a new header field called Event:, which is used to identify an event package including those events of interest to the subscriber. The content of the Event: header field might explicitly or implicitly indicate the events of interest, or the content of the Event: header field might indicate that the details of the events of interest are carried within the SUBSCRIBE message body.

For example, if a user wants to know about the presence and state of Instant Messaging buddy list members, then it would send a SUBSCRIBE request to the Instant Messaging server. The Event: header would simply need to indicate that the user wants to be informed about his buddy list. The server would know of the buddy list contents and, provided that the

Figure 5-18
SIP event notification

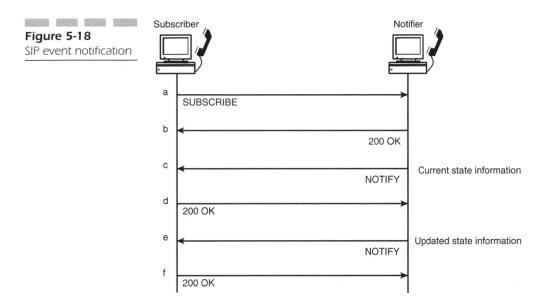

subscriber is authenticated, would have enough information in the Event: header to be able to send the required notifications to the subscriber. On the other hand, if a subscriber wants to be informed of stock prices for an arbitrary set of stocks, then the list of stock ticker symbols might be carried as text within the message body, and the Event: header would simply provide information to enable the server to understand the meaning of the message body.

Upon receipt of a SUBSCRIBE request and assuming that the subscriber is authorized to receive the requested information, the server will respond with a 200 (OK) response. The server will then send a NOTIFY message to indicate the current state of affairs for the requested information (the current prices for the stocks in question, for example). The server will then send a new NOTIFY when state changes occur, such as if a stock price changes or if a buddy list member becomes active. The NOTIFY request is likely to include the requested information within a message body. Provided that no errors occur in a NOTIFY message (such as information about an unsubscribed event), then the subscriber will issue a 200 (OK) response to each NOTIFY.

As mentioned, the SUBSCRIBE-NOTIFY mechanism is a framework for enabling users to subscribe to specific information. Those who want to use the mechanism in support of event notification for some specific application must develop detailed specifications for the manner in which the mechanism is to be used. Such detailed specifications must include the content of the Event: header and the format of the message bodies to be used in the SUBSCRIBE and NOTIFY requests. One example of such specification details can be found in an Internet draft that details how the SUBSCRIBE-NOTIFY technique can be used for presence information.

SIP for Instant Messaging

We have already discussed the usage of SIP event notifications in relation to presence information for instant messaging applications. In fact, SIP has been proposed by many as a unifying protocol to replace the several incompatible protocols that have existed to date. To that end, a separate IETF working group has been established to develop a specification for the use of SIP for instant messaging. That working group is titled *SIP for Instant Messaging and Presence Leveraging Extensions* (SIMPLE).

The SIMPLE working group has already developed an Internet draft for presence functions, an important aspect of instant messaging. Another important aspect is the mechanism by which the instant messages themselves are passed between users. One solution is a new SIP method, the MESSAGE request, which is proposed in an Internet draft.

Instant messaging typically involves the exchange of content between a set of participants in near real time. Generally, the content is short text messages, although that need not be the case, and other content (such as images or other binary data) can also be transferred. Generally, the messages that are exchanged are not stored, but this also need not be the case. Instant messaging differs from e-mail in common usage in that instant messages are usually grouped together into brief live conversations. Although some instant messaging protocols do establish a session between users, most often individual messages are stand-alone, and no explicit relationship exists between messages. Although there can be the appearance of a session between users, that appearance is often a function of the client-user interface or just a function of the conversation that takes place.

SIP is well suited to a session-based instant messaging solution, where the session would be instigated with an INVITE request and terminated with a BYE request. Supporting an instant messaging regime where each message is stand-alone, however, needs an extension to the basic SIP protocol. The proposed extension is the new MESSAGE method.

A user that wants to send an instant message to another user does so through the issuance of a MESSAGE request. This request carries the actual message in a message body. The message body will typically be of the form *text/plain* or may be in the form *message/cpim*, where cpim stands for *common presence and instant message* format, an IETF format designed for instant message data.

Unlike an INVITE, a MESSAGE request does not establish a SIP dialog. MESSAGE requests can, however, be associated with an existing SIP dialog. An instant message sent between two parties during an existing session would be associated with the dialog of that session. In many cases, however, a MESSAGE request will simply be sent between two users and will not be associated with any session. Therefore, if a MESSAGE request is sent outside an existing dialog, it is not required that any tag value be included in the To: header.

Figure 5-19 shows an example of the MESSAGE method as a means for passing instant messages. Two messages are shown. In the first case, Boss sends a message to Daniel via a proxy. When the message is delivered, a 200 (OK) response is returned. Subsequently, Daniel wants to respond to the instant message and constructs a MESSAGE request of his own. This request is also sent via a proxy and a 200 (OK) response is returned. Note that both of the messages are sent via the same proxy in Figure 5-19. From a protocol perspective, the two MESSAGE requests are totally independent. Therefore, the MESSAGE request from Daniel to Boss might traverse a different proxy (or proxies) than the MESSAGE request from Boss to Daniel. In Figure 5-19, we see that none of the Contact:, Record-Route:, or Route: headers are included. The Contact: header is strictly forbidden with the MESSAGE method. The Record-Route: and Route: headers are not relevant to the MESSAGE method because those headers are related to a specific dialog, and the MESSAGE method does not establish a SIP dialog.

The SIP REFER Method

Like many of the proposed SIP extensions, the REFER method is currently documented within the IETF in the form of an Internet draft. This particular SIP extension, however, is extremely popular and has been implemented in many SIP devices.

A REFER request enables the sender of the request to instruct the receiver to contact a third party, with the contact details for the third party included within the REFER request. A common application of the REFER method is in Call Transfer applications.

Imagine, for example, that Joe and Mary are having a conversation. Mary determines that Joe really needs to talk to Susan and decides to transfer the call to Susan. Mary would send a REFER message to Joe, with Susan's contact details contained with a new Refer-to: header field, and Joe

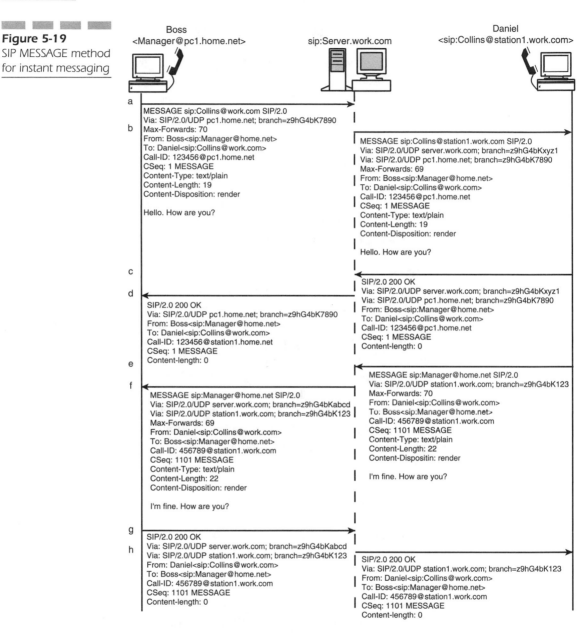

Figure 5-19
SIP MESSAGE method
for instant messaging

can then contact Susan. Moreover, the REFER method can be interpreted as an implicit SUBSCRIBE request, so that Joe can subsequently send a NOTIFY request to Mary to indicate whether he was able to successfully contact Susan. Figure 5-20 shows an example of the use of the REFER method.

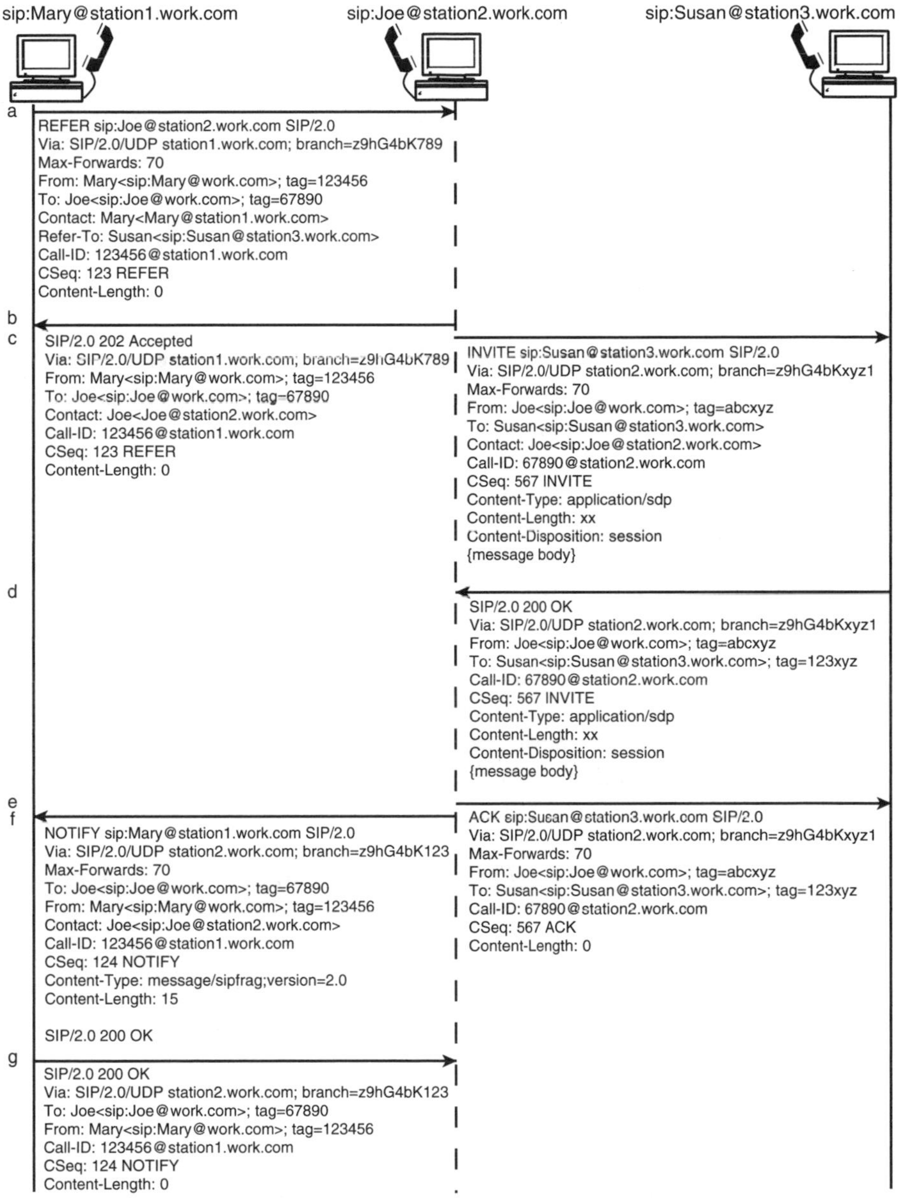

sip:Mary@station1.work.com   sip:Joe@station2.work.com   sip:Susan@station3.work.com

Figure 5-20

Usage of the SIP
REFER method

a

```
REFER sip:Joe@station2.work.com SIP/2.0
Via: SIP/2.0/UDP station1.work.com; branch=z9hG4bK789
Max-Forwards: 70
From: Mary<sip:Mary@work.com>; tag=123456
To: Joe<sip:Joe@work.com>; tag=67890
Contact: Mary<Mary@station1.work.com>
Refer-To: Susan<sip:Susan@station3.work.com>
Call-ID: 123456@station1.work.com
CSeq: 123 REFER
Content-Length: 0
```

b

c

```
SIP/2.0 202 Accepted
Via: SIP/2.0/UDP station1.work.com; branch=z9hG4bK789
From: Mary<sip:Mary@work.com>; tag=123456
To: Joe<sip:Joe@work.com>; tag=67890
Contact: Joe<Joe@station2.work.com>
Call-ID: 123456@station1.work.com
CSeq: 123 REFER
Content-Length: 0
```

```
INVITE sip:Susan@station3.work.com SIP/2.0
Via: SIP/2.0/UDP station2.work.com; branch=z9hG4bKxyz1
Max-Forwards: 70
From: Joe<sip:Joe@work.com>; tag=abcxyz
To: Susan<sip:Susan@station3.work.com>
Contact: Joe<sip:Joe@station2.work.com>
Call-ID: 67890@station2.work.com
CSeq: 567 INVITE
Content-Type: application/sdp
Content-Length: xx
Content-Disposition: session
{message body}
```

d

```
SIP/2.0 200 OK
Via: SIP/2.0/UDP station2.work.com; branch=z9hG4bKxyz1
From: Joe<sip:Joe@work.com>; tag=abcxyz
To: Susan<sip:Susan@station3.work.com>; tag=123xyz
Call-ID: 67890@station2.work.com
CSeq: 567 INVITE
Content-Type: application/sdp
Content-Length: xx
Content-Disposition: session
{message body}
```

e

f

```
NOTIFY sip:Mary@station1.work.com SIP/2.0
Via: SIP/2.0/UDP station2.work.com; branch=z9hG4bK123
Max-Forwards: 70
To: Joe<sip:Joe@work.com>; tag=67890
From: Mary<sip:Mary@work.com>; tag=123456
Contact: Joe<sip:Joe@station2.work.com>
Call-ID: 123456@station1.work.com
CSeq: 124 NOTIFY
Content-Type: message/sipfrag;version=2.0
Content-Length: 15
```

```
ACK sip:Susan@station3.work.com SIP/2.0
Via: SIP/2.0/UDP station2.work.com; branch=z9hG4bKxyz1
Max-Forwards: 70
From: Joe<sip:Joe@work.com>; tag=abcxyz
To: Susan<sip:Susan@station3.work.com>; tag=123xyz
Call-ID: 67890@station2.work.com
CSeq: 567 ACK
Content-Length: 0
```

```
SIP/2.0 200 OK
```

g

```
SIP/2.0 200 OK
Via: SIP/2.0/UDP station2.work.com; branch=z9hG4bK123
To: Joe<sip:Joe@work.com>; tag=67890
From: Mary<sip:Mary@work.com>; tag=123456
Call-ID: 123456@station1.work.com
CSeq: 124 NOTIFY
Content-Length: 0
```

In Figure 5-20, the REFER and NOTIFY requests between Mary and Joe
are sent within an established dialog. Once Joe receives the REFER request
from Mary, he issues a 202 (accepted) response. This is a new status code
that is not defined in the base SIP specification, but it is specified in several

SIP extensions. After sending the 202 response, Joe proceeds to send an INVITE request to Susan.

Once Susan accepts the INVITE request, Joe sends an ACK as normal. He (or his device) also sends a NOTIFY request to Mary to indicate that the actions performed as a result of the referral have been successful. The NOTIFY message has a message body that contains part of the SIP response received from Susan. The inclusion of a SIP message fragment is indicated by the "Content-Type: message/sipfrag; version=2.0" line of the NOTIFY request. This form of message content type is defined within a separate Internet draft, and it enables a SIP message part to be tunneled within a SIP message. If the content type had been "Content-Type: message/sip, version=2.0," then the whole of the 200 (OK) response would be included in the message body.

Although not shown in Figure 5-20, yet another Internet draft defines the Referred-By: header, which can be included in a REFER message and can optionally be included in an INVITE that occurs as a result of a REFER request. In the example of Figure 5-20, the Referred-By: header would enable Joe to indicate to Susan that he is contacting her at Mary's suggestion (which might make Susan more likely to accept his call).

In Figure 5-20, the dialog between Mary and Joe remains established, and Joe could return to the dialog after consultation with Susan. If either Mary or Joe wants to terminate the dialog, then they can do so in the normal manner through the use of the BYE request.

Reliability of Provisional Responses

A SIP session is instigated with an INVITE message. One or more provisional responses might occur, such as 100 (trying) or 183 (session progress) prior to a final response to the INVITE. SIP can use UDP transport, which is unreliable. Therefore, it is possible that messages sent from one end might never reach the other end. SIP addresses some of these issues by specifying rules for the retransmission of a request or a response. Moreover, SIP includes the ACK as a means to confirm that a final response has been received. Provisional responses do not generate an ACK, however. Therefore, a provisional response could be lost forever.

Imagine, for example, an INVITE sent from A to B. B responds with the provisional response such as 180 (ringing), but this response is not received at A. The called user then picks up the phone and B issues a final response of 200 (OK), which is received at A prior to expiry of the retransmission timer for the INVITE. A is happy that a final response has been received and the call is established. However, there might be a problem, particularly in the case of interworking with other networks.

In the case of interworking, the 180 (ringing) response or the 183 (session progress) response can be mapped to the Q.931 Alerting message or the ISUP *Address Complete message* (ACM). External network elements might be expecting such messages/responses to drive state machines, or certain applications might be relying upon guaranteed delivery to ensure that a particular feature or service operates correctly. A simple example would be a call to an unassigned number. The SS7 ACM message is used to open a one-way path from the called end to the calling end in order to relay an announcement such as "The number you have called has been changed." If the ACM is mapped to a 180 or 183 response and that response is not delivered to the calling end, then the caller might be left in the dark and not understand why the call did not connect. Therefore, a SIP extension has been specified to provide reliability for SIP provisional responses.

The document describing this SIP extension is titled "Reliability of Provisional Responses in SIP" (RFC 3262). The extension involves the addition of two new headers: *response sequence* (RSeq) and *response ACK* (RAck). RAck is a request header field, and RSeq is a response header field. The extension also defines an option tag to be used with the "supported" SIP header. This option tag is 100rel. The tag may also be included in the "unsupported" header to indicate that a server does not understand the option. The extension also defines a new SIP method, *Provisional Response ACK* (PRACK).

The extension works according to the following example. A client that supports the reliable delivery of provisional responses sends an INVITE, which includes the supported header with the option tag 100rel and the Require: header with the option tag 100rel. When a server receives the INVITE and determines that a provisional response should be delivered reliably, it sends the response and includes the RSeq header. It then waits for the client to send a PRACK message to confirm that the response has been received. If a PRACK message is received, then it is certain that the provisional response was received. Upon reception of the PRACK, the server returns 200 (OK) to indicate that the PRACK has been received. If the server does not receive the PRACK message, then it shall retransmit the response T1 seconds later (the default T1 value is 0.5 seconds or 500 milliseconds). If the server still does not receive a PRACK, then it retransmits the response every T1 seconds, up to a maximum of 64 times, after which the server assumes that a network failure has occurred.

The initial provisional response contains the RSeq header, which is defined to carry a 32-bit integer value between 1 and 4,294,967,295. The initial value of RSeq chosen the first time that a given response is returned is a random number between 1 and 2,147,483,647. Each time the response is retransmitted, the value of RSeq is incremented by 1. The provisional response also includes the Require: header containing the 100rel option tag.

The PRACK sent by the client contains the RAck header, which contains two components. The first component is the value of the RSeq copied from the provisional response for which the PRACK is being sent. The second is a CSeq number and method token copied from the response for which the PRACK is being issued.

The process is shown in Figure 5-21. In this example, the calling client uses the Supported: header to indicate in the INVITE that it supports the acknowledgment of provisional responses. The first 180 (ringing) response to the INVITE is lost. After a timeout, the server retransmits the response. Note that both instances of the response include the RSeq: header, but that the value has been incremented in the second instance of the response.

Upon receipt of the retransmitted response, the calling client acknowledges the response through the use of the PRACK method. This contains the RAck: header field. The content of the field is the value received in the RSeq: header of the response, plus the content of the CSeq: header in the response. Finally, the server confirms receipt of the PRACK by issuing a 200 (OK) response. Note that this 200 (OK) response relates to the PRACK and not to the initial INVITE.

Although reliability may be applied to provisional responses, it must not be applied to the 100 (trying) response and the RSeq: header should not be included in such a response. The 100 (trying) response is sent on a hop-by-hop basis, whereas the reliability mechanism described in this extension is end to end and does not apply to the 100 (trying) response.

The SIP UPDATE Method

As already described in this chapter, a SIP session is established with an INVITE. The far-end response to the INVITE will include a tag parameter and will lead to the establishment of a dialog. Depending upon whether the response is a provisional (1XX) response or a final (2XX) response, the dialog is either early or confirmed.

If something about the session needs to change (the codecs being used, for example), then a re-INVITE can be issued. One problem with this approach, however, is that the reissuance of an INVITE changes the state of the dialog. A re-INVITE within an established session (one that has a confirmed dialog) will generally work well as a means of changing session

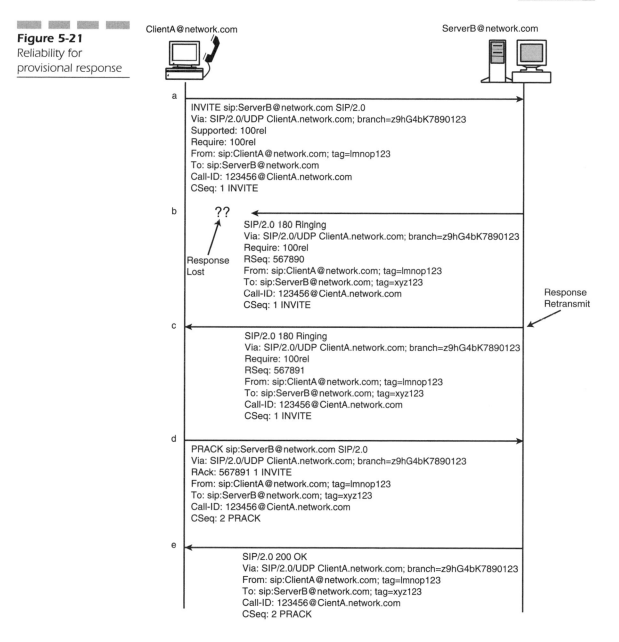

Figure 5-21
Reliability for
provisional response

data such as media parameters. A re-INVITE for an early dialog, however, cannot be used for the same purpose. Consider, for example, an INVITE that receives a 183 (session progress) response that includes a message body. That message body might establish a media stream from callee to caller for sending a ring tone or music while the called party is alerted. At this stage, the dialog is in the early state. If the caller wants to change any media parameters, such as the codec to be used, or wants to put the early media on hold, then a re-INVITE will not fill the need. Something else is required.

The solution is the new UPDATE method, which is specifically designed to enable the modification of session information before a final response to an INVITE is received. Operation of this extension is simple. The caller begins with an INVITE transaction, which proceeds normally. Once a dialog is established, either early or confirmed, the caller can send an UPDATE request method that contains an SDP offer for the purposes of updating the session. The response to the UPDATE method contains the answer.

Similarly, once a dialog is established, the callee can send an UPDATE with an offer, and the caller places its answer in the 2XX response to the UPDATE. The Allow header (in an INVITE, for example) can indicate support for the UPDATE method, so that one party can know if the other party supports the method.

One important usage of the UPDATE method is when reserving network resources as part of a SIP session establishment. We address this issue next.

Integration of SIP Signaling and Resource Management

Ensuring that sufficient resources are available to handle a media stream is a very important aspect of providing a high-quality service and is therefore a necessity for a carrier-grade network. For example, when a called party answers the phone and attempts to speak, it is imperative that the necessary resources are available to carry that speech back to the caller. Equally, resources must be available in the direction from caller to callee. Although customers may expect to experience occasional congestion, that experience should only happen while attempting to establish the call. Once the far end phone rings and is answered, a customer expects that conversation can take place. Anything else is considered defective operation.

In our discussion so far, we have mainly focused on the signaling related to call setup and teardown. The bandwidth required for signaling is miniscule compared to that required for handling the media to be exchanged, however. Furthermore, the signaling may take a very different path from the media. Therefore, we cannot assume that the successful transfer of signaling messages will automatically lead to a successful transfer of media. If carrier-grade operation is desired, the signaling must accommodate some mechanism to ensure that resources are available in the network for the transport of the required media.

As we shall describe in Chapter 8, "Quality of Service," resource reservation can be achieved either on a per-session basis or on an aggregate basis. On a per-session basis, end-to-end network resources are reserved in real time as part of session establishment. On an aggregate basis, however, a certain amount of network resources are reserved in advance for a certain type of usage, and network resource utilization is constantly monitored. If a session is to be established under such circumstances, it is simply necessary to check that the resource demands of the new session will not cause the total usage for sessions of that type to exceed any pre-arranged reservation.

For example, if an IP network were used to carry voice, user data, and management traffic, then it would be common to reserve a certain percentage of resources for voice, a certain percentage of resources for user data, and a certain percentage of resources for management traffic. The management of access to those resources takes place at the edge of the network. Therefore, if a new voice session is to be established, policing functions at the edge of the network are invoked to verify that the new session would not cause the total usage to exceed any prereserved maximum.

Since two main approaches to network resource reservation exist, any SIP solution for resource reservation must be compatible with both approaches. In other words, SIP-related resource reservation must work in a regime where resources are reserved on a session-by-session basis and must also work in a regime where resources are reserved in advance and where new sessions are allowed only if total usage does not exceed a prereserved maximum. The remainder of this section addresses a SIP extension that supports both regimes.

An Internet draft was prepared in 1999 describing a means for reserving network resources in advance of alerting the called user. The draft specified the use of a new SIP header to be used in the INVITE, indicating to the called end that resource reservation is needed and that the user should not yet be alerted. After successful resource reservation, a second INVITE would be issued to set up the call. This approach, however, had some incompatibilities with standard SIP. If new headers were not supported by the called end, a call could be established without sufficient resources, since SIP states that new unrecognized headers should be ignored. Consequently, the extension was rejected. The basic need was recognized, however, and several groups have worked for some time to define a new Internet draft to address the issue in a different way. That draft is titled "Integration of

Resource Management and SIP." As for all Internet drafts, this specification is subject to change.

The new approach utilizes reliable signaling for provisional responses and also utilizes the UPDATE method described previously. The technique also involves extensions to SDP. In fact, the details of the session preconditions are specified through extensions to SDP. This approach is reasonable, since SDP describes the media to be exchanged. Given that the *quality of service* (QoS) requirements are related to the media, it is logical that they should be included as part of the session description rather than within SIP itself. Moreover, if a given session is to include multiple media streams, then different QoS requirements might be applied to the different streams. For example, a session that includes both audio and video might have stronger QoS requirements for the audio in the event that it is considered more important. If different resource requirements are to be applied to different media streams, then there is no choice but to specify those requirements within the message body, rather than within SIP.

Of course, since SIP is used to carry the session description, it is not possible to completely divorce the SIP signaling from the media to be exchanged. In fact, resource reservation is one of the main reasons for the introduction of the UPDATE method described previously. SIP, however, should not care what the resource requirements happen to be. By specifying the requirements at the media level, it is possible to define any number of requirements in SDP in the future without having to revise SIP every time.

The basic methodology is depicted in Figure 5-22. An INVITE carries a session description that describes the offered media and preconditions to be met. The called entity returns a 183 (session progress) response containing a session description that requests the caller to reserve resources in the caller-to-callee direction. This response is sent using reliable signaling (and therefore causes a PRACK message from the caller to callee) because it contains a message body requesting resource reservation and we cannot take a risk that the response might be lost. Once the 183 (session progress) message has been sent, both caller and callee reserve resources in their respective directions. Once the calling entity has reserved resources, it sends an UPDATE request to the called end. The UPDATE indicates that resources have been reserved from the calling end to the called end. The called entity responds to the UPDATE to indicate that resources have been reserved from the called end to the calling end. The called user is now alerted and the call proceeds as normal.

As mentioned, the preconditions/requirements are specified in the SDP description. Let's take a closer look at how those preconditions are specified. A given precondition is defined to have three statuses: desired, current, and

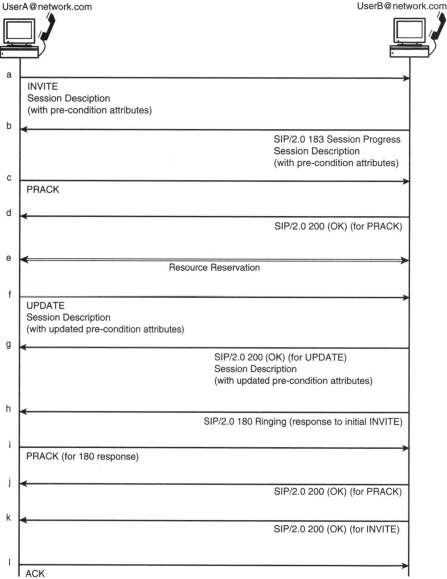

Figure 5-22
Session establishment with end-to-end resource reservation

confirmed. The current status reflects the existing state of affairs for a given resource reservation (which might be that no reservation has yet taken place). The desired status reflects a minimum threshold that must be achieved for a call to proceed. The confirm status indicates that the minimum threshold has been reached.

For each of the status types, we can indicate whether the status relates to the *end-to-end* (e2e) resource reservation, local resource reservation, or remote resource reservation. Recall that resource reservation can be on a session-by-session basis or on an aggregate basis. In the case of per-session resource reservation, resources must be reserved end to end as part of the session establishment. In the case of aggregate reservation, each end must check that it is allowed to access the reserved aggregate resources at its end (local). Moreover, each end must ensure that the far end has also been granted access (remote). In addition to the status type (e2e, local, or remote), we must also specify whether the reservation is for sending media, receiving media, or both. Finally, when requesting that the far end reserve resources, we can use a strength tag to indicate whether the reservation is optional or mandatory.

Example SDP Description for the Initial INVITE Referring again to Figure 5-22, let's look at examples of the media portions of the SDP descriptions that would be associated with the signaling exchange. Here is an example of an SDP description for the initial INVITE:

```
v=0
0=userA 45678 001 IN IP4 stationA.network.com
s=
c = IN IP4 stationA.network.com
t=0 0
m=audio 4444 RTP/AVP 0
a=curr: qos e2e none
a=des: qos mandatory e2e sendrecv
```

This SDP description indicates that User A wants to use G.711 μ-law voice. The description includes QoS-related preconditions, which specify that the current QoS is such that no end-to-end reservation has taken place. The desired precondition is that mandatory end-to-end resource reservation be in place for both the sending and receiving of media.

Example SDP Description for the 183 (Session Progress) Response This description is almost identical to that of the initial INVITE, with the exception of User-B-specific address and port information as well as the addition of the confirmation attribute. The confirmation attribute indicates that User B wants User A to reserve resources in the direction from User A to User B (which is the receive direction from User B's perspective). User B will separately reserve resources in the User-B-to-User-A direction.

```
v=0
0=userB 12345 001 IN IP4 stationB.network.com
```

```
s=
c = IN IP4 stationB.network.com
t=0 0
m=audio 6666 RTP/AVP 0
a=curr: qos e2e none
a=des: qos mandatory e2e sendrecv
a=conf: qos e2e recv
```

Example SDP Description for the UPDATE Request This SDP description indicates that User A has performed his part of the task; resources are reserved in the direction from User A to User B.

```
v=0
0=userA 45678 001 IN IP4 stationA.network.com
s=
c = IN IP4 stationA.network.com
t=0 0
m=audio 4444 RTP/AVP 0
a=curr: qos e2e send
a=des: qos mandatory e2e sendrecv
```

Example SDP Description for the 200 (OK) Response to the UPDATE Request This SDP description indicates that the current QoS status and desired QoS status are the same. In other words, resource reservation has taken place in both directions. The call can now proceed.

```
v=0
0=userB 12345 001 IN IP4 stationB.network.com
s=
c = IN IP4 stationB.network.com
t=0 0
m=audio 6666 RTP/AVP 0
a=curr: qos e2e sendrecv
a=des: qos mandatory e2e sendrecv
```

The foregoing description looks at resource reservation on an end-to-end basis. On the other hand, network resources might have already been reserved on an aggregate basis. In such a case, each end simply needs to verify with a local policing function that it is allowed access to available resources. The approach is to use status types that are segmented (local and remote) as opposed to end to end, which means that each participant deals with network access permission at its own end. The signaling flow for such a scenario would be as shown in Figure 5-23.

In Figure 5-23, User A requests access to network resources in advance of sending the INVITE. Assuming that access permission has been granted, then the INVITE message body indicates that local send and receive

Figure 5-23
Session establishment
with local and
remote resource
verification

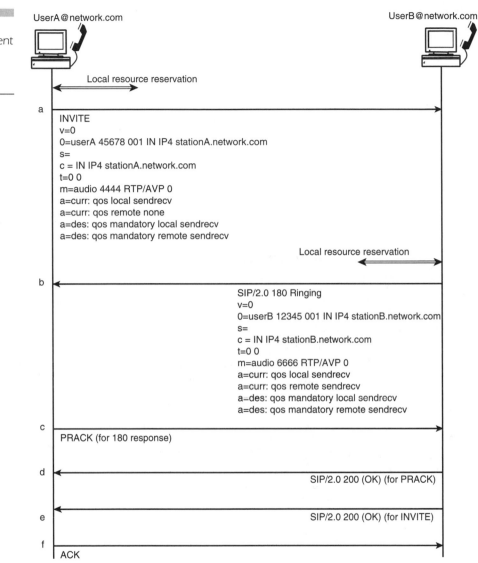

Figure 5-23
Session establishment with local and remote resource verification

resources are available, but that remote resources are not yet known to be available. Upon receipt of the INVITE, User B requests access to resources at the called end. Once access is granted, User B can proceed to alert the called party and returns a 180 (ringing) response to User A. That response contains a message body that indicates that the necessary resources are available at User B's end. The 180 (ringing) response is sent reliably. Since both ends have verified access to network resources, the call can proceed.

In the foregoing descriptions, mandatory strength means that the session cannot continue unless the required resources are definitely available. Optional strength means that the receiver of the session description should reserve resources and proceed with the call, but the resource reservation is not an absolute requirement. Strength of *none* is the initial situation and indicates that no effort to reserve resources has yet taken place. The strength value of *failure* also exists and would be included in a message body in the event that resource reservation did not succeed. This value would be returned in a 580 (precondition failure) response, which is a new response that is included as part of the resource reservation extension to SIP.

Note that, in the event that resources cannot be reserved, the decision not to complete a call might be made only as a last resort, even if the reservation is mandatory. If sufficient resources cannot be used, the option to select a lower-bandwidth codec to reduce the resource requirements is always available. This avenue should be explored before refusing to establish a call. However, it can only be explored if each end supports a lower-bandwidth codec than their first choice. Moreover, a lower-bandwidth codec often means lower speech quality, something that needs to be seriously considered before assuming that a lower-bandwidth codec is really an option.

Usage of SIP for Features and Services

A number of SIP messaging scenarios have already been described in this chapter, and most of them have described fairly straightforward call scenarios. We have also seen a number of extensions to SIP. In fact, one of the great advantages of SIP is the fact that it enables the easy addition of backwards-compatible extensions. Another great advantage of SIP is the fact that it can be used to provide a number of advanced features and services. For example, we have already seen how the REFER method can be used for a call-transfer application. We have also seen that SIP inherently supports personal mobility through the use of registration. We have seen how a forking proxy can enable a one-number service and we have also seen how the Retry-After: header can be used to support call-completion services. In many ways, these are just the tip of the iceberg when it comes to the features that SIP can offer.

It is possible for SIP messages to carry MIME content as well as an SDP description. Therefore, a response to an INVITE could include a piece of text, an HTML document, an image, and so on. Equally, because of the fact that a SIP address is a URL, it can easily be included in web content for click-to-call applications. Basically, SIP is well suited to integration with existing IP-based applications and can leverage those applications in the creation of new services. SIP not only supports new and exciting services, but it can also be used to implement the existing supplementary (CLASS) services that exist in traditional telephony today, including call waiting, call forwarding, multiparty calling, call screening, and so on.

In many cases, the signaling for a SIP call will be routed through a proxy. This approach is useful in many ways because it enables the proxy to invoke various types of advanced feature logic. The feature logic may be resident at the proxy or located in a separate feature server or database. In fact, the proxy may also have access to many other functions such as a policy server, an authentication server, and so on, as depicted in Figure 5-24.

The policy server could hold call-routing information and/or QoS-related information, while the feature server would hold subscriber-specific feature data, such as screening lists, forwarding information, and so on. It is perfectly possible for the proxy to use the services of an *Intelligent Network* (IN) *Service Control Point* (SCP), with which communication might be done over the *IN Application Part* (INAP), as described in Chapter 7, "VoIP and SS7," or the network might use the Parlay *Open Service Access* (OSA) approach, utilizing *application programming interfaces* (APIs) between the nodes. Although a distributed architecture has several advantages, it is also

Figure 5-24
SIP service
architecture

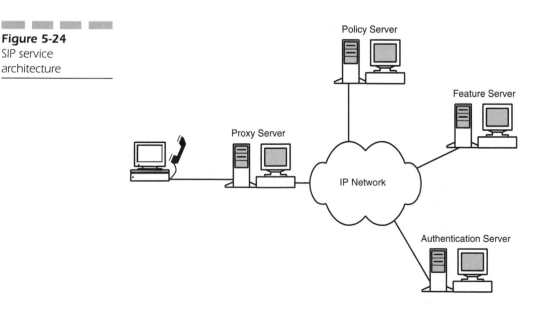

possible, of course, for the various functions depicted in Figure 5-24 to be integrated in the proxy in one monolithic implementation.

The following section describes just a few examples of calling features using SIP. These are simple examples and are certainly not meant as a complete picture of the features that SIP can enable.

Call Forwarding

Figure 5-25 depicts a call-forwarding-on-busy scenario. Note that, in order to avoid unnecessary clutter, the message body and many of the message headers have been omitted from the figure. These items would, of course, be included in a real signaling implementation.

User 1 calls User 2 via a proxy. User 2 wants calls to be forwarded to User 3 whenever User 2 is busy. Since User 2 happens to be busy, a 486 (busy here) response is returned. Upon receiving the response, the proxy forwards the call to User 3. Note that the INVITE request from the proxy to User 3 contains the same To: field as in the INVITE to User 2. This enables User 3 to recognize that this call is a forwarded call, originally sent to User 2. The 200 (OK) response from User 3 contains a Contact: header, indicating User 3. This enables subsequent messages for this call to pass directly from User 1 to User 3. The proxy could, however, have chosen to insert a Record-route: header in the INVITE. This would ensure that future messages and responses are sent through the proxy.

A call-forwarding-on-no-answer scenario would be quite similar to the call-forwarding-on-busy scenario. The difference is that the INVITE from the proxy to User 2 would timeout. The proxy would then send a CANCEL to User 2 and then forward the call by sending an INVITE to User 3.

The proxy would independently take care of a call-forwarding unconditional service. The proxy would not bother to send an INVITE to User 2 at all. It would simply proceed to send an INVITE to User 3. The To: field of the INVITE to user 3 would still indicate user 2, as is the case for all call-forwarding scenarios.

Consultation Hold

Figure 5-26 depicts a scenario where User B calls User A and, after speaking for a while, User B places User A on hold and places a call to User C. Once the call to User C is finished, User B reopens communication with User A. Again, for the avoidance of clutter, many of the SIP headers have been omitted.

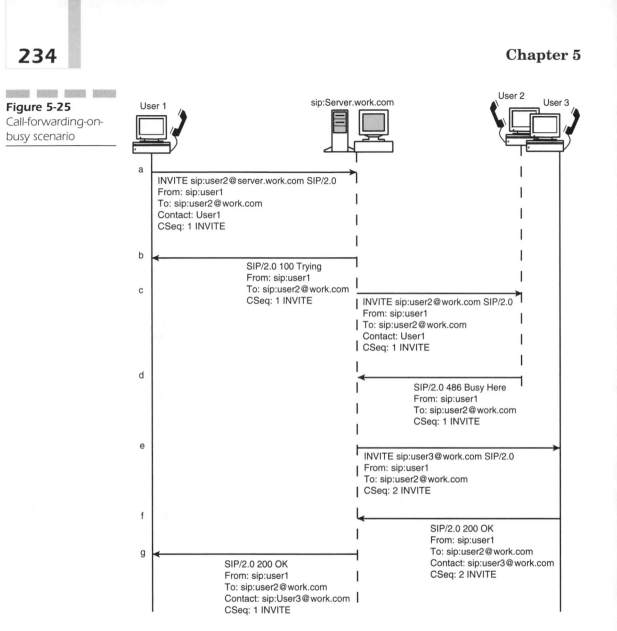

Note the manner in which a call is placed on hold. A SIP UPDATE is issued with the same Call-ID and other parameters as the original INVITE, but the SDP description includes a media attribute that specifies inactive (neither send nor receive). Depending on the situation, the issuer of the UPDATE could specify either inactive or sendonly. When sendonly is used, it means that the receiver of the UPDATE should send audio, but not receive any. The issuer of the UPDATE might also locally mute any received audio stream. The 200 OK response to the UPDATE can also include an updated session description, in accordance with the offer/answer model of SIP.

Figure 5-26
Usage of the UPDATE method to place a call on hold

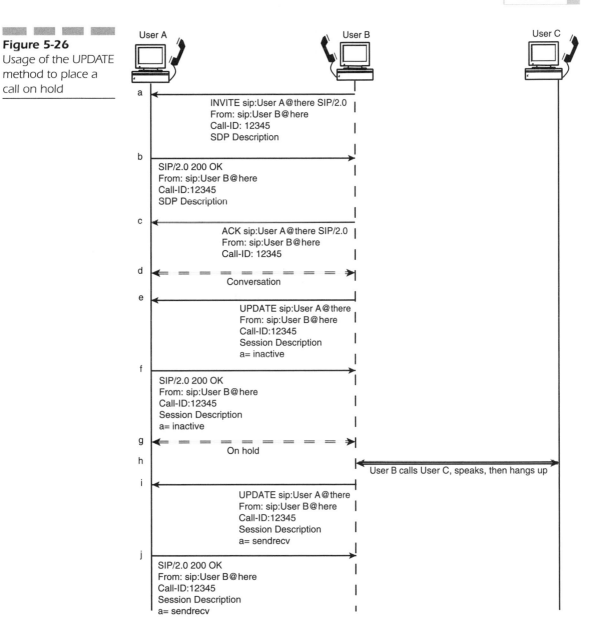

In the example, User B places a call to User C in a normal manner. After that call is finished, User B reopens communication with User A by issuing another UPDATE, again with the same Call-ID as the original call (message i in the figure). The message body in this UPDATE has attribute of sendrecv. Equally, the 200 (OK) response to the UPDATE (message j) has a sendrecv attribute in the SDP description. The 200 response is acknowledged (not shown) and communication between User A and User B is open again.

A real-life scenario for this type of signaling sequence is one where User A asks User B a question, and User B needs to check with User C for the correct answer. In such a situation, User C might need to talk to User A directly, in which case User B could use the REFER method to transfer the call to User C.

Interworking

Obviously, SIP-based networks will never be the only types of networks in existence. To begin with, circuit-switched networks will continue to be with us for a very long time. Therefore, an obvious need exists for SIP-based networks to be able to interwork with the circuit-switched networks of the PSTN. Furthermore, although SIP is viewed by many as the future of IP telephony, an embedded base of H.323 systems already exists, and more H.323 systems are being deployed. Therefore, a need also exists for SIP-based networks to interwork with H.323-based networks. The SIP specification, however, does not specifically address how such interworking should be achieved.

PSTN Interworking

For interworking with the PSTN, gateways will be required to provide the conversion from circuit-switched media to the packet and vice versa. In fact, we have already seen that the H.323 architecture places great emphasis on the role of gateways. Not only must there be media interworking, but there must also be signaling interworking. After all, SIP is a signaling protocol. If calls are to be established between a SIP-based network and the PSTN, then the SIP network must be able to communicate with the PSTN according to the signaling protocol used in the PSTN. In most cases, this protocol is SS7. For the establishment, maintenance, and teardown of speech calls, the applicable SS7 protocol is the *ISDN User Part* (ISUP).

The usage of gateways and the control of those gateways in a SIP-based network are discussed in some detail in Chapter 6. Furthermore, interworking with the PSTN through the use of signaling gateways is discussed in detail in Chapter 7. For the moment, however, it is sufficient to assume that a gateway entity exists on the network for interworking between SIP and the PSTN. Let us call that entity a *network gateway* (NGW). In such a case, let us consider a call from Manager (at work) to Collins (at home), where Collins has a standard residential phone connected to a PSTN switch. Such a scenario would appear as in Figure 5-27. Note that most of the fields within the SIP messages and responses are omitted to avoid clutter.

Recall that SIP enables a SIP URI to have the form of a telephone number. The INVITE includes such a URI in the To: field, which is mapped to the calling party number of the *initial address message* (IAM). The From: field of the INVITE could also have a URI in the form of a telephone for the calling party. This would enable the called party to still avail itself of the PSTN Caller-ID service.

The PSTN switch responds with an ACM, which is mapped to the SIP 183 (session progress) response. This response contains a session description and enables in-band information to be returned from the called switch to the SIP caller. Once the call is answered, an ISUP ANM is returned, which is mapped to the 200 (OK) response. This response will have an updated session description, since the session description in the 183 response is considered temporary and is part of an early dialog.

If the call were to be in the opposite direction, from the PSTN to a SIP user, then it would appear as in Figure 5-28. In this case, the 180 ringing response is used instead of the 183 session progress response, but this is for example purposes only. In this example, the 180 ringing response is used by the NGW to trigger a ringing tone towards the calling party locally. Alternatively, the called SIP user agent might return a 183 (session progress) response, with an early media description (such as a ringing tone) to be played to the caller. As is the case in the previous example, not every field is shown for the SIP messages and responses in Figure 5-28. Furthermore, for the avoidance of clutter, the usage of reliable signaling for SIP provisional responses is not shown in either example. It can be assumed that the reliability of these responses would be applied, so that PRACK methods and corresponding 200 (OK) responses would occur.

Although these two interworking examples might seem quite straightforward, seamless interworking between two different protocols is not often quite that easy, because rarely does a one-to-one mapping of messages and parameters take place between one protocol and the other. Sometimes the mapping from protocol A to B means that what is sent in protocol B is only an approximation of what was received in protocol A. If dual interworking is used (from protocol A to protocol B and back to protocol A), what comes out at one end may not be the same as what went in at the other end. Given

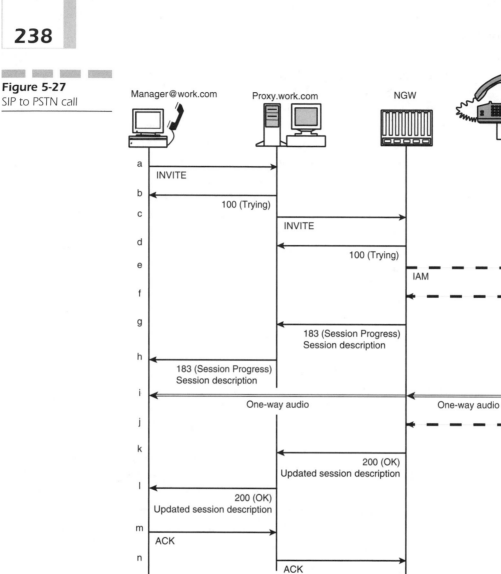

Figure 5-27
SIP to PSTN call

that certain VoIP networks provide a transit service from PSTN at one end to PSTN at the other end, this lack of transparency is a concern.

The solution to this issue is known as *SIP for Telephony* (SIP-T), which is documented in an Internet draft. SIP-T is not so much an extension to SIP, but rather it is a specification of how best to use SIP when interworking with traditional telephony networks. As such, the SIP-T draft is categorized as a *Best Current Practice* (BCP) rather than a potential technical standard.

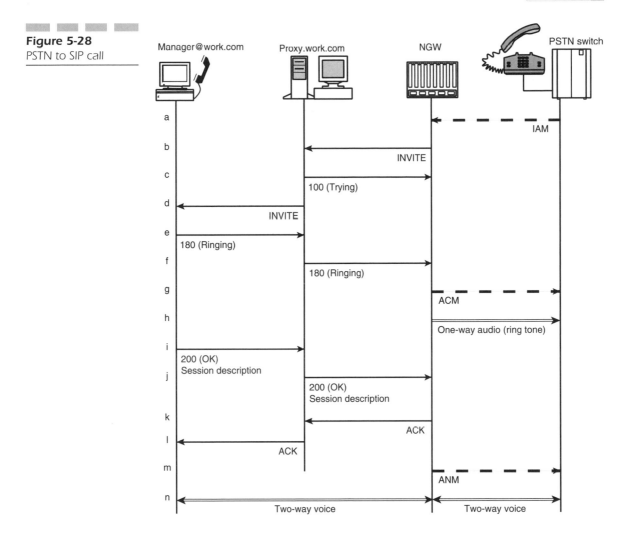

Figure 5-28
PSTN to SIP call

A significant aspect of SIP-T is the capability to carry ISUP or QSIG message contents within a SIP message body. This capability is described in RFC 3204 (MIME media types for ISUP and QSIG objects). The approach is to carry an ISUP or QSIG message within a SIP message body in hex format. In such a case, the SIP message body could be a multipart message body, including both an SDP session description as one part and an ISUP or QSIG message as another. The SIP INFO method is another component of SIP-T, as it provides for the transfer of midcall information within a SIP session. Such midcall information would typically be generated within the PSTN.

The whole issue of interworking with SS7 is fundamental to the success of VoIP in the real world. The issue is a major concern not just to enable SIP-based networks to interwork seamlessly with the PSTN for regular phone calls, but also to enable SIP-based systems to avail themselves of services in the PSTN that rely on SS7 signaling, including IN services such as toll-free calling. Therefore, Chapter 7 of this book is devoted to the topic and addresses the transit issue just described as well as many other aspects.

Interworking with H.323

Interworking with H.323 is also a major consideration, particularly because H.323 is already widely deployed. Several Internet drafts have been prepared to address the issue. One in particular, titled "SIP-H.323 Interworking Requirements," is under consideration by the IESG as a potential informational RFC.

The proposed approach for interworking between a SIP network and an H.323 network involves the use of a gateway, as depicted in Figure 5-29. Depending on the function to be performed, the gateway may appear to the SIP network as a user-agent client or user-agent server. To the H.323 network, the gateway will appear as an H.323 endpoint. On the SIP side, the gateway might also include a SIP registrar function and/or proxy function, and on the H.323 side, it could contain a gatekeeper function.

Figure 5-30 depicts an example scenario of a call from a SIP caller to an H.323 terminal. The H.323 Fast Connect procedure is used, and we assume that a gatekeeper is not involved in the call. If a gatekeeper were to be used, then the RAS signaling would occur on the H.323 side of the gateway and would not be seen by the SIP network in any case. The SDP information in the SIP INVITE is mapped to the logical channel information included in

Figure 5-29
SIP-H.323
interworking
gateway

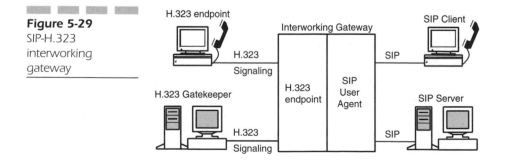

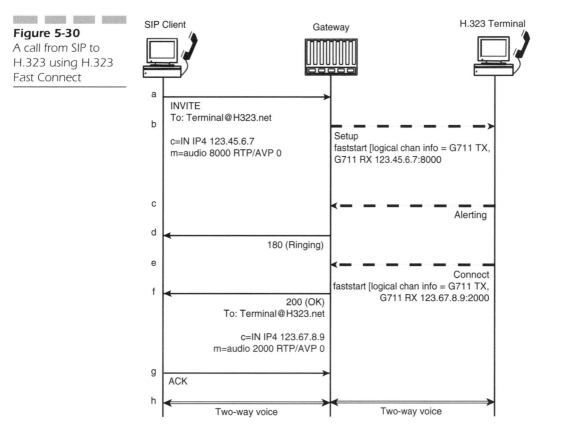

Figure 5-30

A call from SIP to H.323 using H.323 Fast Connect

the faststart element of the Q.931 SETUP message from the gateway to the H.323 terminal. In the example, a faststart element is conveniently returned from the called terminal in the Connect message rather than the Alerting. In this case, the logical channel information within the Connect message can be mapped directly to the session description within the SIP 200 (OK) response.

If the faststart element had been included in the Alerting message instead or some other messages (such as Call Proceeding), then the gateway would simply have had to store the information until it was ready to send it to the SIP client in the 200 (OK) response. Note that the text included with the messages in the figure is for explanatory purposes only and does not match the actual syntax of the messages as they should appear in a real implementation.

If we assume that the called H.323 terminal network does not support the Fast Connect procedure, then the sequence becomes more long-winded, as shown in Figure 5-31. In this case, the gateway supports Fast Connect, but the called terminal does not. Thus, the Setup message includes the fast-start element, but the messages returned from the called H.323 terminal do not. Therefore, upon receipt of the Connect message, the gateway must perform H.245 logical channel signaling to and from the called terminal before returning the 200 (OK) to the calling SIP client. The H.245 messaging includes capability exchange and the establishment of logical channels. For

Figure 5-31
A call from SIP to H.323 without H.323 Fast Connect at the destination

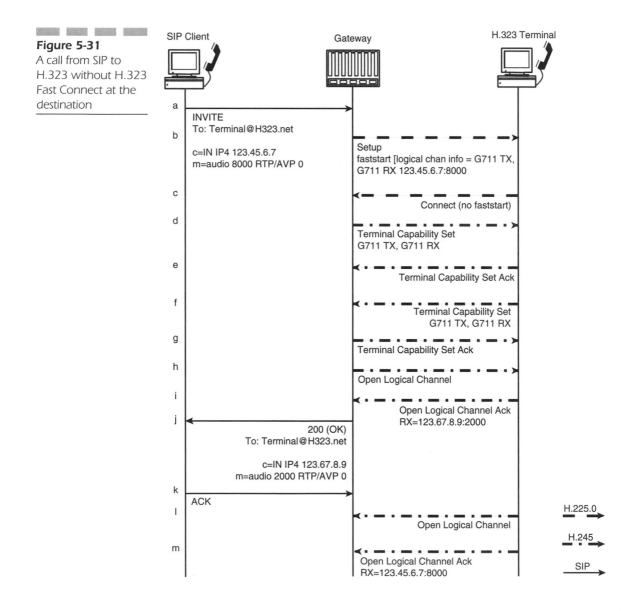

SIP Client

Gateway

H.323 Terminal

a

INVITE
To: Terminal@H323.net

b

c=IN IP4 123.45.6.7
m=audio 8000 RTP/AVP 0

Setup
faststart [logical chan info = G711 TX,
G711 RX 123.45.6.7:8000

c

Connect (no faststart)

d

Terminal Capability Set
G711 TX, G711 RX

e

Terminal Capability Set Ack

f

Terminal Capability Set
G711 TX, G711 RX

g

Terminal Capability Set Ack

h

Open Logical Channel

i

Open Logical Channel Ack
RX=123.67.8.9:2000

j

200 (OK)
To: Terminal@H323.net

c=IN IP4 123.67.8.9
m=audio 2000 RTP/AVP 0

k

ACK

l

Open Logical Channel

H.225.0

H.245

m

Open Logical Channel Ack
RX=123.45.6.7:8000

SIP

clarity, interim responses from the called terminal (Call Progress, Alerting, and so on) are not shown in the figure.

A call in the opposite direction, from H.323 to SIP, is relatively straight-forward if the H.323 terminal and the gateway both support the H.323 Fast Connect procedure. Such a scenario is depicted in Figure 5-32. In this case, the session information is included in the H.225/Q.931 SETUP message and this can be mapped to the SDP information within the SIP INVITE. Equally, the session description accompanying the 200 (OK) response from the called SIP user agent can be mapped to the information within the faststart element of the Connect message back to the H.323 terminal.

If, on the other hand, the calling H.323 terminal does not support the Fast Connect procedure, then the interworking becomes more complex for two reasons. First, the fact that Fast Connect is not used means that separate H.245 logical-channel signaling is required on the H.323 side of the gateway. Second, no logical channel information is carried within the Setup message. Therefore, nothing can be mapped to a session description in the SIP INVITE.

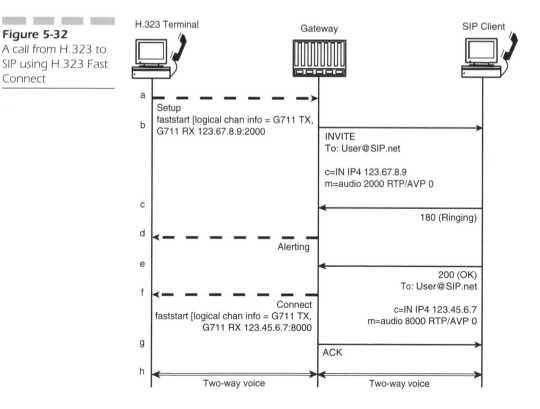

Figure 5-32
A call from H.323 to SIP using H.323 Fast Connect

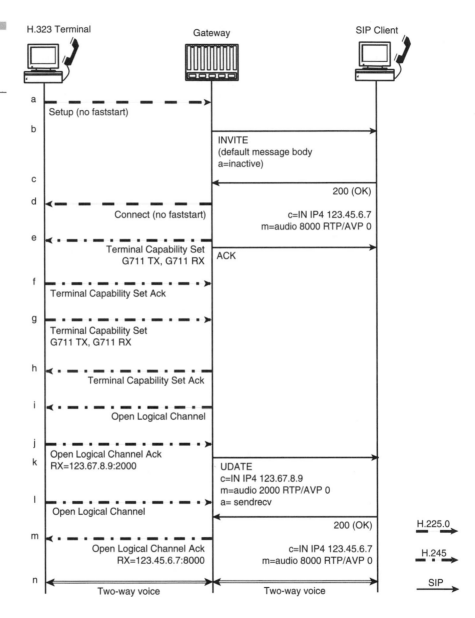

In that event, the INVITE could contain a default session description. The IP address in the connection (c=) line of the session description could contain a default value, while the media (m) line would indicate a port number of 0 and an attribute of inactive, as shown in Figure 5-33. Once the appropriate H.245 signaling has taken place between the gateway and the

H.323 terminal, the gateway can issue a SIP UPDATE to the SIP client, including a revised session description.

Summary

Although quite new, SIP has already established itself as the way of the future for signaling in VoIP networks. Its strengths include the fact that it is far simpler than H.323 and is therefore easier to implement. Its simplicity, however, belies its flexibility and power. It can support all the features that are already supported by traditional telephony networks and more besides. In fact, SIP is the protocol of choice for the evolution of third-generation wireless networks, where it is expected that SIP-based mobile devices will become available and SIP-based network elements will be introduced within mobile networks.

As we shall see in the next chapter, SIP also fits very well with those protocols used for media gateway control and, as such, forms part of the overall architecture known as softswitch.

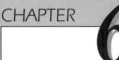

Media Gateway Control and the Softswitch Architecture

Previous chapters described some of the advantages that *Voice over IP* (VoIP) can offer over traditional circuit-switched telephony, such as lower cost of network implementation, integration of voice and data applications, new features, and potentially reduced bandwidth for voice calls. It would be nice if all telephony were to be carried over the *Internet Protocol* (IP) so that these advantages could be made available globally. Unfortunately, that day will not arrive for a long time. Replacing all traditional circuit-switched networks is not feasible, and perhaps not even desirable. Among other reasons, the cost would simply be exorbitant. Rather, one can expect a gradual evolution from circuit-switched to IP-based networks such that both types of networks will exist side by side for a very long time.

Given that VoIP networks and traditional circuit-switched networks will have to coexist for many years, they will need to interwork as seamlessly as possible. In other words, users of existing circuit-switched systems should be able to place calls to VoIP users, and vice versa. In each case, the user should not have to do anything radically different from simply calling another user on the same type of network. The challenge therefore is to develop solutions that make such seamless interworking possible.

Separation of Media and Call Control

Previous chapters discussed the interworking between VoIP networks and circuit-switched networks through the use of gateways. The primary function of such gateways is to make the VoIP network appear to the circuit-switched network as a native circuit-switched system and vice versa. In other words, VoIP networks must accept signaling and media from circuit-switched networks in native format and convert both to the format used within the IP network.

We also saw in Chapter 4, "H.323," that the signaling involved in a VoIP call can take a very different path from the media path itself. The media might go directly from end to end, while the signaling might pass through one or more intermediate entities, such as gatekeepers or *Session Initiation Protocol* (SIP) proxies. In fact, the logical separation of signaling and media is not a new concept; it is something that has existed for many years in the *Signaling System 7* (SS7) architecture.

Putting these two considerations together means that a network gateway has two related but separate functions: signaling conversion and media conversion. The signaling aspect is coupled to whatever call-control functions are operating on the network. In fact, signaling can be considered the language that call-control entities use to communicate. The media conversion and transport can be considered a slave function, invoked and manipulated to meet the needs dictated by call control and signaling.

In terms of network architecture, this leads to a configuration of the kind depicted in Figure 6-1. This figure shows a scenario where a VoIP network sits between two external networks, as might be the case for a VoIP network that provides long-distance service. Figure 6-1 illustrates the separation of call control and signaling from the media path. Moreover, it illustrates the master-slave relationship between call control and media, even within the network gateway itself. Even if the gateway is just a single box handling both signaling and media, some internal control system will be used between the call-control functions and the media-handling functions.

Given that such a logical separation exists, there is no compelling reason why the separation could not also be made physical. In fact, such a physical separation has some significant advantages. First, a physical separation enables media conversion to take place as close as possible to the traffic source or sink within the circuit-switched network while centralizing the call-handling functions. Second, a smaller number of gateway controllers or call agents located more centrally can control multiple gateways placed in various locations. Such a network architecture leads to another benefit, which is that new features could be rolled out more quickly as they would need to be implemented only in the centralized call-control nodes, rather than at every node in the network. All of this leads us to the type of architecture shown in Figure 6-2.

Of course, the physical separation that could bring about these benefits requires a well-standardized control protocol between the *media gateway*

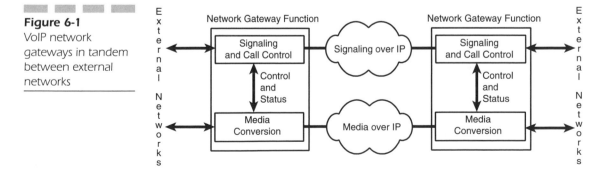

Figure 6-1

VoIP network gateways in tandem between external networks

Figure 6-2
Distributed system
architecture

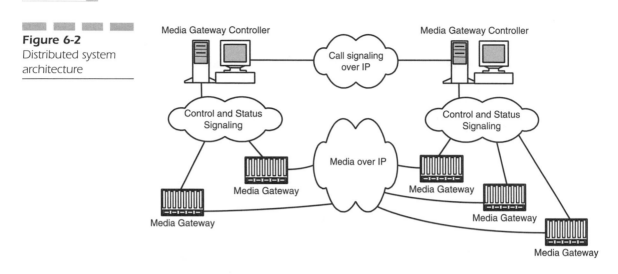

controller (MGC) or the call agent and *media gateway* (MG). Numerous such standardized protocols have been developed. The two most common protocols are the *Media Gateway Control Protocol* (MGCP or MEGACO) and MEGACO/H.248. By using one or another of these protocols for media gateway control and SIP as the call-signaling protocol, the architecture shown in Figure 6-2 is a reality and is known as the *softswitch architecture*.

Softswitch Architecture

The softswitch architecture involves the separation of the media path and media-conversion functions from the call control and signaling functions, as depicted in Figure 6-2. The entities that handle call control are known as call agents or MGCs, while the entities that perform the media conversion are known as MGs. The name softswitch is used because many of the switching functions traditionally handled by large monolithic systems in the circuit-switched world are instead emulated by software systems. In fact, in many quarters, the term softswitch is used to refer to a call agent or MGC as opposed to the overall architecture.

The softswitch architecture has many supporters within the VoIP industry. In fact, many of the leaders in the industry have come together in an organization known as the *International Softswitch Consortium*. This organization promotes the softswitch concept and related technologies, it allows interested parties to share ideas and cooperate in the ongoing development and deployment of the technology, and it facilitates interwork testing between different implementations.

One of the reasons why the softswitch approach is so popular is the fact that it is a distributed architecture. For a network operator, it is possible to use different network components from different vendors, such that the best in each class may be chosen in each area. For equipment vendors, it is possible to focus efforts on one area and not have to develop or acquire expertise in all areas. For example, a company can focus on the development of an MGC without having expertise in the voice-coding techniques and related *digital signal processors* (DSPs) used within MGs. Equally, a company that does have the necessary voice-coding expertise can develop an MG without having the call management expertise needed to develop a call agent. Another distinct advantage is the rapid rollout of new features described previously.

In the softswitch architecture, SIP is often used as the signaling protocol between the MGCs, so that an MGC appears as a SIP user agent in the SIP network. Therefore, when interworking between SIP and the *Public Switched Telephone Network* (PSTN), the MGC handles the functional interworking between SIP and PSTN signaling such as *ISDN User Part* (ISUP), while the gateways take care of the media conversion functions, as shown in Figure 6-3.

As mentioned, two popular protocols exist for media gateway control. MGCP was developed within the *Internet Engineering Task Force* (IETF) and MEGACO/H.248 has been developed as a cooperative effort between

Figure 6-3
Softswitch/PSTN
interworking

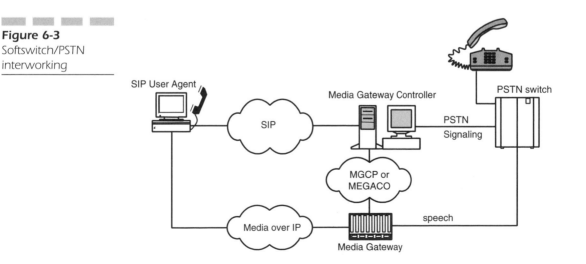

the IETF and the *International Telecommunication Union Telecommunications Standardization Sector* (ITU-T) Study Group 16. This reason is why MEGACO/H.248 has such a strange name. Within the IETF, it is known as MEGACO (for media gateway control protocol) and within the ITU it is known as H.248. Although MGCP is still popular in many implementations, MEGACO/H.248 is the media gateway control protocol of the future, and can be considered the successor to MGCP. Most new implementations use MEGACO/H.248.

Requirements for Media Gateway Control

Before delving deeply into the details of MGCP or MEGACO, it is first worth considering the functions that the protocols should support and what requirements exist for interworking with other protocols. This background information will clarify the reasons behind the protocols in the first place.

RFC 2805, "Media Gateway Control Protocol Architecture and Requirements," is an informational *Request for Comments* (RFC), describing the functions to be supported by a distributed gateway system, as depicted in Figure 6-3. Although the focus in this chapter is interworking between IP-based networks and circuit-switched networks, the document addresses generic requirements to be met for interworking with any type of network, not just circuit-switched networks. At a very high level, the requirements for the control protocol are that it should

- Enable the creation, modification, and deletion of media streams across an MG, including the capability to negotiate the media formats to be used.

- Allow the specification of the transformations to be applied to media streams as they pass through an MG.

- Allow the MGC to request and the MG to report the occurrence of specified events within the media stream (such as *Dual Tone Multifrequency* [DTMF] tones or call-associated signaling), and allow the MGC to specify the actions that should be automatically taken by the MG when such events are detected.

- Allow the MGC to request that the MG apply tones or announcements either by direct instruction or upon the occurrence of certain events, such as applying a dial tone to an analog line when an off-hook event is detected.

- Enable the establishment of media streams according to certain *quality of service* (QoS) requirements and to change QoS during a call.
- Enable the reporting of QoS statistics from an MG to an MGC. Since the MGC is not involved in the media transfer, it has no visibility of quality considerations such as delay, jitter, and packet loss and will need to be informed for network management purposes.
- Enable the reporting of billing/accounting information such as call start and stop times, the volume of content carried by media flow (such as the number of packets/bytes), and other statistics.
- Support the management of associations between an MG and an MGC, such as the capability for a given gateway to be controlled by a different MGC in the case of failure of a primary MGC.
- Support a flexible and scalable architecture, whereby different MGs with different capacities and different interfaces may be controlled by an MGC.
- Facilitate the independent upgrade of MGs and MGCs, including the use of a version indicator in message exchanges.

Protocols for Media Gateway Control

As mentioned earlier, several protocols have been written for media gateway control. Of these, two have been widely embraced. The first of these is MGCP (RFC 2705). This is an interesting document in that it is purely an informational RFC (rather than a standards-track protocol), even though it includes a protocol specification. The reason why it is categorized as informational is because it is being succeeded by MEGACO/H.248. However, work on MEGACO started much later than MGCP. Even while MGCP was still an Internet draft, many systems developers proceeded to include it within their product development rather than wait for MEGACO. Therefore, the decision was taken to release the MGCP specification as an informational RFC.

RFC 2885 was the first release of MEGACO/H.248 as a standards-track protocol. It was subsequently discovered that the protocol had some errors, and a separate RFC (RFC 2886) was prepared to document the errata and propose appropriate corrections. Subsequently, the original RFC was updated, and a new RFC (RFC 3015) was published and included the necessary corrections to RFC 2885. Thus, RFC 3015 is now the official version of MEGACO version 1. As of this writing, version 2 of the MEGACO specification is being prepared at the IETF and ITU. The second version adds some new capabilities and modifies some aspects of the version 1 specification.

Although MEGACO is clearly the way forward for media gateway control, many systems developers continue to implement MGCP within their systems, and MGCP specification enhancement work continues at the IETF. In this chapter the description of MGCP is based on the latest draft update to RFC 2705. Given that MGCP and MEGACO are both popular, we describe both in this chapter.

MGCP

MGCP is a master-slave protocol in which controllers, known as call agents, control the operation of MGs. The call control intelligence and related call signaling is taken care of by the call agent, while the MG takes instruction from a call agent and basically does what the call agent commands. The commands from the call agent generally relate to the establishment of connections and the teardown of connections from one side of the gateway to another. In most cases, the call agent tells the MG to make a connection from a line or trunk on the circuit-switched side of the gateway to a *Real-Time Transport Protocol* (RTP) port on the IP side of the gateway, as depicted in Figure 6-4.

MGCP only addresses the communication between a call agent and an MG and does not address communication from one call agent to another.

Figure 6-4
The basic approach to MGCP connection establishment

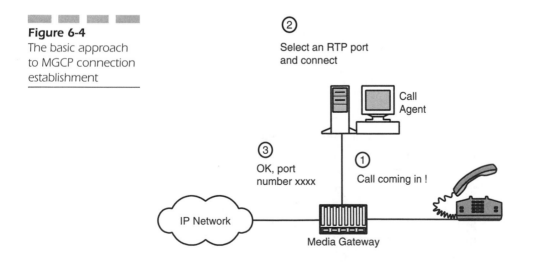

Since instances will occur where media is to be exchanged between gateways controlled by different call agents, signaling communication will be needed between call agents. MGCP assumes that such a signaling protocol exists, but does not specify what it is. In reality, the protocol between call agents is likely to be SIP.

The MGCP Model

MGCP describes a model containing endpoints and connections. Endpoints are sources or sinks of media and include such elements as trunk interfaces or *plain old telephone service* (POTS) line interfaces. Endpoints exist within MGs, and depending on the type of endpoint, it might or might not have one or more external channel or line interfaces. For example, an analog line endpoint would be connected to a physical analog line, which would in turn connect to an analog telephone. On the other hand, endpoints also exist that have no external connections. One example is an announcement endpoint that would reside internally in the gateway. Other entities in the IP network would connect to it for the purpose of retrieving an announcement.

Connections are the allocation of IP resources to an endpoint such that the endpoint may be accessed from the IP network. For example, when a call is to be established across a gateway from a circuit-switched line to an IP interface, an ad hoc relationship is established between the circuit-switched line and an RTP port on the IP side. This relationship is called a connection, and a single endpoint can have several connections. A general representation of endpoints and connections is provided in Figure 6-5.

MGCP Endpoints

MGCP describes a number of different endpoints as follows:

- **DS0 channel** A digital channel operating at 64 Kbps. Such a channel will typically be multiplexed within a larger transmission facility such as a DS1 (1.544 Mbps) or E1 (2.048 Mbps). In most cases,

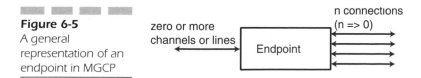

Figure 6-5
A general
representation of an
endpoint in MGCP

zero or more
channels or lines

Endpoint

n connections
(n => 0)

the DS0 will carry voice coded according to either G.711 μ-law or G.711 A-law. In some cases, the DS0 will carry signaling instead of voice, such as an *Integrated Services Digital Network* (ISDN) D-channel, in which case it is necessary that the signaling information received be passed from the MG to the call agent for processing.

■ **Analog line**   Typically, an endpoint that interfaces to a standard telephone line and provides service to a traditional telephone. In most cases, the media to be received on the analog line will be an analog voice stream, though it could also be audio-encoded data from a modem, in which case the gateway shall be required to extract the data and forward it as IP packets.

■ **Announcement server access point**   An endpoint providing access to a single announcement. Normally, a connection to this type of endpoint will be one-way, since the requirement is for media to be sent from the announcement rather than to it. We assume that the announcement server is within the IP network and IP-based connections are to be made to it. Thus, the endpoint does not have any external circuit-switched channels or lines associated with it.

■ *Interactive Voice Response* **(IVR) access point**   An endpoint providing access to an IVR system, where prompts or announcements may be played, and responses from the listener may be acted upon.

■ **Conference bridge access point**   An endpoint where media streams from multiple callers may be mixed and provided back to some or all of the callers.

■ **Packet relay**   A specific form of conference bridge supporting only two connections. This type of endpoint will use packet transport on both connections. A typical example would be a firewall between an open and a protected network, where the media stream must pass through such a relay point rather than directly from end to end.

■ **Wiretap access point**   A point from which a connection is made to another endpoint for the purposes of listening to the media transmitted or received at that endpoint. Connections to a wiretap access point will be one-way only.

■ **ATM trunk-side interface**   An endpoint corresponding to the termination of an *Asynchronous Transfer Mode* (ATM) trunk, where such a trunk might be an ATM virtual circuit. Such an endpoint might be used in a gateway that provides an interface between a VoIP network on one side and a Voice over ATM network on the other.

Each endpoint is identified with an endpoint identifier. This identifier will be comprised of the domain name of the gateway to which it belongs and a local name within the gateway. The local name within the gateway will obviously depend on the gateway in question, but it will generally have a hierarchical form such as X/Y or X/Y/Z, where Y is an entity within X, and Z is an entity within Y. An example would be a gateway that supports multiple channelized T3 interfaces. If one wanted to identify a single DS0 channel on the gateway, then one would have to identify the particular T3 (X), the particular DS1 within the T3 (Y), and the particular DS0 (Z) within that DS1. If we wanted to identify DS0 number 7 within DS1 number 12 on DS3 number 4, then the identifier could be something like trunk4/12/7@ gateway.somenetwork.net.

We can issue a wildcard to certain components of the identifier using either $ (any) or * (all). Thus, if we wanted to refer to any DS0 within the fifth DS1 within the first T3 in our example, then we could do so through the term trunk1/5/$@gateway.somenetwork.ne. This type of wildcarding is useful when a call agent wants to create a connection on an endpoint in a gateway and does not really care which endpoint is used. Through the use of the $ wildcard, the choice of a particular endpoint can be left to the gateway itself. Through the use of the * wildcard, the call agent can instruct the gateway to perform some action related to a number of endpoints, such as a situation where the call agent requests statistical information related to all endpoints on a gateway.

MGCP Calls and Connections

A connection is the relationship established between a given endpoint and an RTP/IP session. Consider, for example, a DS0 endpoint. If the DS0 is idle, then no connection is associated with the endpoint and no resources are allocated to the endpoint on the IP side of the gateway. When the DS0 is to be used to carry voice traffic, however, then IP resources are allocated to the endpoint, and a link is created from the DS0 on the line side of the gateway to an IP session on the IP side of the gateway. In MGCP terminology, this link is called a connection. If two endpoints are involved in a call, then two connections are used.

A call is a grouping of connections such that the endpoints associated with those connections may send media to each other and/or receive media from each other. Figure 6-6 shows a scenario where a call is established between two analog phones, such that they communicate across an IP

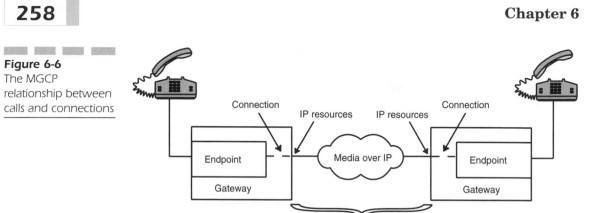

Figure 6-6
The MGCP
relationship between
calls and connections

network using two gateways. On each gateway, a connection is created between an endpoint on the line side of the gateway and IP resources (such as the IP address and RTP port number) on the IP side of the gateway. Once a connection is created on each of the two endpoints, then the call can be considered the linking of these two connections such that media may be passed between the endpoints.

When a gateway creates a connection for an endpoint and allocates IP resources, those resources are described in terms of a *Session Description Protocol* (SDP) session description. The establishment of a call involves the endpoints sharing their respective session descriptions with each other so that they may exchange media streams. The primary function of MGCP is to enable those connections to be created and to enable the session descriptions to be exchanged. MGCP does so through the use of commands and responses between call agents and MGs.

Overview of MGCP Commands

MGCP defines nine commands, some of which are sent from call agent to gateway and some from gateway to call agent. MGCP commands are composed of a command line, a number of parameter lines, and, optionally, a session description. The command line and parameters are text, using the US-ASCII character set. The session description uses SDP, also a text-based protocol.

The command line and the various parameters are separated from each other by a *carriage return and line feed character* (CRLF). The session description is separated from the command line and the various parameter lines by a single empty line, similar to the way things are done in SIP.

The command line is composed of the requested verb (the name of the command), a transaction identifier, the endpoint for which the command applies (or from which the command has been issued), and the protocol version (MGCP 1.0). These four items are separated from each other by a *space* (SP) character. Thus, the command appears as

```
CommandVerb SP TransactionID SP EndpointID SP MGCP 1.0
```

MGCP defines the nine commands in the following list. For each command, the requested verb is coded as a four-letter uppercase ASCII string:

- **EndpointConfiguration (EPCF)** Issued by a call agent to a gateway to inform the gateway about bearer/coding characteristics on the line side of one or more endpoints. The only characteristic currently specified is whether the encoding on the line side is A-law or μ-law.

- **CreateConnection (CRCX)** Issued by a call agent to a gateway and used for the creation of a connection at an endpoint.

- **ModifyConnection (MDCX)** Issued by a call agent to a gateway and used to change some of the characteristics of an existing connection. One of the primary functions is to provide information to the endpoint regarding the far end of the call.

- **DeleteConnection (DLCX)** Issued by a call agent to a gateway to instruct the gateway to terminate a connection (or multiple connections). The command can also be issued by a gateway to a call agent (for example, when the gateway has detected that line-side connectivity has been lost).

- **NotificationRequest (RQNT)** Issued by a call agent to a gateway to request that the gateway notify the call agent when certain events occur. The gateway could be instructed, for example, to detect an off-hook signal from an analog line.

- **Notify (NTFY)** Issued by a gateway to a call agent to report the occurrence of certain events.

- **AuditEndpoint (AUEP)** Issued by a call agent to a gateway to query the status of a given endpoint and/or other data regarding the endpoint. The gateway will respond with the requested information.

- **AuditConnection (AUCX)** Issued by a call agent to a gateway to query information regarding a specific connection.

- **RestartInProgress (RSIP)** Issued by a gateway to a call agent to inform the call agent when one or more endpoints are taken in or out of service.

Command Encapsulation   MGCP supports the concept of encapsulation, where one command can be included within another. For example, when instructing a gateway to create a connection (a CRCX command), a call agent can simultaneously instruct the gateway to notify the call agent of certain events. Thus, we can find a NotificationRequest encapsulated within a CreateConnection command. This feature is particularly useful in cases when certain actions need to be taken or events need to be detected only in conjunction with other activities. For example, a call agent might require an MG to detect and report DTMF tones, but only when a call is in progress. In such a case, a NotificationRequest command instructing the gateway to detect and report DTMF tones would be encapsulated within a CreateConnection command, so that the tone detection would only occur in the context of the connection created.

MGCP enables only one level of encapsulation. In other words, one command cannot be encapsulated with a second command, if that second command is already encapsulated within a third command. However, MGCP does enable multiple commands to be sent at the same time within the same *User Datagram Protocol* (UDP) packet.

MGCP Parameters   Each parameter is identified by a code that comprises one or two uppercase characters. A parameter code is followed by a single colon, one space character, and the parameter value. In some cases, the parameter value might be a single numeric value. In other cases, the parameter value might be a single hexadecimal string. In still other cases, it might be a comma-separated list of values or might comprise a list of subfields. The same set of parameters is available for use in commands and responses.

MGCP defines the following parameters, where the parameter code is indicated within parentheses:

- **BearerInformation (B)**   Specifies the line side encoding as A- or μ-law. For example, B:e:mu means μ-law encoding.

- **CallId (C)**   The unique identifier for a single call, comprised of hexadecimal digits.

- **Capabilities (A)**   Coded in the same way as LocalConnectionOptions, this is used in response to an audit to indicate the capabilities supported by an endpoint.

- **ConnectionId (I)**   Unique identifier for a connection on an endpoint, comprised of hexadecimal digits.

■ **ConnectionMode (M)**   Includes options such as send-only, receive-only, and send-receive. These choices enable a given endpoint to be configured for one-way or two-way media transfers.

■ **ConnectionParameters (P)**   Connection-related statistical information such as average latency, average jitter, the number of packets sent/received, and the number of lost packets. These statistics are reported to a call agent at the end of a call, within the response to the DeleteConnection command. They may also be returned to the call agent as a response to the AuditConnection command.

■ **DetectEvents (T)**—A list of events that an endpoint should detect. This list might include the detection of such items as off-hook, on-hook, hook-flash, DTMF digits, and so on.

■ **DigitMap (D)**—A digit map is a representation of a dialing plan and specifies rules for which combinations of digits can be dialed. The purpose of the digit map is to enable the gateway to understand what types of digit strings might be received from a caller. When equipped with a digit map, the gateway is able to collect a set of digits and send them en bloc to the call agent, rather than sending each digit individually. The digit map is sent in a *Notification Request* (RQNT) command from the call agent to the gateway.

■ **EventStates (ES)**   Used in response to an audit command, this parameter is a comma-separated list of events associated with the current state of the endpoint and can indicate which events caused the endpoint to be in that particular state.

■ **LocalConnectionDescriptor (LC)**   An SDP session description for the current connection on an endpoint, indicating at which IP address and port number the endpoint is expecting to receive media and in what format. In a two-party call, the LocalConnectionDescriptor for the connection at one end corresponds to the RemoteConnectionDescriptor for the connection at the other end.

■ **LocalConnectionOptions (L)**   A list of connection options such as bandwidth, packetization period, silence suppression, gain control, echo cancellation, and encryption. These are connection-related characteristics that a call agent suggests to a gateway when a connection is to be established. In the syntax, the parameter is identified by L:. The individual components are separated by commas, and each is identified by one or two characters, followed by a colon and the specific value. For example, to turn echo cancellation off and to turn silence suppression on, the parameter would be L: e:off, s:on.

- **MaxMGCPDatagram (MD)** Indicates the maximum size MGCP datagram (MGCP packet) supported by an MG. The parameter can be included in the response to an AUEP command.

- **NotifiedEntity (N)** An address for a call agent to which notify commands should be sent from a gateway.

- **ObservedEvents (O)** A list of events detected by an endpoint. This parameter is used to report the events to the call agent. Generally, the events reported will be a subset of those requested by the RequestedEvents in a command previously received from the call agent. The RequestIdentifier (described later) is used to correlate events requested by the call agent with events reported by the gateway.

- **Package List (PL)** A list of packages supported by an endpoint. Different types of endpoints have different characteristics. A given endpoint can support only certain types of events, tones, or actions. For example, a dial tone is an event that would be associated only with an analog line endpoint. To ease protocol handling, events and signals are grouped into packages. For example, the line (L) package includes events and signals (such as the detection of off-hook or on-hook) that are associated only with analog line endpoints. The PL parameter can be included in the response to multiple commands.

- **QuarantineHandling (Q)** Indicates how an endpoint should handle events that occur during the period while the gateway is waiting for a response to a Notify command sent to a call agent (the quarantine period). The choices are that the endpoint should go ahead and process the events or simply discard them.

- **ReasonCode (E)** Used by a gateway when deleting a connection to indicate the cause of the deletion or after a restart of an endpoint to indicate the reason for the restart.

- **RemoteConnectionDescriptor (RC)** An SDP session description related to the far end of a call. This parameter indicates an address and port to which media should be sent and the format of that media.

- **RequestedEvents (R)** A list of events that an endpoint is to watch for. Associated with each event, the endpoint may be instructed to perform one or more actions, such as notify the call agent immediately, collect digits, or apply a signal to the line. For example, an endpoint might be instructed to detect an off-hook evet on an analog line and, upon detection of the event, to apply a dial tone to the line.

- **RequestedInfo (F)** Information to be provided by a gateway to a call agent in response to either of the two audit requests. The requested information can include such items as the current value of RequestedEvents, the current value of DigitMap, or the current value of NotifiedEntity. This parameter should not be confused with the RequestedEvents parameter.

■ **RequestIdentifier (X)**   Used in RQNT and NTFY commands to identify a particular request for information. The purpose of the parameter is to enable a call agent to correlate a given notification from a gateway with a particular notification request previously sent from the call agent to the gateway.

■ **ResponseAck (K)**   A list of one or more transaction ID ranges.

■ **RestartDelay (RD)**   A number of seconds indicating when an endpoint will be brought back into service.

■ **RestartMethod (RM)**   Used to indicate whether a restart at an endpoint is graceful (existing connections are not terminated) or forced (existing connections are terminated). This parameter is also used to indicate that an endpoint is about to be brought back into service after a specified delay.

■ **SecondConnectionId (I2)**   Connection identifier associated with the connection on a second endpoint. When a command, such as CreateConnection, specifies a second endpoint, then actually two connections are created. Each has its own identifier and each may be manipulated through commands such as ModifyConnection.

■ **SecondEndpointID (Z2)**   Used to indicate a second endpoint on the same gateway. This would apply when a call agent wants to create a connection between two endpoints on the same gateway. It can be used in the command to specify the second endpoint to be included in the connection. Its use in a command may include the use of the any ($) wildcard. In such a case, the parameter will also be used in the response to the command to indicate the specific endpoint chosen as the second endpoint.

■ **SignalRequests (S)**   Signals to be applied by an endpoint to the media stream, such as a ring tone or dial tone.

■ **SpecificEndpointID (Z)**   Used to indicate a single endpoint, particularly in a response to a command where a wildcard was used by the call agent. For example, when creating a connection using CRCX, the call agent might not care exactly which endpoint is used for the connection and might let the gateway make the choice. In such a case, the gateway would respond to the CRCX command and would use this parameter in that response to indicate the identity of the exact endpoint chosen.

Not all parameters will appear in all commands. Rather, for each command, certain parameters are mandatory, certain parameters are optional, and certain parameters are forbidden. Table 6-1 shows the mapping of command parameters to commands, where M = mandatory, O = optional, and F = Forbidden.

Overview of MGCP Responses

For each MGCP command, a response is returned. Responses are comprised of a response line, optionally followed by a number of response parameters. The response line is made up of a return code, followed by the transaction identifier, optionally followed by a commentary or reason phrase. Each of these is separated by a *single space* (SP) character and the response line is terminated by a *carriage return line feed* (CRLF) and may be represented as

```
Return code SP TransactionID SP Commentary CRLF
```

A return code is an integer value. Return codes are divided into the following categories:

- **0XX (000 to 099)**   This indicates an acknowledgement to a response.
- **1XX (100 to 199)**   Provisional responses. A final response will follow later.
- **2XX (200 to 299)**   The command successfully completed.
- **4XX (400 to 499)**   Failure due to a transient error
- **5XX (500 to 599)**   Failure due to a permanent error.
- **8XX (800 to 899)**   Package-specific responses.

Table 6-2 gives a complete listing of the return codes defined in MGCP and their meanings.

A response must be correlated with the command that triggered the response. Therefore, the TransactionId appears in both commands and responses. The response to a command will use the same TransactionId as the command that generated the response.

The same parameters that are available for use in MGCP commands are also available for use in MGCP responses. The allowed usage in responses differs from the allowed usage in commands, however. For example, the LocalConnectionOptions parameter is an optional parameter in a CRCX command, but is forbidden in a response to a CRCX command. Table 6-3 provides a mapping of response parameters to the commands that generate

Table 6-1

Command parameters and commands (as specified in proposed update to RFC 2705)

Parameter name	EPCF	CRCX	MDCX	DLCX	RQNT	NTFY	AUEP	AUCX	RS IP
BearerInformation	O[1]	O	O	O	O	F	F	F	F
CallId	F	M	M	O	F	F	F	F	F
Capabilities	F	F	F	F	F	F	F	F	F
ConnectionId	F	F	M	O	F	F	F	M	F
ConnectionMode	F	M	M	F	F	F	F	F	F
ConnectionParameters	F	F	F	O[2]	F	F	F	F	F
DetectEvents	F	O	O	O	O	F	F	F	F
DigitMap	F	O	O	O	O	F	F	F	F
EventStates	F	F	F	F	F	F	F	F	F
LocalConnectionOptions	F	O	O	F	F	F	F	F	F
MaxMGCPDatagram	F	F	F	F	F	F	F	F	F
NotifiedEntity	F	O	O	O	O	O	F	F	F
ObservedEvents	F	F	F	F	F	M	F	F	F
PackageList	F	F	F	F	F	F	F	F	F
QuarantineHandling	F	O	O	O	O	F	F	F	F
ReasonCode	F	F	F	O	F	F	F	F	O
RequestedEvents	F	O	O	O	O[3]	F	F	F	F
RequestIdentifier	F	O[4]	O[4]	O[4]	M	M	F	F	F
RequestedInfo	F	F	F	F	F	F	O	M	F
ResponseAck	O	O	O	O	O	O	O	O	O
RestartDelay	F	F	F	F	F	F	F	F	O
RestartMethod	F	F	F	F	F	F	F	F	M
SecondConnectionId	F	F	F	F	F	F	F	F	F

265

Table 6-1 cont.

Command parameters and commands (as specified in proposed update to RFC 2705)

Parameter name	EPCF	CRCX	MDCX	DLCX	RQNT	NTFY	AUEP	AUCX	RS IP
SecondEndpointId	F	O	F	F	F	F	F	F	F
SignalRequests	F	O	O	O	O*	F	F	F	F
SpecificEndpointId	F	F	F	F	F	F	F	F	F
RemoteConnectionDescriptor	F	O	O	F	F	F	F	F	F
LocalConnectionDescriptor	F	F	F	F	F	F	F	F	F

[1]The BearerInformation parameter is only conditionally optional and must be included under certain circumstances.

[2]The ConnectionParameters parameter is only valid in a DeleteConnection request sent by the gateway.

[3]The RequestedEvents and SignalRequests parameters are optional in the NotificationRequest. If these parameters are omitted, the corresponding lists will be considered empty.

[4]The RequestIdentifier parameter is optional in connection creation, modification, and deletion commands. The parameter is required, however, if the command contains an encapsulated notification request.

Table 6-2

MGCP return codes

Return Code	Description
000	Response acknowledgement.
100	The transaction is currently being executed. An actual completion message will follow later.
101	The transaction is being queued for later execution. A completion response will follow later.
200	The requested transaction was completed normally.
250	The connection was deleted.
400	The transaction could not be executed due to a transient error.
401	The phone is already off-hook.
402	The phone is already on-hook.
403	Transaction not executed because the endpoint does not have sufficient resources at this time.
404	Insufficient bandwidth at this time.
405	The transaction could not be executed because the endpoint is restarting.
406	Transaction timeout. The transaction did not complete in a reasonable time and has been aborted.
407	Transaction aborted. The transaction has been aborted due to some external (from management, for example) action.
409	The transaction has not been executed because of internal overload.
410	No endpoint available. An any ($) wildcard was used in the request, but no suitable endpoint is available to satisfy the request.
500	The transaction was not executed because the endpoint is unknown.
501	The transaction was not executed because the endpoint is not ready.
502	The transaction was not executed because the endpoint does not have sufficient resources.
503	All of (*) wildcard too complicated.
504	Unknown or unsupported command.
505	Unsupported RemoteConnectionDescriptor.
506	Unable to satisfy both LocalConnectionOptions and RemoteConnectionDescriptor.
508	Unknown or unsupported quarantine handling.
509	Error in RemoteConnectionDescriptor.
510	The transaction was not executed because a protocol error was encountered.
511	The transaction was not executed because the command contained an unrecognized extension.
512	The transaction was not executed because the gateway is not equipped to detect one of the requested events.

Table 6-2 cont.

MGCP return codes

Return Code	Description
513	The transaction was not executed because the gateway is not equipped to generate one of the requested signals.
514	The transaction was not executed because the gateway cannot send the requested announcement.
515	The transaction refers to an incorrect ConnectionID (may already have been deleted).
516	The transaction refers to an unknown CallID.
517	Unsupported or invalid mode.
518	Unsupported or invalid package.
519	Endpoint does not have a digit map.
520	Transacting could not be executed because the endpoint is restarting.
521	Endpoint redirected to another call agent.
522	No such event or signal.
523	Unknown action or illegal combination of actions.
524	Internal inconsistency in LocalConnectionOptions.
525	Unknown extension in LocalConnectionOptions.
526	Insufficient bandwidth.
527	Missing RemoteConnectionDescriptor.
528	Incompatible protocol version.
529	Internal hardware failure.
530	CAS signaling protocol error.
531	Failure of a grouping of trunks (such as facility failure).
532	Unsupported value(s) in LocalConnectionOptions.
533	Response too large.
534	Codec negotiation failure.
535	Packetization period not supported.
536	Unknown or unsupported RestartMethod.
537	Unknown or unsupported digit map extension.
538	Event/signal parameter error (missing, erroneous, unsupported, unknown, and so on) package or vendor extension parameter.
539	Invalid or unsupported command parameter. This code should only be used when the parameter is neither a package or vendor extension parameter.
540	Per-endpoint connection limit exceeded.
541	Invalid or unsupported LocalConnectionOptions. This code should only be used when the LocalConnectionOptions is neither a package nor a vendor extension LocalConnectionOptions.

the response, where M is mandatory, O is optional, and F is forbidden. We can see from the table that responses generally contain fewer parameters than commands.

Table 6-3

Response parameters

Parameter name	EPCF	CRCX	MDCX	DLCX	RQNT	NTFY	AUEP	AUCX	RSIP
BearerInformation	F	F	F	F	F	F	O	F	F
CallId	F	F	F	F	F	F	F	O	F
Capabilities	F	F	F	F	F	F	O[1]	F	F
ConnectionId	F	O[2]	F	F	F	F	O[1]	F	F
ConnectionMode	F	F	F	F	F	F	F	O	F
ConnectionParameters	F	F	F	O[3]	F	F	F	O	F
DetectEvents	F	F	F	F	F	F	O	F	F
DigitMap	F	F	F	F	F	F	O	F	F
EventStates	F	F	F	F	F	F	O	F	F
LocalConnectionOptions	F	F	F	F	F	F	F	O	F
MaxMGCPDatagram	F	F	F	F	F	F	O	F	F
NotifiedEntity	F	F	F	F	F	F	O	O	O
ObservedEvents	F	F	F	F	F	F	O	F	F
QuarantineHandling	F	F	F	F	F	F	O	F	F
PackageList	O[4]	O[4]	O[4]	O[4]	O[4]	O[4]	O	O[4]	O[4]
ReasonCode	F	F	F	F	F	F	O	F	F
RequestIdentifier	F	F	F	F	F	F	O	F	F
ResponseAck	O[5]	O[5]	O[5]	O[5]	O[5]	O[5]	O[5]	O[5]	O[5]
RestartDelay	F	F	F	F	F	F	O	F	F
RestartMethod	F	F	F	F	F	F	O	F	F
RequestedEvents	F	F	F	F	F	F	O	F	F
RequestedInfo	F	F	F	F	F	F	F	F	F
SecondConnectionId	F	O	F	F	F	F	F	F	F
SecondEndpointId	F	O	F	F	F	F	F	F	F

Table 6-3 cont.

Response
parameters

Parameter name	EPCF	CRCX	MDCX	DLCX	RQNT	NTFY	AUEP	AUCX	RSIP
SignalRequests	F	F	F	F	F	F	O	F	F
SpecificEndpointId	F	O	F	F	F	F	O[1]	F	F
LocalConnection Descriptor	F	O[6]	O	F	F	F	F	O[7]	F
RemoteConnection Descriptor	F	F	F	F	F	F	F	O[7]	F

[1]Multiple ConnectionId, SpecificEndpointId, and Capabilities parameters may be present in the response to an AuditEndpoint command.

[2]In the case of a CreateConnection message, the response line is followed by a ConnectionId parameter and a LocalConnectionDescriptor. It may also be followed a SpecificEndpointId parameter if the creation request was sent to a wildcarded EndpointId. The ConnectionId and LocalConnectionDescriptor parameter are marked as optional in the table. In fact, they are mandatory with all positive responses when a connection was created, and forbidden when the response is negative and no connection was created.

[3]Connection parameters are only valid in a response to a non-wildcarded DeleteConnection command sent by the call agent.

[4]The PackageList parameter is only allowed with return code 518 (unsupported package), except for AuditEndpoint, where it can also be returned if audited.

[5]The ResponseAck parameter must not be used with any other responses than a final response issued after a provisional response for the transaction in question. In that case, the presence of the ResponseAck parameter should trigger a Response Acknowledgement; any ResponseAck values provided will be ignored.

[6]A LocalConnectionDescriptor *must* be transmitted with a positive response (code 200) to a CreateConnection. The parameter must also be transmitted in response to a ModifyConnection command, if the modification resulted in a modification of the session parameters. The LocalConnectionDescriptor is encoded as an SDP session description.

[7]When several session descriptors are encoded in the same response, they are encoded one after each other, separated by an empty line.

Command and Response Details

From the information in Tables 6-1 and 6-3, we can identify the structure of each command and the corresponding response. All that remains is to indicate where a given command can be encapsulated within another command. This information is included in the following command and response details, where optional parameters and optional encapsulated commands are shown within square brackets.

Note that the representation of the command and response formats here is slightly different than that in certain parts of the MGCP specification. That is because the MGCP specification defines both a protocol and an *application programming interface* (API). The format here is based on the protocol and shows the complete command line with the EndpointId and MGCP version included. The format here also shows the complete response line with the ReturnCode, TransactionId, and Commentary all included. The information is presented in this way in order to enable a better visualization of the commands and responses as actually transmitted by a call agent or gateway.

EndpointConfiguration (EPCF) Command

```
EPCF   TransactionId EndpointId MGCP 1.0
            [BearerInformation]
```

EndPointConfiguration (EPCF) Response

```
ReturnCode TransactionId Commentary
            [PackageList]
```

NotificationRequest (RQNT) Command

```
RQNT TransactionId EndpointId MGCP 1.0
            [NotifiedEntity]
            [RequestedEvents]
            RequestIdentifier
            [DigitMap]
            [SignalRequests]
            [QuarantineHandling]
            [DetectEvents]
            [encapsulated EndpointConfiguration]
```

NotificationRequest (RQNT) Response

```
ReturnCode TransactionId Commentary
            [PackageList]
```

Notify (NTFY) Command

```
NTFY TransactionId EndpointId MGCP 1.0
            [NotifiedEntity]
            RequestIdentifier
            ObservedEvents
```

Notify (NTFY) Response

```
ReturnCode TransactionId Commentary
            [PackageList]
```

CreateConnection (CRCX) Command

```
CRCX TransactionID EndpointID MGCP 1.0
            CallId
            [NotifiedEntity]
            [LocalConnectionOptions]
            Mode
            [RemoteConnectionDescriptor | SecondEndpointId]
            [Encapsulated NotificationRequest]
            [Encapsulated EndpointConfiguration]
```

CreateConnection (CRCX) Response

```
ReturnCode TransactionId Commentary
                  ConnectionId
                  [SpecificEndpointId]
                  [LocalConnectionDescriptor]
                  [SecondEndpointId]
                  [SecondConnectionId]
                  [PackageList]
```

ModifyConnection (MDCX) Command

```
MDCX TransactionId EndpointId MGCP 1.0
            CallId
            ConnectionId
            [NotifiedEntity]
            [LocalConnectionOptions}
            [Mode]
            [RemoteConnectionDescriptor]
            [encapsulated NotificationRequest]
            [encapsulated EndpointConfiguration]
```

ModifyConnection (MDCX) Response

```
ReturnCode TransactionId Commentary
                  [LocalConnectionDescriptor]
                  [PackageList]
```

DeleteConnection (DLCX) Command (from Call agent to Gateway)

```
DLCX TransactionId EndpointId MGCP 1.0
            CallId
            [ConnectionId]*
            [encapsulated NotificationRequest]
            [encapsulated EndpointConfiguration]
```

---

*A call agent might want to delete all connections for a given endpoint and a given call, rather than just a single connection. In such a case, the DLCX command from the call agent would not include the ConnectionId parameter.

DeleteConnection (DLCX) Response (from Gateway to Call agent)

```
ReturnCode TransactionId Commentary
                    ConnectionParameters
                    [PackageList]
```

DeleteConnection (DLCX) Command (from Gateway to Call Agent)

```
DLCX TransactionId EndpointId MGCP 1.0
        CallId
        ConnectionId
        ReasonCode
        Connectionparameters
```

DeleteConnection (DLCX) Response (from Call Agent to Gateway)

```
ReturnCode TransactionId Commentary
                    [PackageList]
```

AuditEndpoint (AUEP) Command

```
AUEP TransactionId EndpointId MGCP 1.0
        [RequestedInfo]
```

AuditEndpoint (AUEP) Response

```
ReturnCode TransactionId Commentary
                        EndpointIdList | {
                        [RequestedEvents]
                        [QuarantineHandling]
                        [DigitMap]
                        [SignalRequests]
                        [RequestIdentifier]
                        [NotifiedEntity]
                        [ConnectionIdentifiers]
                        [DetectEvents]
                        [ObservedEvents]
                        [EventStates]
                        [BearerInformation]
                        [RestartMethod]
                        [RestartDelay]
                        [ReasonCode]
                        [MaxMGCPDatagram]
                        [Capabilities]
                        [PackageList]
```

AuditConnection (AUCX) Command

```
AUCX TransactionId EndpointId MGCP 1.0
        ConnectionId
        RequestedInfo
```

AuditConnection (AUCX) Response

```
ReturnCode TransactionId Commentary
                        [CallId]
                        [NotifiedEntity]
                        [LocalConnectionOptions]
                        [Mode]
                        [RemoteConnectionDescriptor]
                        [LocalConnectionDescriptor]
                        [ConnectionParameters]
                        [PackageList]
```

RestartInProgress (RSIP) Command

```
RSIP TransactionId EndpointId MGCP 1.0
            RestartMethod
            [RestartDelay]
            [ReasonCode]
```

RestartInProgress (RSIP) Response

```
ReturnCode TransactionId Commentary
                        [NotifiedEntity]
                        [PackageList]
```

Call Setup Using MGCP

Utilizing the foregoing descriptions of commands and responses, it is possible to describe MGCP-based call setup in some detail. The following description of the process assumes that two endpoints are to be used on two different gateways, but that both gateways are connected to the same call agent.

Call setup using MGCP involves the creation of at least two connections, as shown in Figure 6-7. First, a call agent uses the CRCX command to instruct a gateway to create a connection on an endpoint. The call agent may specifically identify the endpoint to be used or may use a wildcard notation to enable the gateway to choose the endpoint. The gateway creates the connection and responds to the call agent with a session description (LocalConnectionDescriptor) applicable to the connection. The LocalConnectionDescriptor is an SDP description of the session as it applies to the connection just created. Therefore, it contains information such as an IP address, port number, and media coding.

Upon receipt of the response from the first gateway, the call agent instructs the second gateway to create a connection on an endpoint. The call agent has more information available for inclusion in the CRCX command

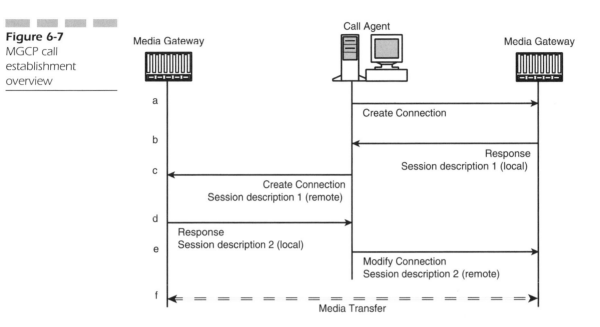

Figure 6-7
MGCP call
establishment
overview

to the second gateway than it had for the CRCX command for the first gateway. Specifically, the call agent is armed with knowledge of the session description applicable to the connection on the first endpoint, and it can pass this information to the second endpoint. Therefore, the call agent issues a CRCX command to the second gateway and includes the optional RemoteConnectionDescriptor parameter, which contains the session description obtained for the connection made on the first endpoint. Thus, the second endpoint in the call now has a session description related to the first endpoint and knows where to send media. For the moment, the first endpoint in the call is still in the dark.

The second endpoint sends a response to the CRCX command and includes the LocalConnectionDescriptor parameter, which is a session description related to the connection created on the second endpoint.

Third, the call agent issues an MDCX command to the first gateway. The command contains a RemoteConnectionDescriptor parameter, corresponding to the session description received from the second gateway for the second endpoint. At this stage, both endpoints now hold a session description for the other and know exactly where to send the media stream and in what format. Media can now be exchanged.

Figure 6-8 shows a detailed example of call establishment between two gateways, with each supporting DS0 endpoints and managed by the same call agent, including the session descriptions involved.

Figure 6-8
MGCP call setup
between two
gateways

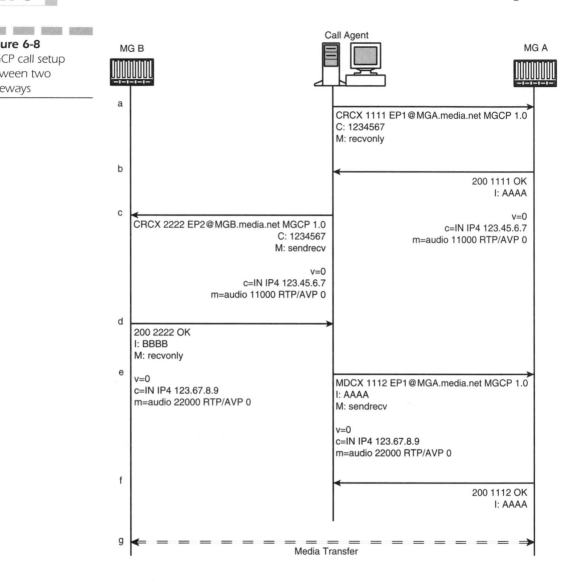

The following text describes the messages of Figure 6-8 in more detail.

a. The call agent sends a CRCX command to MGA, specifying that a connection is to be created on *endpoint 1* (EP1). The command also specifies the TransactionId (111) and the CallId (1234567). No BearerInformation is sent in this case, because the endpoint is specifically identified, and we assume that the bearer-related information for the line side of the endpoint is already known. If that information were not known or if the selection of endpoint were wildcarded, then the bearer information would specify either A-law or μ-law. The CRCX command specifies the mode to be receive-only for now. This means that the endpoint can receive media from the IP network, but cannot send media to the IP network. The reason for this choice is the fact that the endpoint does not yet have a session description for the remote end and hence cannot know where to send a media stream. The remote end is about to be presented with a session description for the local end, however, and could start to send media before the local endpoint receives a session description for the remote end. Therefore, the local endpoint should be prepared to receive media from the remote endpoint.

b. MG A responds to the CRCX command with the same TransactionId, a ConnectionId (AAAA), and a session description indicating voice coded according to G.711 μ-law (RTP payload format 0). Note that this is the voice coding chosen on the IP side of the endpoint as opposed to the line side of the endpoint. It could equally well be some other coding choice such as G.728 or G.729.

c. The call agent sends a CRCX command to the second gateway (MG B). Again a specific endpoint (EP2) is indicated in the command. The TransactionId is 2222 and the same CallId (1234567) is used. The call agent includes a RemoteConnectionDescriptor, which is simply the session description received earlier from MG A. This time the CRCX command specifies the mode as send-receive, since a session description for the remote end is available, which means that the endpoint has an address and port to which it can send media.

d. MG B sends a positive response to the CRCX, including a ConnectionId (BBBB) and a session description corresponding to that connection. The session description in this case also happens to be G.711 μ-law coded voice.

e. The call agent sends an MDCX command to the command to the first gateway. The command specifically specifies the ConnectionId and includes a RemoteConnectionDescriptor, which is the session description received from MG B. Given that a session description is available for the remote end, MG A now has an address and port to which it may send media. Therefore, the MDCX command specifies the mode to be send-receive.

f. MG A sends a positive response to the MDCX command.

g. Media can now be exchanged between the two endpoints.

MGCP Events, Signals, and Packages

The example of Figure 6-8 was a simple scenario between two trunking gateways. The example also assumes that the call agent has some prior knowledge of the requirements for the call. Specifically, in Figure 6-8, the call agents know *a priori* that a call is to be established between the two gateways. Such a situation implies that the call agent is exchanging signaling with an external network, such as the SS7 network.

The situation can be more complex if the call is triggered in another way. If, for example, MGA were to be a residential gateway that supports analog lines, then the call would be instigated and controlled by events and signals on the line itself. Such events would include a transition from on-hook to off-hook, the application of dial tone to the line, the receipt of dialed digits, and so on. In other words, the MG would detect an off-hook event and would notify the call agent. The MG would also apply a dial tone and wait for digits. Upon receipt of those digits, the MG would pass them to the call agent, which would make the necessary call-routing decisions. The MG would pass the dialed digits to the call agent only if those digits match with a digit map previously downloaded from the call agent. If the dialed digits were invalid (did not match a digit map stored at the MG), then the MG would be required to apply an appropriate tone to the line.

Alternatively, one could imagine a situation where one of the gateways uses *multifrequency* (MF) signaling. In such a case, the gateway would need to detect and send MF signals, including digits for a calling party number and a called party number. We can see that the events that a gateway must detect and the signals that a gateway must apply depend on the function of the gateway and its role in the network.

Because different types of endpoints will need to detect different events and apply different signals, MGCP groups events and signals into packages, and a given endpoint can implement one or more packages. These packages have package names and a given event or signal is identified by a package name and an event or signal name, with the format "package name/event name." For example, a gateway that terminates analog lines will implement the line package (L). The off-hook event has the abbreviation hd. Therefore, when referring to an off-hook event detected by an analog line endpoint, the formal identification is L/hd. Incidentally, the package name is actually optional when it comes to identifying an event or signal, since each event or signal has a unique identifier.

The use of package names is useful with wildcarding, however. If, for example, a gateway were required to detect every event in L, then the events to be detected can be represented by L/*. This format is a lot easier than listing every event that might occur. Wildcarding can also be applied to the package name. If for example, the event "surprise" happened to be included in more than one package, then the syntax */surprise would refer to the surprise event in every package.

When the SignalRequests and the RequestedEvents parameters are both specified in an RQNT command, then the signals that are requested are synchronized with the events that are to be detected. For example, if a ringing signal is to be applied to a line and an off-hook event is to be detected, then the ringing signal will be stopped as soon as the off-hook event occurs.

A number of signals and events are defined in the MGCP specification and are grouped into various packages. Whenever new events or packages need to be defined, then these should be documented and the names of the events or packages should be registered with the *Internet Assigned Numbers Authority* (IANA). Prior to registration with the IANA, developers may want to test new packages. Such experimental packages should have names beginning with the two characters "x-." These characters indicate an experimental extension to the protocol, and packages starting with these characters do not need to be registered.

Interworking Between MGCP and SIP

The example of Figure 6-8 assumes that two gateways are used, but that a single call agent controls both gateways. It is equally likely that separate call agents control the gateways. In such a case, a need exists for signaling between the two call agents. In the softswitch architecture, we commonly use SIP.

Figure 6-9 is an adaptation of Figure 6-8, with the difference that separate call agents control the gateways. The figure shows how MGCP and SIP fit very well together, particularly since they both utilize SDP for describing sessions. Once the connection is created on the first endpoint (line b of Figure 6-10), the returned session description becomes the message body of a SIP INVITE sent to the other call agent (line c). Note that most of the SIP headers and other fields have been omitted, and to avoid clutter, the names of the endpoints and the TransactionIds have been shortened compared to Figure 6-8.

Once the second connection is created, on the second gateway (line e), the corresponding session description becomes the message body of the SIP 200

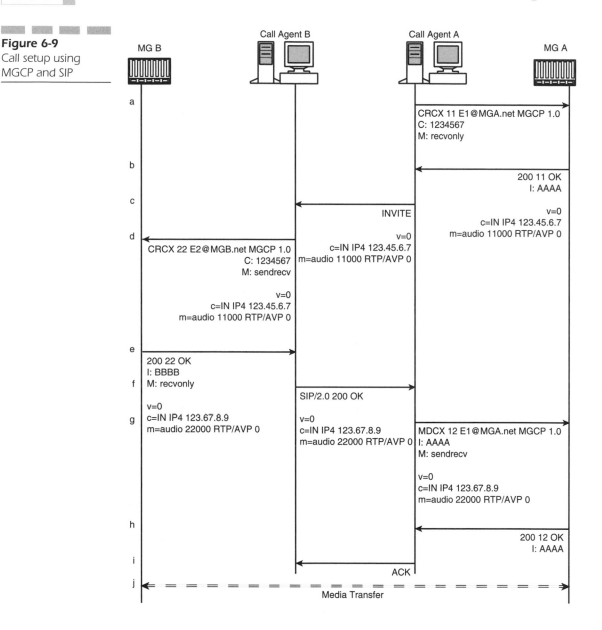

Figure 6-9
Call setup using
MGCP and SIP

(OK) response (line f). In turn, the session description is mapped to the MDCX command sent to the first endpoint (line g). Finally, the positive acknowledgment to the MDCX command (line h) is mapped to the SIP ACK (line i), thereby completing the MGCP and SIP signaling sequences.

MEGACO/H.248

Known as MEGACO within the IETF and H.248 within the ITU, this protocol has widespread backing within both the Internet and telecommunications community. In fact, the cooperation between the two industry organizations reflects the convergence that has occurred between traditional telecommunications and the Internet.

Given the fact that the term MEGACO/H.248 is rather a mouthful, for the purpose of brevity, the remainder of the book uses only the term MEGACO.

MEGACO Architecture

MEGACO has a number of architectural similarities to MGCP. Similar to MGCP, it defines MGs, which perform the conversion of media from the format required in one network to the format required in another. MEGACO defines MGCs, which control call establishment and teardown within MGs and are akin to call agents as defined by MGCP. The similarities are hardly that surprising, given that fact that they have both been designed to address a similar set of requirements.

The MEGACO protocol involves a series of transactions between MGCs and MGs. Each transaction involves the sending of a TransactionRequest by the initiator of the transaction and the sending of a TransactionReply by the responder. A TransactionRequest comprises a number of commands and the TransactionReply comprises a corresponding number of responses. For the most part, transactions are requested by an MGC and the corresponding actions are executed within an MG. As in MGCP, however, a number of cases occur when an MG initiates the transaction request.

The MEGACO specification is written such that it provides both a text encoding and a binary encoding of the protocol. The text encoding is written according to *Augmented Backus-Naur Form* (ABNF), described in RFC 2234. The binary encoding is written according to *Abstract Syntax Notation One* (ASN.1), described in ITU-T Rec. X.680, 1997. This reflects one difference between the IETF, which normally uses ABNF, and the ITU where ASN.1 is the format of choice. MGCs should support both types of encoding, while MGs may support both types of encoding but are not required to do so. Within this book, the ABNF form is used when offering examples of MEGACO usage as it is more human-readable than ASN.1.

Terminations   MEGACO defines terminations, which are logical entities on an MG that act as sources or sinks of media streams. Certain terminations are physical. They have a semipermanent existence and may be associated with external physical facilities or resources. These would include a termination connected to an analog line or a termination connected to a DS0 channel. Physical terminations exist as long as they are provisioned within the MG and are akin to endpoints as defined in MGCP.

Other terminations have a more transient existence and only exist for the duration of a call or media flow. These are known as ephemeral terminations and represent media flows such as an RTP session. These terminations are created as a result of a MEGACO Add command, which is somewhat similar in function to the CreateConnection command in MGCP. Ephemeral terminations are destroyed by means of the Subtract command, which is similar to the MGCP DeleteConnection command.

Terminations have specific properties and the properties of a given termination will vary according to the type of termination. Clearly, for example, a termination connected to an analog line will have different characteristics to a termination connected to a TDM channel (such as a DS0). The properties associated with a termination are grouped into a set of descriptors. These descriptors can be included in MEGACO commands, thereby enabling termination properties to be changed according to instruction from MGC to MG.

A TerminationID references a termination, and the MG chooses the TerminationID schema to be used. Similar to MGCP, MEGACO enables the use of the wildcards all (*) and choose ($). A special TerminationID called Root can also be used. This TerminationID is used to refer to the gateway as a whole rather than to any specific terminations within the gateway.

Contexts   A context is an association between a number of terminations for the purposes of sharing media between those terminations. Terminations may be added to contexts, removed from contexts, or moved from one context to another. A termination can exist in only one context at any time, and terminations in a given gateway can exchange media only if they are in the same context.

A termination is added to a context through the use of the Add command. If the Add command does not specify a context to which the termination should be added, then a new context is created as a result of the execution of the Add command. This is the only mechanism for creating a new context. A termination is moved from one context to another through the use of the Move command and is removed from a context through the use of the Subtract command. If the execution of a Subtract command results in the removal of the last termination from a given context, then that context is deleted.

The relationship between terminations and contexts is illustrated in Figure 6-10, where a gateway is depicted with four active contexts. In context C1, we see a simple two-way call across the MG. In Context C2, we see a three-way call across the MG. In contexts C3 and C4, we see a possible implementation of call waiting. In context C3, terminations T6 and T7 are involved in a call. Another call arrives from termination T8 for termination T7. If the user wanted to accept this waiting call and place the existing call on hold, then this could be achieved by moving termination T7 from context C3 to context C4.

The existence of several terminations within the same context means that they have the potential to exchange media. The existence of terminations in the same context does not necessarily mean that they can all send data to each other and receive data from each other at any given time, however. The context itself has certain attributes, including the topology, which

Figure 6-10

MEGACO contexts and terminations, showing possible Move command

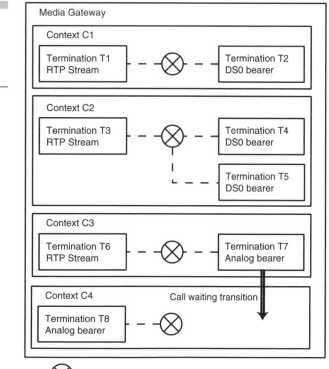

indicates the flow of media between terminations (which terminations may send media to others or receive media from others). Also, the priority attribute indicates the precedence applied to a context when an MGC must handle many contexts simultaneously, and an emergency attribute is used to give preferential handling to emergency calls.

A context is identified by a ContextID, which is assigned by the MG and is unique within a single MG. As is the case for terminations, MECAGO enables wildcarding when referring to contexts, such that the all (*) and choose ($) wildcards may be used. The all wildcard may be used by an MGC to refer to every context on a gateway. The choose wildcard ($) is used when an MGC requires the MG to create a new context.

A special context is also available, known as the null context. This context contains all terminations that do not exist in any other context. Idle terminations normally exist in the null context. The ContextID for the null context is simply "-".

Transactions   MEGACO transactions involve the passing of commands and the responses to those commands. Commands are directed towards terminations within contexts. In other words, every command specifies a ContextID and one or more TerminationIDs to which the command applies. This is the case even for a command that requires some action by an idle termination that does not exist in any specific context. In such a case, the null context is applicable.

Commands between an MGC and an MG are grouped together in a transaction structure where a set of commands related to one context may be followed by a set of commands related to another context. The grouped commands are sent together in a single TransactionRequest. This can be represented as

```
TransactionRequest (TransactionID {
        ContextID1 {Command, Command, ... Command},
        ContextID2 {Command, Command, ... Command},
        ContextID3 {Command, Command, ...Command}  } )
```

A TransactionRequest does not need to contain commands for more than one context or even contain more than one command. A TransactionRequest containing just a single command for a single context is perfectly valid.

Upon receipt of a TransactionRequest, the recipient executes the enclosed commands. The commands are executed sequentially in the order specified in the TransactionRequest. Upon completed execution of the commands, a TransactionReply is issued. This has a similar structure to the

TransactionRequest in that it contains a number of responses for a number of contexts. A TransactionReply may be represented as

```
TransactionReply (TransactionID {
        ContextID1 {Response, Response, ... Response},
        ContextID2 {Response, Response, ... Response},
        ContextID3 {Response, Response, ...Response}  } )
```

As mentioned, commands within a transaction are executed in sequence. If a given command in the transaction happens to fail, then the subsequent commands in the transaction are not processed. One exception is when a transaction contains one or more optional commands. If an optional command fails, then the processing of the remaining commands will continue. Optional commands in a transaction are indicated by inserting the string O- immediately prior to the command name.

If the recipient of a TransactionRequest might take some time to execute the request, the recipient can return an interim reply so that the sender of the request does not assume that it has been lost. This interim reply is simply an indication that the TransactionRequest has been received and is being handled. The term applied to this response is TransactionPending. It simply returns the TransactionID received and no parameters. The TransactionID is a 32-bit integer.

```
TransactionPending (TransactionID { })
```

Messages   Not only is it possible to combine multiple commands within a single transaction, but it is also possible to concatenate multiple transactions within a message. In ABNF syntax, a message begins with the word MEGACO followed by a slash, the protocol version number (1), a message ID (mId), and finally the message body. The mId can be the domain name or IP address (and optionally the port) of the entity transmitting the message. As an alternative, it can be the MTP point code and network indicator of the entity issuing the message.

The following example combines the concept of a message, transactions, and commands and it shows the text format of a message:

```
MEGACO/1 [111.111.222.222]:34567
Transaction = 12345 {
        Context = 1111 {
                        Add = A5555,
                        Add = A6666
        }
          Context = $ {
                        Add = A7777
          }
}
```

In this example, the message is issued by an MGC at address 111.111.222.222 and port 34567. One transaction is identified by TransactionID 12345 and two commands are related to context 1111. Both commands are Add commands, causing terminations A5555 and A6666 to be added to context 1111. The transaction includes one command related to any context that the MG may choose. This command is an Add command, whereby termination A7777 is added to the context. The processing of this command would result in the MG creating a new context, for which the ContextID would be returned in the response.

Transactions within a message are treated independently. The order of the transactions within the message does not imply an order in which the recipient of the message should handle the transactions. Note that this is quite different from the handling of commands within a transaction, where the ordering of the commands is of significance.

Overview of MEGACO Commands

MEGACO defines eight commands that provide the ability to control and manipulate contexts and terminations. Most of the commands are sent from an MGC to an MG. The exceptions are the exception of the Notify command, which is always sent from MG to MGC and the ServiceChange command, which may be sent from either an MG or an MGC. The eight commands are as follows:

- **Add** The Add command adds a termination to a context. If the command does not specify a particular context to which to add the termination, then a new context is created. If the command does not indicate a specific TerminationID but instead uses the choose ($) wildcard, the MG will create a new ephemeral termination and add it to the context.

- **Modify** The Modify command changes the property values of a termination, instructs a termination to issue one or more signals, or instructs a termination to detect and report specific events.

- **Subtract** The Subtract command removes a termination from a context. The response to the command can provide statistics related to the termination's participation in the context. These reported statistics depend upon the type of termination in question. For an RTP termination, the statistics can include items such as packets sent, packets received, jitter, and so on. In this respect, the command is similar to the MGCP DLCX command. If the result of a Subtract command is the removal of the last termination from a context, then the context itself is deleted.

- **Move** The Move command moves a termination from one context to another. The command is not used to move a termination from or to the null context as these operations must be performed with the Add and Subtract commands respectively. The capability to move a termination from one context to another provides a useful tool for accomplishing the call-waiting service.

- **AuditValue** The AuditValue command is used by the MGC to retrieve current values for properties, events, and signals associated with one or more terminations.

- **AuditCapabilities** The AuditCapabilities command is used by an MGC to retrieve the possible values of properties, signals, and events associated with one or more terminations. At first glance, this command may appear very similar to the AuditValue command. The difference between them is that the AuditValue command is used to determine the current status of a termination, whereas the AuditCapabilities command is used to determine the possible statuses that a termination might assume. For example, AuditValue would indicate any signals that are currently being applied by a termination, where AuditCapabilities could indicate all the possible signals that the termination could apply if required.

- **Notify** The Notify command is issued by an MG to inform the MGC of events that have occurred within the MG. The events to be reported will typically have previously been requested as part of a command from the MGC to the MG, such as a Modify command. The events reported will be accompanied by a RequestID parameter to enable the MGC to correlate reported events with previous requests.

- **ServiceChange** The ServiceChange command enables an MG to inform an MGC that a group of terminations is about to be taken out of service or is being returned to service. The command is also used in a situation where an MGC is handing over control of an MG to another MGC. In that case, the command is first issued from the controlling MGC to the MG to instigate the transfer of control. Subsequently, the MG issues the ServiceChange command to the new MGC as a means of establishing the new relationship.

Descriptors

MECAGO defines a number of descriptors available for use with commands and responses. These descriptors form the parameters of the command and/or response and provide additional information to qualify a given command or response. Depending on the command or response in question, a given descriptor will be mandatory, forbidden, or optional. In most cases, when a descriptor is not mandatory, it is optional. Relatively few cases occur when a descriptor is forbidden.

The general format of a descriptor is

```
Descriptorname=<someID>{parm=value, parm=value, . . . }
```

The parameters of a descriptor may be fully specified, underspecified, or overspecified. Fully specified means that the parameter is assigned a unique and unambiguous value, which the command recipient must use in processing the command. Underspecified involves the use of the choose wildcard ($), allowing the command recipient to select any value it can support. Overspecified means that the sender of the command provides a list of acceptable parameter values in order of preference. The command recipient selects one of the offered choices for use. The response to the command will indicate which of the choices was selected.

Modem Descriptor   This descriptor specifies the modem type and associated parameters to be used in modem connections for audio, video, or data communications or for text conversion. The following types of modems are included: V.18 (text telephony), V.22 (1,200 bps), V.22bis (2,400 bps), V.32 (9,600 bps), V.32bis (14,400 bps), V.34 (33,600 bps), V.90 (56 Kbps), V.91 (64 Kbps), and synchronous ISDN.

By default, a termination does not have a modem descriptor. In other words, at start up no modem-related properties will be specified for a termination. They will be applied later as a result of a command (Add or Modify) from an MGC to an MG.

The modem descriptor was included in the first version of the MEGACO specification (RFC 3015). The descriptor should not be used in subsequent versions of the protocol and should be ignored if received. Later in this chapter we describe the syntax of the MEGACO protocol. The modem descriptor is listed as a parameter in various commands and responses, but it is included in the MEGACO version 2 syntax only for backwards-compatibility reasons.

Multiplex Descriptor   In multimedia communications, different media streams can be carried on different bearer channels. The multiplex descriptor specifies the association between media streams and bearers. The following multiplex types are supported: H.221. H.223, H.226, V.76, and Nx64K.

Media Descriptor   The media descriptor describes the various media streams. This is a hierarchical descriptor in that it contains an overall descriptor, known as the *termination state descriptor*, which is applicable to the termination in general and it contains a number of *stream descriptors* applicable to each media stream. Furthermore, each stream descriptor also contains up to three subordinate descriptors, known respectively as the *local control descriptor*, the *local descriptor*, and the *remote descriptor*. This type of hierarchy may be represented by the following list:

Media descriptor
   Termination state descriptor
    Stream descriptor
      Local control descriptor
      Local descriptor
      Remote descriptor

Termination State Descriptor   This descriptor contains the two properties ServiceStates and EventBufferControl, plus other properties of a termination that are not specific to any media stream.

The ServiceStates property indicates whether the termination is available for use. Three values are applicable: test, out of service, and in service. In service does not mean that the termination is currently involved in a call. Rather it means that the termination is either actively involved in a call or available for use in a call. In service is the default state of a termination.

The EventBufferControl property specifies whether events detected by the termination are to be buffered following detection or processed immediately. To begin, a termination will report events that the MGC has asked it to report, which the MGC will do through the issuance of a command that contains an EventsDescriptor. Whether the termination will actually report the events indicated in the EventsDescriptor depends on whether the EventBufferControl is set to off or lockstep. If set to off, then the termination will report detected events immediately. If the value is set to lockstep, then events will be buffered in a *first-in, first-out* (FIFO) buffer, along with the time that the event occurred. The contents of the buffer will be examined upon receipt of a new EventsDescriptor specifying the events that the termination is to detect. Inclusion of the EventBufferControl within the TerminationStateDescriptor enables the MGC to turn on or off the immediate notification of events, without having to send a separate EventsDescriptor each time.

The termination state descriptor is optional within a media descriptor.

Stream Descriptor   A stream is defined by its subordinate descriptors LocalControlDescriptor, LocalDescriptor, and RemoteDescriptor, and it is identified by a StreamID.

StreamID values are used between an MG and MGC to indicate which media streams are interconnected. Within a given context, streams with the same StreamID are connected. A stream is created by specifying a new StreamID on a particular termination in a context. A stream is deleted by setting empty local and remote descriptors and setting the values of ReserveGroup and ReserveValue in the subordinate LocalControlDescriptor to false.

LocalControlDescriptor   This descriptor contains properties specific to the media stream, in particular the Mode property, the ReserveGroup property, and the ReserveValue property.

Mode can take one of the following values: sendonly, receiveonly, send/receive, inactive, or loopback. Send and receive directions are with respect to the exterior of the context. If mode is set to sendonly, then a termination can only send media to entities outside the context. The termination cannot pass media to other entities within the same context. If mode is set to receiveonly, then a termination can receive media from outside the context and pass it to other terminations in the context, but it cannot receive media from other terminations and pass it to destinations outside of the context.

When an MGC wants to add a termination to a context, it can specify a set of choices for the session (using local and remote descriptors) that the MGC would prefer the MG to use, and the MGC specifies these in order of preference. The MG is not obligated to select the first choice offered by the MGC, but it might be obligated to reserve resources for the sessions indicated by the MGC. The properties ReserveValue and ReserveGroup indicate which resources should be reserved for the alternatives specified by the MGC in the LocalDescriptor and RemoteDescriptor.

The LocalDescriptor and RemoteDescriptor may specify several properties and/or property groups. For example, a given SDP description could indicate two groups of properties: one for G.711 A-law and one for G.729. If the value of ReserveGroup is true, then the MG should reserve resources for one of these property groups. The use of ReserveValue is similar, but it is applied to reserve resources for a particular property value as opposed to a group of properties.

LocalDescriptor and RemoteDescriptor When used with a text encoding of the protocol, the LocalDescriptor and RemoteDescriptor contain zero or more SDP session descriptions describing the session at the local end of a connection and the remote end of a connection respectively.

The MEGACO usage of SDP includes some variations from the strict syntax of SDP as specified in RFC 2327. In particular, the s=, t=, and o= lines are optional, the choose wildcard ($) is permitted, and it is permissible to indicate several alternatives in places where a single parameter value should normally be used.

Each of the LocalDescriptors and RemoteDescriptors may contain several session descriptions, with the beginning of the session description indicated with the SDP v= line.

Events Descriptor This descriptor contains a RequestIdentifier and a list of events that the MG should detect and report, such as off-hook transition, fax tone, and so on. The purpose of the RequestIdentifier is to correlate requests and the consequent notifications. Normally, detected events will be reported immediately to the MGC. However, events may be buffered depending on the value of the EventControlBuffer property (which is a property specified in the TerminationStateDescriptor).

If events are buffered and subsequently need to be reported to the MGC, then the information related to those events is stored in the EventBuffer-Descriptor.

SignalsDescriptor This descriptor contains a list of signals that a termination is to apply. Signals can be applied to just a single stream or to all streams in a termination. Typical signals will include, for example, tones applied to a subscriber line such as a ring tone or a dial tone. Three types of signals exist:

- **On/off** The signal remains on until it is explicitly turned off.
- **Timeout** The signal persists until a defined period of time has elapsed or until it is turned off, whichever occurs first.
- **Brief** The signal is to be applied only for a very limited time (such as would be the case for an MF signal on R1 MF trunks).

A SignalsDescriptor may include a sequential sequence of signals that are to be played one after another.

Audit Descriptor   The audit descriptor specifies a list of information to be returned from an MG to an MGC. The audit descriptor is simply a list of other descriptors that should be returned in the response. These can include the following descriptors: multiplex, events, signals, Observed-Events, DigitMap, statistics, packages, and EventBuffer.

ServiceChange Descriptor   This descriptor is used only in association with the ServiceChange command and includes information such as the type of service change that has occurred or is about to occur, the reason for the service change, and a new address to be used after the service change.

The type of service change is defined by the parameter ServiceChangeMethod, which can take one of the defined values: Graceful, Forced, Restart, Disconnected, Handoff, or Failover. Graceful indicates the removal of terminations from service after a specified delay and without interrupting existing connections. Forced indicates an abrupt removal from service including the loss of existing connections. Restart indicates that restoration of service will begin after a specified delay. Disconnected applies to the whole MG and indicates that connection with the MGC has been restored after a period of lost contact. The MGC can issue an Audit command to verify that termination characteristics have not changed during the period of lost contact. Handoff is used from an MGC to an MG to indicate that the MGC is about to go out of service and that a new MGC is taking over. The handoff value is also used from an MG to a new MGC when trying to establish contact after receiving a handoff from the previous MGC. A failover is sent from an MG to an MGC if an MG has detected a fault and a backup MG is taking over.

The ServiceChangeDelay specifies a number of seconds indicating the delay applicable. It would be used, for example, in conjunction with a Graceful service change. If the parameter is absent or set to zero, then no delay is applicable. In the case of a graceful service change, then a value of zero indicates that the MGC should wait for the natural removal of terminations from their contexts; that is, the MGC should wait for calls on the specified terminations to come to an end on their own.

As suggested by its name, the ServiceChangeReason parameter indicates the reason for the service change. Table 6-4 shows the reasons currently defined in MEGACO.

Table 6-4

Service change reasons

Reason Code	Interpretation
900	Service restored
901	Cold boot
902	Warm boot
903	MGC-directed change
904	Termination malfunctioning
905	Termination taken out of service
906	Loss of lower-layer connectivity
907	Transmission failure
908	MG impending failure
909	MGC impending failure
910	Media capability failure
911	Modem capability failure
912	Mux capability failure
913	Signal capability failure
914	Event capability failure
915	State loss
916	Packages change
917	Capability change

DigitMap Descriptor The DigitMap is a description of a dialing plan. In other words, a DigitMap specifies a set of valid dialed digit combinations, and it is stored in the MG so that the MG can send received digits en bloc to the MGC rather than one by one. A digit map can be loaded into an MG through operation and maintenance action or it can be loaded from the MGC by MECAGO commands. If it is loaded from the MGC, then the Digit-Map descriptor is used to convey the information.

From a syntax perspective, a digit map is a string or a list of strings, with each string comprising a number of symbols representing the digits 0 through 9 and the letters A through K. The letter X can be used as a wild-card, designating any digit from 0 through 9, and the dot (.) symbol is used to indicate zero or more repetitions of the immediately preceding digit or digit string. This symbol can be used to indicate an indeterminate number length in the dial plan. For example, in the United States, domestic calls use 10-digit numbers. International calls, however, use numbers of variable lengths, such that a call to Germany may require more digits than a call to Iceland. To indicate calls to international numbers, the string "011x." could be used. The 011 indicates an international call and the "x." indicates some number of digits, each in the range 0 to 9.

In addition, the string can contain three symbols indicating a start timer (T), a short timer (S), and a long timer (L). To understand the usage of these timers, consider what happens when someone picks up a standard residential telephone to make a call. First, the off-hook is detected and the system (for example, the telephone exchange) applies a dial tone and gets ready to collect digits. The system will only wait for a certain amount of time for the person to press a button on the phone. If the person does not press a button in time, the system will timeout and take some other action, such as prompting the user.

This initial timeout value corresponds to the start timer. Once the person starts to dial digits, an interdigit timer will be applied. This timer will either be a short timer or a long timer depending on the dial plan. If the person has dialed a certain number of digits and the system still needs more digits to be able to route the call, a short timer will be applied while waiting for the next digit. On the other hand, if the person has dialed sufficient digits to route the call, but further digits could still be received and would cause a different routing, then a long timer will be applied. The purpose of the long timer is for the system to be confident that the person has finished dialing.

An example in the Unites States would be where dialing 0 causes routing to an operator from the local exchange carrier and dialing 00 causes routing to a long-distance operator. After the system has received the first 0, it must wait before routing the call to make sure that the person does not intend to dial a second 0 in the expectation of being connected to a long distance operator. Table 6-5 gives an example of a dial plan.

This dialing plan could be represented in a digit map as

(0 | 00 | [1-7]xxx | 8xxxxxxx | Fxxxxxxx | Exx | 91xxxxxxxxxx | 9011x.)

Note that the timers have been omitted from this digit map. This omission does not mean that timing does not occur at the MG; rather, it just means that the timers are applied by default according to the usage rules

Table 6-5

Dial plan

Dialed digits	Interpretation
0	Local operator
00	Long-distance operator
Xxxx	Local extension number (start with 1 through 7)
8xxxxxxx	Local number
#xxxxxxx	Off-site extension
*xx	Star services
91xxxxxxxxxx	Long-distance number
9011 + up to 15 digits	International number

just described. For example, each one of the patterns in this digit map could be preceded by the start timer symbol (T). The fact that the symbol is not listed in the digit map means that the MG acts as if it were placed at the start of each pattern in any case.

StatisticsDescriptor   This descriptor provides statistical information regarding the usage of a termination within a given context. The specific statistics to be returned depend upon the type of termination in question.

Strictly speaking, this descriptor is always optional and can be returned as a result of several different commands. The descriptor should be returned when a termination is removed from a context through the Subtract command and should also be returned in response to the AuditValue command.

ObservedEvents Descriptor   This descriptor is a mandatory parameter in the Notify command, where it is used to inform the MGC of events that have been detected. This descriptor is optional in most other command responses, except for ServiceChange. When used in a response to the Audit-Value command, it is used to report events stored in the event buffer and that have not yet been reported to the MGC. The descriptor contains a RequestIdentifier with a value matching that received in the events descriptor that listed the events to be detected in the first place. This enables a correlation between the events requested and the events reported.

The descriptor enables the optional inclusion of a timestamp with each observed event indicating the time at which the event was detected. If the timestamp is to be used, it is structured in the form yyyymmddThhmmssss, which includes hundredths of a second. The T separates the year, month, and day from the hour, minute, and seconds. When used, the timestamp itself is separated from the specific event description by a colon (:).

Error Descriptor   The error descriptor is returned in a transaction reply when the requested command could not be executed. The descriptor can also be included in a Notify command from MG to MGC. The error descriptor consists of an error code, optionally accompanied by a textual description of the error. The error codes are registered at the IANA. Table 6-6 lists the various MEGACO error codes and related descriptions. New error codes can be specified, in which case they must be registered with the IANA.

Table 6-6

MEGACO error
codes

Error code	Error description
400	Bad request
401	Protocol error
402	Unauthorized
403	Syntax error in transaction
404	Syntax error in TransactionReply
405	Syntax error in TransactionPending
406	Version not supported
410	Incorrect identifier
411	The transaction refers to an unknown ContextId
412	No ContextIds available
421	Unknown action or illegal combination of actions
422	Syntax error in action
430	Unknown TerminationID
431	No TerminationID matched a wildcard
432	Out of TerminationIDs or no TerminationID available
433	TerminationID is already in a context
440	Unsupported or unknown package
441	Missing remote descriptor
442	Syntax error in command
443	Unsupported or unknown command
444	Unsupported or unknown descriptor
445	Unsupported or unknown property
446	Unsupported or unknown parameter
447	Descriptor not legal in this command

Table 6-6 cont.

MEGACO error
codes

Error code	Error description
448	Descriptor appears twice in a command
450	No such property in this package
451	No such event in this package
452	No such signal in this package
453	No such statistic in this package
454	No such parameter value in this package
455	Parameter illegal in this descriptor
456	Parameter or property appears twice in this descriptor
461	TransactionIds in reply do not match request
462	Commands in transaction reply do not match commands in request
463	TerminationID of transaction reply does not match request
464	Missing reply in transaction reply
465	TransactionId in transaction pending does not match any open request
466	Illegal duplicate transaction request
467	Illegal duplicate transaction reply
471	Implied add for multiplex failure
500	Internal gateway error
501	Not implemented
502	Not ready
503	Service unavailable
504	Command received from unauthorized entity
505	Command received before restart response
510	Insufficient resources
512	MG unequipped to detect requested event
513	MG unequipped to generate requested signals
514	MG cannot send the specified announcement
515	Unsupported media type
517	Unsupported or invalid mode
518	Event buffer full
519	Out of space to store digit map
520	MG does not have a digit map
521	Termination is "ServiceChanging"
526	Insufficient bandwidth
529	Internal hardware failure
530	Temporary network failure
531	Permanent network failure
581	Does not exist

Topology Descriptor   The topology descriptor is somewhat different from other descriptors in that it is relevant to a context only, rather than to a specific termination within a context. The purpose of the descriptor is to indicate how media streams should flow within a context (who can hear or see whom). By default, all terminations in a context can send and receive media to and from each other. If a different situation is desired, then the topology descriptor is used.

The descriptor comprises a sequence of triples in the form Termination1, Termination 2, association. Such a triple indicates whether there is to be no media flow, one-way media flow, or two-way media flow between the two terminations. These three alternatives are indicated by *isolate*, *oneway*, and *bothway* respectively. The order in which the terminations appear in the triple is important. For example, the triple (T1, T2, oneway) means the T1 can send media to T2, but T2 must not send media to T1.

If more than two terminations are used in a context, then more than one triple is required. For example, if three terminations (T1, T2 and T3) are used, then three triples are needed to describe the association between T1 and T2, the association between T1 and T3, and the association between T2 and T3. Figure 6-11 provides an example of several topologies between three terminations. It is assumed that the topologies are applied in sequence.

Figure 6-11

Context topologies

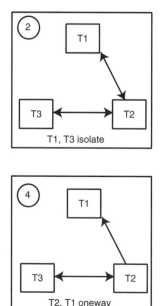
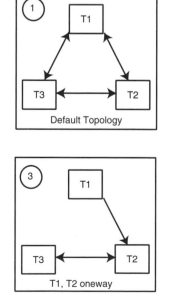

The topology descriptor can be a useful tool for implementing services such as call waiting, where the controlling party needs to swap between two other parties. In a conference call situation, this descriptor can facilitate features such as the ability for one or more parties to the call to have a private discussion before returning to the whole group.

The topology descriptor will not be found as a parameter to any of the MEGACO commands. This is because the commands are directed at specific terminations within a context and the topology is a characteristic of a context as a whole, rather than any specific termination. When the descriptor is to appear in a transaction, it appears before any commands for terminations in a particular context. The topology descriptor can only appear in a transaction for a context that already exists.

Packages

In the previous section on the various descriptors, instances occur where a reference is understandably made to specifics that depend on the type of termination in question. Clearly, different types of terminations exist, and the properties and capabilities of those terminations will vary according to type. For example, it is obvious that a termination connected to an analog telephone line will be quite different from a termination connected to a digital trunk. Consequently, it is not practical to define within the protocol itself all the possible termination properties that could exist, all the possible signals that could possibly be sent, all the possible statistics that might apply, and all the possible events that might be detected.

To address the variations between different types of terminations, MEGACO incorporates the concept of packages. Packages are groups of properties, signals, events, and statistics. Within a package, these items are defined and are given identifiers, and the parameters associated with them are specified. A given termination implements a set of packages, and the MGC can audit a termination to determine which packages it supports. Several basic packages are described in the MEGACO specification. Among these are the Tone Detection Package, the DTMF Generator Package, and the Analog Line Supervision Package, to name just a few. The MEGACO spec also provides guidance for the specification of new packages, which should be registered with the IANA.

Packages Descriptor   In the previous discussion of descriptors, one descriptor is purposely not mentioned. This is the packages descriptor. It was left out until the concept of packages could be introduced. The packages descriptor is used only in responses, and it provides the MGC with a list of the packages implemented in a termination.

MEGACO Command and Response Details

Having described the purpose of the various commands and descriptors, it is worth examining the details of how they are put together into transactions and messages. This is done here prior to specifying which descriptors are applicable to each command so that the reader may have a good understanding of how commands and responses are syntactically structured. This knowledge can help to better visualize the various command/descriptor combinations.

Syntax for Commands and Responses
The following sections specify the syntax for each command and response.

Command Syntax   As described earlier, commands are applied to terminations within a specific context (or a choose [$] context in some cases). Commands applicable to the terminations within the same context are grouped. Several such groups form a transaction and several transactions form a message. In the text syntax of the protocol, groupings of commands are enclosed by the brackets { and }. Similarly, the contents of the different descriptors are enclosed in brackets and individual descriptors are separated by commas. This structure leads to a syntax that appears as in the following example. The line numbers A to K are not part of the syntax, but they are used just for reference in describing the various components:

```
A.  MEGACO/1 [111.111.222.222]:3456
B.  Transaction = 12345 {
C.       Context = - {
D.          Modify = * {
E.              Events = 7777 {al/of},
F.              DigitMap = Map1 {
G.                  (0|00|[1-7]xxx|8xxxxxxx|Fxxxxxxx|Exx|91xxxxxxxxxx|9011x.) }
H.          }
I.       }
J.  }
```

This example shows a message, using MEGACO version 1, from an entity at IP address 111.111.222.222. The message indicates that the MGC is using port number 3456 for MEGACO messages (line A). This transaction contains only one message, and it has a TransactionID of 12345 (line B).

The transaction relates to only one context: the null context identified by the symbol "-" (line C). Only one command is specified, a Modify command, and it applies to all terminations in this context as indicated by the all (*) wildcard (line D).

The command has two descriptors. The first is an events descriptor with a RequestID value of 7777 (line E). Only one event is to be detected: the off-hook (of) event, which belongs to the Analog Line Supervision package (al) (line E). Note that line E, by having both a right and left bracket, totally describes this descriptor. Note also the comma, which indicates that another descriptor is to be expected.

The next descriptor is the DigitMap descriptor, which specifies a Dig-itMap name of Map1 (line F). The details of the digit map are given on line G. This line also contains a right bracket, which terminates the DigitMap descriptor. The fact that no subsequent comma is used indicates that no more descriptors are included for this command.

Line H just contains a right bracket, terminating the Modify command. Line I just contains a right bracket, terminating the set of commands applied to this context. Line J contains a right bracket terminating the transaction.

Within the syntax, we know that this is a transaction request, as opposed to a transaction reply or a transaction pending. This is due to the use of the string Transaction on line B. A transaction reply would utilize the string Reply and a transaction pending would utilize the string Pending.

Response Syntax The previous example command leads to a fairly simple response as follows:

```
A2.   MEGACO/1 [111.111.333.333]:5678
B2.   Reply = 12345 {
C2.        Context = - { Modify = *}
D2.   }
```

The same basic structure is applied to the response. In this case, the message is issued from the MG, and consequently the IP address and port numbers are different (line A2). The TransactionID is the same as in the transaction request; however the Reply token is used to indicate that this is a reply (line B2). Line C2 specifies the context, the null context, as in the transaction request. Line C2 also lists the commands the response is issued to (Modify) and the terminations the commands apply to (all of them in this case). No descriptors are used in the response. Line D2 uses a right bracket to indicate the end of the reply.

Responses in the Case of Failure The previous example showed a reply where the command was carried out successfully. Cases will occur, however, when one or more commands cannot be carried out. Situations might also arise when a complete transaction might fail. To accommodate such situations, MEGACO uses the error descriptor described earlier to indicate failure of a command and the reason for the failure.

Let's assume that the MG was completely unable to execute the foregoing command due to an internal error. In that case, the MG would respond as follows:

```
MEGACO/1 [111.111.333.333]:5678
Reply = 12345 {
    Context = - { Modify = * {
                     Error = 500 { Internal Gateway Error}
                  }
    }
}
```

By examining this response, we can see that the error being reported is applicable to the Modify command. Equally, the error descriptor might be applied higher up, say, at the level of a transaction if a complete transaction could not be executed. This could arise, for example, in the case of some internal error in the MG, a syntax error in the transaction request, or for any number of reasons.

Add, Modify, and Move Commands The Add, Modify, and Move commands are practically identical in terms of the syntax used. The only syntactical difference between them is in the names of the commands themselves. All three commands include the descriptors media, modem, mux, events, EventBuffer, signals, DigitMap, and audit —all of which are optional. Version 2 of the specification does not actually use the modem descriptor, but it is retained in the command syntax for backwards compatibility with MEGACO version 1. To indicate that items are optional in the ABNF syntax, we enclose them in square brackets ([,]). Therefore, the syntax of the commands can be represented as follows, where the command is Add, Move, or Modify.

```
Command ( TerminationID
          [, MediaDescriptor]
          [, ModemDescriptor]
          [, MuxDescriptor]
          [, EventsDescriptor]
          [, EventBufferDescriptor]
          [, SignalsDescriptor]
```

```
          [, DigitMapDescriptor]
          [, AuditDescriptor]
    )
```

Note that this code is a representation of the command and is not exactly how the command will appear as it crosses the interface between the MGC and MG. The exact protocol structure as it crosses the interface is described in the "Command Syntax" section earlier.

Although the three commands are identical in terms of the descriptors that may be used, they obviously have different functions. This fact has an impact on whether wildcards can be used and where. In the Add command, the ContextID can be choose ($), in which case a new context is created. Equally, the TerminationID can take choose ($), in which case a new ephemeral termination is created as a result of the execution of the Add command. The context cannot be the null context, however, since one does not add a termination to the null context. Rather, a termination exists in the null context when it has been subtracted from whatever other context it was in.

For the Modify command, the choose ($) option is not permitted if used either for the TerminationID or for the ContextID. It is not possible to make modifications to terminations that do not exist, and it is not possible to create a new context except with the Add command. It is possible, however, to refer to the null context when issuing a modify command, since it is perfectly valid to modify a termination that exists in the null context.

For the Move command, the null context is not permitted, because the Move command is used only to move terminations from one active context to another. The Move command cannot be used to move a termination to or from the null context. We can use the all (*) wildcard as a TerminationID in the Move command, but we cannot use the choose ($) wildcard.

Add, Modify, and Move Commands' Responses   A successful response to a command can also contain a number of descriptors. As is the case for the command itself, all the allowed descriptors in a response to an Add, Modify, or Move command are optional. The response can be represented as follows. Again, the command can be either Add, Modify, or Move. As is the case for the command itself, the modem descriptor is included in version 2 of the protocol only for backwards-compatibility reasons.

```
Command ( TerminationID
          [, MediaDescriptor]
          [, ModemDescriptor]
          [, MuxDescriptor]
          [, EventsDescriptor]
```

```
        [, SignalsDescriptor]
        [, DigitMapDescriptor]
        [, ObservedEventsDescriptor]
        [, EventBufferDescriptor]
        [, StatisticsDescriptor]
        [, PackagesDescriptor]
    )
```

At first glance, with the exception of some changes in the list of allowed descriptors, the response appears very similar to the command, which it is. Within the syntax, however, the two are easily distinguished by the use of the token Transaction when sending a command and the use of the token Reply when responding.

Subtract Command The Subtract command is used to remove a termination from a context. If the termination removed by a Subtract command happens to be the last termination in a given context, then the context itself is deleted. The Subtract command includes only one optional descriptor, the Audit descriptor. The command is represented as follows:

```
Subtract ( TerminationID
            [, AuditDescriptor]
        )
```

Subtract Command Response Although the syntax of the Subtract command differs significantly from that of the Add, Modify, and Move commands, the syntax of the response to a Subtract command is the same as the syntax of a response to an Add, Move, or Modify command:

```
Subtract ( TerminationID
            [, MediaDescriptor]
            [, ModemDescriptor]
            [, MuxDescriptor]
            [, EventsDescriptor]
            [, SignalsDescriptor]
            [, DigitMapDescriptor]
            [, ObservedEventsDescriptor]
            [, EventBufferDescriptor]
            [, StatisticsDescriptor]
            [, PackagesDescriptor]
        )
```

Although all the descriptors in the response to a Subtract command are optional, the statistics descriptor will be returned by default. If the termination happens to represent an RTP session, this provides the MGC with information such as the number of packets sent, the number of packets received, the number of lost packets, the average latency, and so on.

AuditValue Command   The AuditValue command includes just a single mandatory descriptor, the audit descriptor. The command is used to collect the current settings of a termination's attributes, such as properties and signals:

```
AuditValue ( TerminationID
                    AuditDescriptor
            )
```

AuditValue Command Response

```
AuditValue ( TerminationID
                    [, MediaDescriptor]
                    [, ModemDescriptor]
                    [, MuxDescriptor]
                    [, EventsDescriptor]
                    [, SignalsDescriptor]
                    [, DigitMapDescriptor]
                    [, ObservedEventsDescriptor]
                    [, EventBufferDescriptor]
                    [, StatisticsDescriptor]
                    [, PackagesDescriptor]
            )
```

As is to be expected, the response to the AuditValue command contains a significant number of optional descriptors, each of which can provide information regarding the current state of play at the termination. As is the case for all commands and responses, the modem descriptor should be supported only for backwards-compatibility reasons.

AuditCapabilities Command   With the exception of the command name, the syntax of the AuditCapabilities command is identical to that of the AuditValue command. Whereas the AuditValue command returns information regarding the current settings for properties, events, and signals, the AuditCapabilities command returns information on what properties, events, and signals can be supported by a termination:

```
AuditCapabilities ( TerminationID
                            AuditDescriptor
                    )
```

AuditCapabilities Command Response

```
AuditCapabilities ( TerminationID
                            [, MediaDescriptor]
                            [, ModemDescriptor]
                            [, MuxDescriptor]
```

```
                                       [, EventsDescriptor]
                                       [, SignalsDescriptor]
                                       [, ObservedEventsDescriptor]
                                       [, EventBufferDescriptor]
                                       [, StatisticsDescriptor]
                          )
```

Notify Command   The Notify command is issued only by an MG to an MGC, and it provides a list of events detected on a termination. As such, the command contains just a single mandatory descriptor: the Observed-Events descriptor. The command can optionally contain the error descriptor. Within the ObservedEvents descriptor is a RequestID. This matches the RequestID of an Events descriptor previously sent from the MGC to the MG. This parameter enables the MGC to correlate reported events with a particular request to detect and report such events.

```
    Notify ( TerminationID
               ObservedEventsDescriptor
                 [, ErrorDescriptor]
           )
```

Notify Command Response   The only response that will be issued to a Notify command will occur if an erroneous command is made, in which case the error descriptor is used in the response. This approach is the case for all failed commands.

ServiceChange Command   This command, which is used at restart, failover of an MG, or handoff from one MGC to another, carries only one mandatory descriptor — the ServiceChange descriptor. The TerminationID may be specific, may be wildcarded using the all wildcard (*), or may have the value Root to indicate the MG as a whole.

```
    ServiceChange ( TerminationID
                              ServiceChangeDescriptor
                  )
```

ServiceChange Command Response   The response to the ServiceChange also includes the ServiceChange descriptor, but the descriptor is optional in the response.

```
    ServiceChange ( TerminationID
                              [ServiceChangeDescriptor]
                  )
```

Call Setup Using MEGACO

In Figure 6-8, the setup of a call between two gateways using MGCP is examined. Based on the foregoing descriptions of MEGACO commands and responses, we are in a position to describe how the same call establishment would be handled if MEGACO were the control protocol instead of MGCP. The corresponding call flow is shown in Figure 6-12. The overall objective is to establish communication between termination T1 on MG A and termination T4 on MG B, and we assume that the MGC knows from external signaling (such as SIP or SS7) that the connection is required. MG A has an IP address of 311.311.1.1, while MG B has an IP address of 322.322.1.1. The MGC has an IP address of 333.333.1.1 and uses port number 3333 for MEGACO messages.

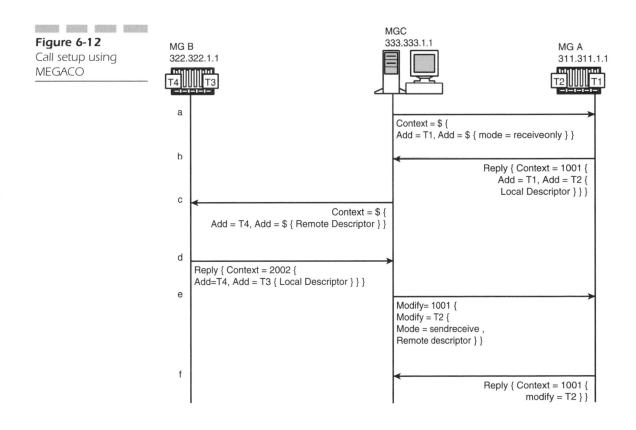

Figure 6-12
Call setup using MEGACO

The figure includes only a representation of the MEGACO messages and commands used, and it does not include a complete syntax of the messages. The complete syntax is given in the following code, along with descriptive text. The syntax assumes the use of MEGACO version 2:

```
Line (a)
MEGACO/2 [333.333.1.1] : 3333
Transaction = 1 {
    Context = $ {
        Add = T1,
        Add = $ {
            Media {
                Stream = 1 {
                    LocalControl {
                        Mode = receiveonly
                    }
                    Local {
                        v=0
                        c=IN IP4 $
                        m=audio $ RTP/AVP 0
                        v=0
                        c=IN IP4 $
                        m=audio $ RTP/AVP 15
                    }
                }
            }
        }
    }
}
```

The MGC issues a transaction request to MG A using a TransactionId value of 1. The transaction is to be executed on a new context created by the MG, as indicated by the choose ($) wildcard as the ContextID. Two Add commands are used. The first is to add termination T1 to the new context. The second is to add a new termination to the same context. This is indicated by the use of the $ as a TerminationID.

The media descriptor for the media associated with the new termination has a StreamID value of 1, assigned by the MGC. The LocalControl is set to receiveonly, which means that the termination may accept media from the remote end, but not send any.

The MGC is also making two suggestions as to the media properties to be applied by the termination. This can be seen in the use of two session descriptions, with the v= line indicating the boundary between each description. The first is the use of RTP payload type 0 (G.711 μ-law). The second is the use of RTP payload type 15 (G.728). Note the use of the choose wildcard within the SDP descriptions. The wildcard is used for both the IP address and the port number to be used in each of the descriptions. Note also that the o=, s=, and t= lines are not included in the SDP descriptions.

Although the SDP specification defines those fields to be mandatory, the MEGACO specification enables the fields to be omitted when SDP is used by MEGACO. The use of a wildcard ($) in the c= line is also a MEGACO-specific modification to SDP and applies only when the MGC is using SDP to suggest media characteristics.

```
Line (b)
MEGACO/2 [311.311.1.1] : 1111
Reply = 1 {
     Context = 1001 {
          Add = T1,
          Add = T2 {
             Media {
                Stream = 1 {
                   Local {
                      v=0
                      c=IN IP4 311.311.1.1
                      m=audio 1199 RTP/AVP 0
                      }
                   }
                }
             }
          }
     }
```

MG A replies to the transaction using the same TransactionID. It provides the ContextID value of 1001 to indicate the newly created context. It indicates that the addition of termination T1 has been successful, and it also indicates the TerminationID of the newly created termination, T2. Associated with this termination, the MG provides a media description, using the same StreamID as specified by the MGC. The MG also provides a session description as the local descriptor. In this case, the session description indicates voice coded according to G.711 μ-law (RTP payload type 0). Furthermore, the IP address and port for the media stream have been filled in by the MG.

```
Line (c)
MEGACO/2 [333.333.1.1] : 3333
Transaction = 2 {
     Context = $ {
          Add = T4,
          Add = $ {
             Media {
                Stream = 2 {
                   LocalControl {
                      Mode = sendreceive
                      }
                   Local {
                      c=IN IP4 $
                      m=audio $ RTP/AVP 0
```

```
                                        v=0
                                     },
                                     Remote {
                                         c=IN IP4 311.311.1.1
                                         m=audio 1199 RTP/AVP 0
                                         v=0
                                               }
                                      }
                           }
                         }
                        }
                       }
                      }
```

The MGC issues a message to MG B, with a TransactionId value of 2. The transaction applies to a new context to be created on MG B, and two terminations are to be added to the new context. The first is termination T4 and the second is a termination that the MG needs to create, as indicated by the use of $ as the TerminationID.

For the new termination, the mode is set to sendreceive, so that the end-point can both send and receive media. The MGC makes a suggestion as to the local session description that should be applied, indicating RTP payload type 0, but leaves the choice of IP address and port number to MG B. The MGC specifies the remote descriptor in detail. This descriptor happens to be the session description received from MG A for termination T2.

```
Line (d)
MEGACO/2 [322.322.1.1] : 2222
Reply = 2 {
     Context = 2002 {
            Add = T4,
            Add = T3 {
                Media {
                      Stream = 2 {
                              Local {
                                  v=0
                                  c=IN IP4 322.322.1.1
                                  m=audio 2299 RTP/AVP 0
                                  }
                                }
                         }
                      }
                    }
             }
```

MG B replies to the transaction request, using the same value of TransactionID. The MG indicates the ContextID of the newly created context (2002) and also indicates that the two terminations, T4 and T3, have been successfully added to the context. T3 is a newly created termination. The reply also specifies media-related information for T3. Specifically, the reply includes a local descriptor in the form of an SDP description, indicating the

IP address of 322.322.1.1, the port number of 2299, and a media format according to RTP payload type 0.

```
Line (e)
MEGACO/2 [333.333.1.1] : 3333
Transaction = 3 {
    Context = 1001 {
        Modify = T2 {
            Media {
                Stream = 1 {
                    LocalControl {
                        Mode = sendreceive
                        }
                    Remote {
                        v=0
                        c=IN IP4 322.322.1.1
                        m=audio 2299 RTP/AVP 0
                            }
                        }
                    }
                }
            }
        }
    }
```

The MGC issues a new message to MG A. The message has one transaction request with a TransactionID value of 3, and it includes a Modify command for termination T2 in context 1001. The Modify command first changes the local control to send-receive so that the termination can send media to the far end. Of course, the termination needs a session description for the far end so that it has an IP address and port number to which to send the media. This is provided in the remote descriptor, which contains the session description received by the MGC from MG B.

```
Line (f)
MEGACO/2 [311.311.1.1] : 1111
Reply = 3 {
    Context = 1001 {
        Modify = T2
            }
        }
```

MG A sends a positive response to the MG to confirm that the modification has been performed. At this stage, termination T1 is in the same context as termination T2 and the two can share media. Terminations T2 and T3 have exchanged session descriptions and can share media. Terminations T3 and T4 are in the same context and can share media. The net effect is that a two-way path is formed between terminations T1 and T4, which means that the original objective has been achieved.

Interworking Between MEGACO and SIP

Interworking between MEGACO and SIP works in very much the same manner as interworking between MGCP and SIP. Once a local descriptor is available at the gateway where the call is being originated, then this can be passed in a SIP INVITE as a SIP message body. Upon acceptance of the call at the far end, the corresponding session description is carried back as a SIP message body within the SIP 200 (OK) response and is passed to the originating side gateway. Once the gateway on the originating side has acknowledged receipt of the remote session description, the MGC can send a SIP ACK to complete the SIP call setup.

No fundamental difference exists between MGCP and MEGACO as far as the manner in which SIP interworking is achieved. This is understandable, since MEGACO and MGCP deal only with MGC-MG control and both handle call setup in quite similar ways.

VoIP and SS7

For a long time, signaling in circuit-switched networks was such that the signaling related to a particular call followed the same path as the speech for that call. This approach is known as *Channel Associated Signaling* (CAS), and the technique is still widely deployed today. Examples include the R1 *Multifrequency* (MF) signaling used in North America and the R2 *Multifrequency Compelled* (MFC) signaling used in many other countries. Although it is still widely used in circuit-switched networks, CAS is considered old technology. The prevailing technology in newer circuit-switched networks is *Common Channel Signaling* (CCS).

CCS involves the use of a separate transmission path for call signaling compared to the bearer path for the call itself, as shown in Figure 7-1. This separation enables the signaling to be handled in a different manner to the call itself. Specifically, other nodes in the network may analyze the signals and take action based on the content of the signals, without needing to be involved in the bearer path. Furthermore, CCS enables signaling messages to be exchanged in cases where no call is to be established at all. Imagine, for example, dialing a star-code to activate some feature. In such a scenario, the local switch might communicate with a service node to activate the feature in the network without actually establishing a call to the service node.

The standard for CCS today is *Signaling System 7* (SS7). Various versions of it are deployed all over the world. In fact, SS7 could be considered the ultimate signaling standard in telecommunications. Anyone who wants to implement a circuit-switched network and who does not implement SS7

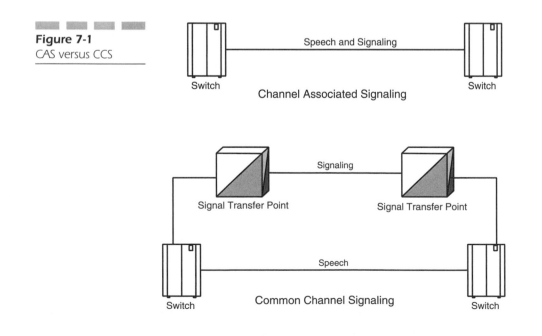

Figure 7-1
CAS versus CCS

Speech and Signaling

Switch   Switch
Channel Associated Signaling

Signaling
Signal Transfer Point   Signal Transfer Point

Speech
Switch   Common Channel Signaling   Switch

is not really playing in the big leagues and cannot really call the network carrier-grade.

Why is SS7 so important? SS7 enables a wide range of services to be provided to the end-user. These services include Caller-ID, toll-free calling, call screening, number portability, and others. In fact, SS7 is the foundation for *Intelligent Network* (IN) services.

Given that SS7 is so prevalent and so important to telecommunications, it follows that any *Voice over IP* (VoIP) carrier that wants to interwork with the *Public Switched Telephone Network* (PSTN) must support SS7 simply in order to communicate with circuit-switched networks. If a VoIP network is to be considered carrier-grade, then it must support SS7. The capability to speak the same language as the PSTN is not the only reason, however. Customers who use VoIP networks will expect the same types of features that are available in existing networks and will expect them to work in the same way. In order to compete effectively, a new carrier has no alternative but to match (if not exceed) the features and functionality already available.

To do so, a carrier can utilize the capabilities of SS7 or try to create a totally new way of providing such services. Utilizing SS7 is the obvious choice. This does not mean that the carrier is limited to what SS7 can provide—far from it. In fact, the *Internet Protocol* (IP) provides the opportunity to offer a wide range of new services. Customers will still expect to see existing services such as toll-free calling, however, and a new carrier has little alternative but to use SS7 to provide many of those services.

This chapter provides a brief introduction to SS7, followed by a description of the technical solutions that can be deployed to support interworking between VoIP networks and SS7. The interworking that needs to be provided has two aspects. First, SS7 needs to be supported by applications in an IP environment even though SS7 was not originally designed for such an environment. Second, different network configuration issues can arise, depending on the architecture of the VoIP network itself. For example, the VoIP network may follow the H.323 model. Alternatively, the VoIP network may use a softswitch architecture based on the *Session Initiation Protocol* (SIP) and the *Media Gateway Control Protocol* (MCGP) or MEGACO. To some degree, how the interworking is done is influenced by the network architectures in question.

The SS7 Protocol Suite

SS7 is not a single protocol per se; rather, it is a suite of protocols operating in concert. In the same way that the *Transmission Control Protocol* (TCP) or the *User Datagram Protocol* (UDP) use the services of the IP in a stack arrangement, SS7 also defines a stack that has certain similarities with the *Open Systems Interconnection* (OSI) seven-layer model. Figure 7-2 shows the SS7 stack.

The Message Transfer Part (MTP)

The three lower levels of the SS7 stack comprise the *Message Transfer Part* (MTP). This is the part of the protocol responsible for getting a particular message from the source to the destination. Level 1 maps closely to layer 1 of the *Open System Interconnection* (OSI) layers. Level 1 handles the issues related to the signals on the physical links between one signaling node and another. These links are known as signaling links and typically operate at 56 Kbps or 64 Kbps.[1] At a node that supports SS7, the physical termination of a signaling link is known as a *signaling terminal*.

MTP level 2 (MTP2) deals with the transfer of messages on a given link from one node to another. Signaling messages from higher-layer protocols are passed down the stack to MTP2, which packages them for transmission on a given signaling link and takes care of getting the message across that link. As well as providing functions for passing messages on behalf of MTP users, MTP2 also has some messages of its own, such as the *Link Status Signal Unit* (LSSU) and the *Fill-In Signal Unit* (FISU). The LSSU is used between two ends of a link to ensure alignment and correct link functioning, and FISUs are sent when nothing else can be sent. The FISU can also carry acknowledgments of received messages.

Figure 7-2
The SS7 protocol stack

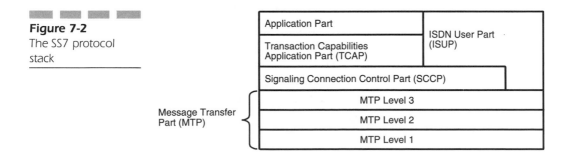

---

[1]This statement applies only to narrowband SS7 links. High-speed links also operate over 1.5 Mbps or 2 Mbps interfaces. High-speed links operate over *Asynchronous Transfer Mode* (ATM).

MTP level 3 (MTP3) deals with the management of messaging on the signaling network as a whole. For a message destined for a particular signaling destination, many different paths could be taken from the source to the destination. These paths could include a direct signaling link if one exists or a path via a number of different intermediate nodes, known as *signal transfer points* (STPs). MTP3 takes care of determining which outgoing link should be used for a particular message. Therefore, MTP3 includes functions for the mapping of signaling destinations to various signaling link sets. MTP3 manages load-sharing across different signaling links and also handles the rerouting of signaling in the case of link failure, congestion, or the failure of another node in the signaling network. In order to help manage the overall signaling network, MTP3 includes a complete signaling network management protocol for ensuring the proper operation of the SS7 network.

Furthermore, MTP3 provides a number of services to the protocol layer above it. These services involve the transfer of messages to and from the upper layer, indicating (for example) the availability or unavailability of a particular destination and signaling network status (such as congestion). These services are provided through the primitives MTP-Transfer request, MTP-Transfer indication, MTP-Pause indication, MTP-Resume indication, and MTP-Status indication.

ISDN User Part (ISUP) and Signaling Connection Control Part (SCCP)

Above MTP, two main alternatives exist: the *ISDN User Part* (ISUP) and the *Signaling Connection Control Part* (SCCP).[2] Let's deal with ISUP first.

Strictly speaking, ISUP is a signaling protocol used to provide services to *Integrated Services Digital Network* (ISDN) applications. In reality, ISUP is most often used as the protocol for setting up and tearing down phone calls between switches and is the most commonly used of the SS7 applications. Examples of ISUP messages include call setup messages such as the *Initial Address message* (IAM) that is used to initiate a call between two switches, the *Answer message* (ANM) that is used indicate that a call has been accepted by the called user, and the *Release message* (REL) that is used to

---

[2]Other alternatives exist, most notably the *Telephony User Part* (TUP). This can be considered a simpler version of ISUP. Numerous national variants of TUP are available.

initiate call disconnection. ISUP is a connection-oriented protocol, which means that it relates to the establishment of connections between users. ISUP messages between the source and destination are related to a bearer path between the source and destination, although the path of the messages and the path of the bearer might be very different. In terms of mapping to the OSI model, ISUP maps to layers 4 through 7.

SCCP provides both connection-oriented and connectionless signaling, although it is most often used to provide connectionless signaling. Connectionless signaling refers to signaling that is passed from one switch or network element to another without the need to establish a signaling connection. In other words, connectionless signaling means that information is simply sent from the source to the destination, without the prior establishment and subsequent release of a dedicated signaling relationship. For example, consider a mobile phone user who roams away from his home network. When the user arrives at the visited network, it is necessary for the visited network and home network to communicate so that the visited network can learn a little about the subscriber and so that the home network can know where the subscriber is. Such signaling is connectionless and it uses the services of SCCP.

SCCP also provides an enhanced addressing mechanism to enable signaling between entities, even when those entities do not know each other's signaling addresses (known as point codes). This addressing is known as *global title addressing*. Basically, it is a means whereby some other address, such as a telephone number, can be mapped to a point code either at the node that initiated the message or some other node between the originator and destination of the message.

The *Transaction Capabilities Application Part* (TCAP) enables the management of transactions and procedures between remote applications, such that operations can be invoked and the results of those operations can be communicated between the applications. TCAP is defined for connectionless signaling only. TCAP provides services to any number of application parts. Common application parts include the *Intelligent Network Application Part* (INAP) and the *Mobile Application Part* (MAP).

SS7 Network Architecture

SS7 is purely a signaling protocol that enables services and features in the telephony network. The SS7 network, however, should be considered as a separate network in its own right for a number of reasons. First, SS7 involves the separation of signaling from the bearer (and in some cases, there might be no bearer at all). Second, SS7 involves a number of network elements that deal only with signaling itself and that are tied together in a network of their own. Consequently, when we think about SS7 in the context of interworking between circuit-switched networks and VoIP networks, we should think in terms of three interworking networks, as shown in Figure 7-3. The VoIP system and the circuit-switched network pass signaling via the SS7 network and pass media directly to and from one another.

Figure 7-4 depicts a typical SS7 network arrangement. The two switches shown do not communicate signaling to each other via direct paths. Instead, the signaling is carried via one or more STPs, typically arranged in a quad. This configuration serves several purposes. To start, it avoids the need for direct signaling links between all endpoints (such as switches), meaning that a fully meshed signaling network is not required. Instead, a separate, more compact signaling network can be created. Second, the quad arrangement ensures great robustness. Each switch will be connected to two STPs, and multiple paths exist between the different STPs. Thus, the failure of a single link will never result in a complete loss of signaling capability between different endpoints.

In fact, SS7 networks in general are extremely robust. The design of SS7 is such that very reliable signaling networks can be built, and the complete failure of a node or of a number of signaling links does not mean catastrophe for the network. As far as the applications are concerned, once information is sent to the SS7 stack, it is as good as delivered to the ultimate destination. SS7 really is that reliable—and that fast.

Figure 7-3
Interworking between IP, telephony, and SS7 networks

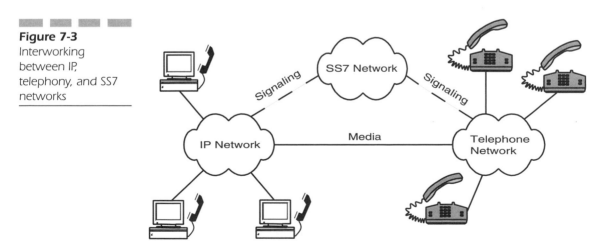

Figure 7-4
SS7 network

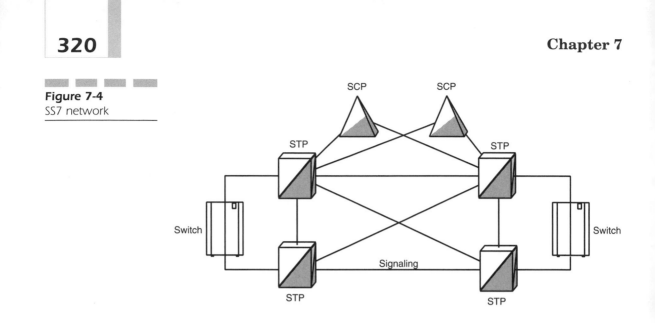

Signaling Points (SPs)

Each node in an SS7 network is a *signaling point* (SP) and has one or more signaling addresses, each of which is known as a *signaling point code* (SPC). SPs that are directly connected are said to be adjacent and the links between such adjacent SPs are known as signaling links. Each of these is typically a 56 Kbps or 64 Kbps link. Several signaling links might exist between two adjacent signaling points for capacity or security reasons, and the set of links is not surprisingly called a linkset. More formally, a linkset is that group of signaling links directly connecting two SPCs.

Signal Transfer Point (STP)

The function of an STP is to transfer messages from one SPC to another. Basically, it works as depicted in Figure 7-5. Let us assume that Switch A, with SPC value 1, wants to send a message known as a *message signal unit* (MSU), to Switch B, which has an SPC value of 2. Let's also assume that no signaling link exists between them, but that they are both connected to an STP with SPC value 3. In that case, the routing tables at Switch A could be set up to send the MSU to the STP. The MSU contains the destination signaling address of Switch B, SPC value 2. When the message is received at the STP, the STP first checks to see if the message is destined for itself by checking the destination address. Given that the message is not, the STP

Figure 7-5
STP function

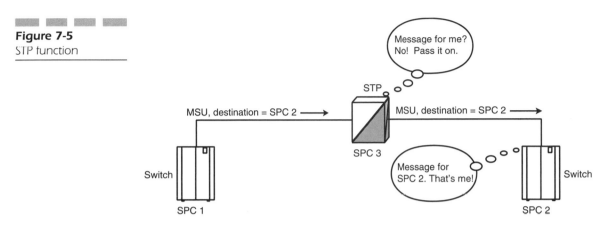

Figure 7-5
STP function

then checks its routing tables to see where messages for SPC 2 should be sent. The STP determines that such MSUs should be sent on the link to Switch B. When the MSU arrives at Switch B, the same process is repeated, except that Switch B recognizes that the MSU is destined for it and sends the content of the message to the appropriate signaling application.

Service Control Point (SCP)

A *service control point* (SCP) is a network entity that contains additional logic and that can be used to offer advanced services. To use the service logic of the SCP, a switch needs to contain functionality that will enable it to act upon instructions from the SCP. In such a case, the switch is known as a *service switching point* (SSP). If a particular service needs to be invoked, the SSP sends a message to the SCP asking for instructions. The SCP, based upon data and service logic that is available, will tell the SSP which actions need to be taken.

A good example of such a service is a call to a toll-free 800 number. When a subscriber dials such a number, the SSP knows that it needs to query the SCP for instructions (which it does). The SCP contains translation information between the dialed 800 number and the real number to which the call should be sent. The SCP responds to the SSP with a routable number, which the SSP uses to route the call to the correct destination. Note that the signaling between the SSP and the SCP is connectionless signaling. Hence, this signaling uses the services of SCCP. In fact, this type of application uses the services of TCAP, which in turn uses the services of SCCP.

Message Signal Units (MSUs)

As mentioned, the messages that are sent in the SS7 signaling network are known as MSUs. Each MSU has the format shown in Figure 7-6. The *Service Information Octet* (SIO) contains the service indicator, which indicates the upper-level protocol to which the message applies (such as SCCP or ISUP). The SIO also contains the Subservice field, which indicates the signaling numbering plan in use. We will discuss this topic in more detail shortly. The *Signaling Information Field* (SIF) contains the actual user information being sent.

If we look at the details of the SIF, then an MSU has one of the formats shown in Figure 7-7, depending on whether the *American National Standards Institute* (ANSI) version of SS7 or the *International Telecommunication Union—Telecommunications Standardization Sector* (ITU-T) version of SS7 is being used. Of particular importance is the routing label, which contains two very important fields: the *destination point code* (DPC) and the *origination point code* (OPC). These are the signaling addresses of the

Figure 7-6

SS7 messages

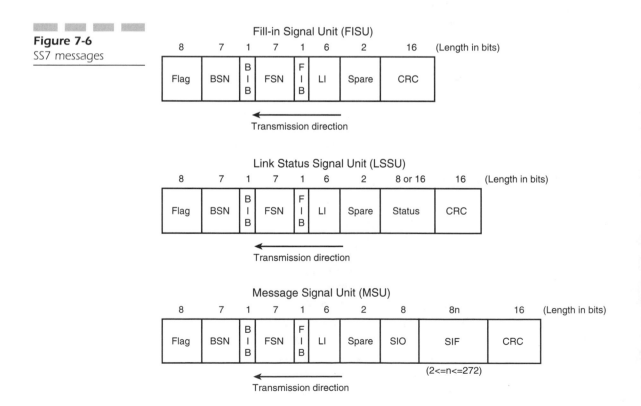

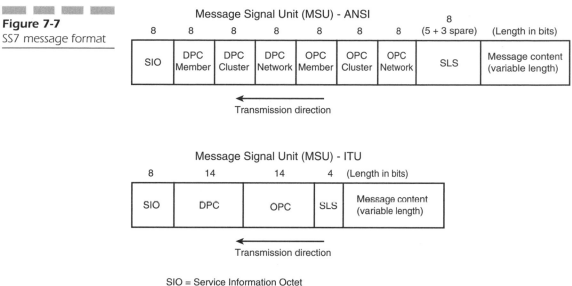

Figure 7-7
SS7 message format

SIO = Service Information Octet
SLS = Signaling link selection
The ANSI SLS is 8 bits, but historically was 5 bits with 3 bits to spare.

destination of the message and the originator of the message respectively. The *Signaling Link Selection* (SLS) field also exists, which indicates the particular signaling link to be used for carrying this message between the two nodes in question.

SS7 Addressing

As mentioned, signaling entities in an SS7 network are addressed by SPCs. A quick look at Figure 7-7 shows that unfortunately different versions of SS7 exist, with different point code structures. For example, in North America, point codes are 24 bits long, whereas in most other countries, point codes are 14 bits long. A notable exception is China, where point codes are 24 bits long.

In North America, point codes have a hierarchical structure, with the format network-cluster-member, each of eight bits. The operator of a very large network would be allocated a network code, enabling that operator to have about 65,000 different SPCs.

Given that different SS7 formats and addressing schemes exist, how do SS7 entities in different countries communicate? The answer is through international *signaling gateways* (SGs). These gateways are signaling points that support the national SS7 variant at one side and the ITU-T international variant at the other. This fact of both national and international variants brings us back to the Subservice field mentioned earlier. Recall that this field is used to indicate the signaling numbering plan in use. Four variants exist: National, National Spare, International, and International Spare. A network element that acts as an international gateway will have at least two point codes, one a national point code and one an international point code. Messages to and from other nodes in the national network will be addressed with the national point code, and messages to and from other international nodes will use the international point code. Thus, a hierarchy exists in the SS7 network whereby messages between countries must go from the national signaling plane to the international signaling plane and back down to the national signaling plane in the destination country, as shown in Figure 7-8.

ISUP

In regular telephony, ISUP is the most often used SS7 application. ISUP is the protocol utilized for the establishment and release of telephone calls. Figure 7-9 shows a typical call establishment and release.

The call begins with the IAM, which contains information about the called number, the calling number, the transmission requirement (typically 64 Kbps), the type of caller (ordinary, operator, payphone, and so on), and other information, such as whether a satellite link has been included in the call so far.

Upon receipt of the IAM at the destination switch, an *Address Complete Message* (ACM) is returned. This message indicates that the call is through-connected to the destination. The ACM causes a one-way audio path to be opened from the destination switch to the originating switch so that the

Figure 7-8
International
signaling

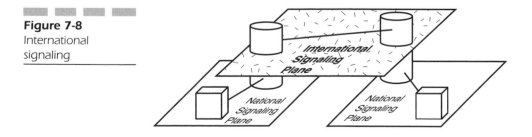

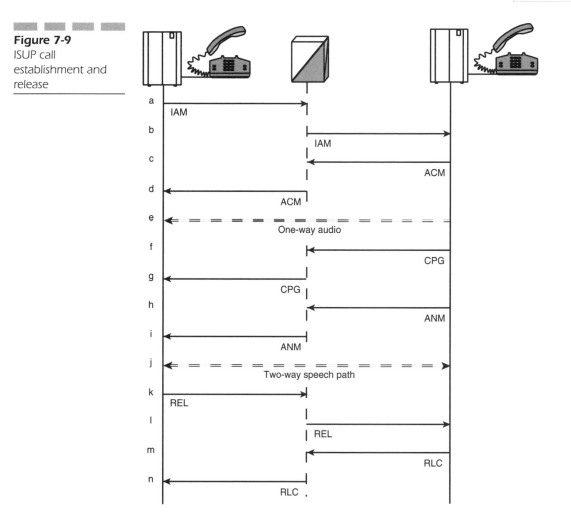

Figure 7-9
ISUP call
establishment and
release

caller can hear a ring-back tone. Note that the tone is generated at the destination end.

Strictly speaking, the ACM is an optional message. Although it is returned in most calls, that is not always the case. Situations can exist where an ACM is not returned. If it is not returned, the caller does not hear a ring-back tone at all, and it appears as though the call is answered immediately. This happens quite frequently on calls to toll-free numbers, particularly if the call is answered automatically.

After the ACM, there might be one or more *Call Progress* (CPG) messages. The CPG is an optional message and is used to provide additional information to the calling switch regarding the handling of the call.

Once the called party answers, an *Answer Message* (ANM) is returned. This message typically has two purposes. The first is to open the transmission path in both directions so that the parties can converse. The second is to instigate charging for the call, since charging for most calls begins when the call is answered.

After conversation, one of the parties hangs up. This causes a *Release* (REL) message to be sent to the other end. The end that receives the REL responds with a *Release Complete* (RLC) message. At this point, the call is complete.

Note that the signaling passes through one or more STPs, whereas the speech takes a more direct transmission path from A to B. That situation leads to one important question. Given that there might be many simultaneous calls between switch A and switch B, how does the ISUP signaling differentiate between calls? In particular, how is a given ISUP message correlated with a given speech path between the switches? The answer is the *Circuit Identification Code* (CIC), which is contained within each ISUP message. For an ISUP message between two switches, the CIC indicates the specific trunk between the switches to which the message applies. In fact, a given circuit between two switches is completely identified by the combination of OPC, DPC, and CIC. The CIC can be seen in the ISUP message formats shown in Figure 7-10. The figure also serves to further highlight the differences between different SS7 variants.

Figure 7-10

The ISUP message format

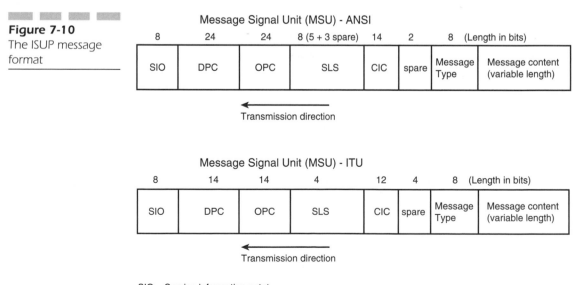

Message Signal Unit (MSU) - ANSI

8	24	24	8 (5 + 3 spare)	14	2	8	(Length in bits)
SIO	DPC	OPC	SLS	CIC	spare	Message Type	Message content (variable length)

← Transmission direction

Message Signal Unit (MSU) - ITU

8	14	14	4	12	4	8	(Length in bits)
SIO	DPC	OPC	SLS	CIC	spare	Message Type	Message content (variable length)

← Transmission direction

SIO = Service information octet
SLS = Signaling link selection

Performance Requirements for SS7

Given that VoIP networks need to interwork with the PSTN, it is clear that VoIP networks need to speak SS7 (at least to the outside world). Not only should VoIP networks support the messages of SS7, however, but they must also support the performance requirements specified for SS7. For example, for MTP, Bellcore[3] specification GR-246-Core states that a given route set should not be out of service for more than 10 minutes per year. Other requirements are that no more than 1×10^{-7} messages should be lost, and no more than 1×10^{-10} messages should be delivered out of sequence. In ISUP, numerous timing requirements must be met. For example, a 2-second timer is initiated when a continuity check request is sent, requiring that a continuity check tone be returned within that time.

Imagine a VoIP network that operates as a long-distance network, as depicted in Figure 7-11. This network receives an IAM from a network at one side of the country, processes the information, and determines that the call should be sent to a network at the other side of the country. The VoIP network sends an IAM to the destination network, which responds with an ACM. The VoIP network must process the ACM and send it to the originating network. This is exactly the same process as is performed by a traditional long-distance network.

If we think about the amount of internal signaling that goes on in an H.323 network or even in a softswitch network, however, we realize that a lot of work is done between the passing of external SS7 messages. Furthermore, if we think about the postdial delay of today's telephone networks (a

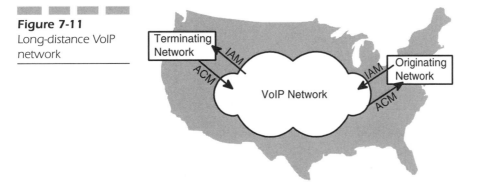

Figure 7-11
Long-distance VoIP network

Terminating Network

Originating Network

VoIP Network

IAM
ACM

---

[3]Bellcore is now known as Telcordia. It is an organization that has written many technical standards, most of which are adopted by North American network operators.

second or two), we realize just how fast this work needs to be done. The point is that a VoIP network using SS7 must meet the stringent requirements that are already met by today's circuit-switched networks. The issue, then, is how to make sure that VoIP networks can emulate the signaling performance of SS7. Fortunately, many groups have been working on this issue. In particular, the *Signaling Transport* (Sigtran) group of the *Internet Engineering Task Force* (IETF) has made great progress in this area. The following sections of this chapter are devoted to describing the architectures and solutions developed within Sigtran.

Sigtran

Sigtran is a working group within the IETF dedicated to addressing the issues regarding the transport of signaling within IP networks. In particular, the group addresses the issues related to signaling performance within IP networks and the interworking with other networks, such as the PSTN.

In order to better understand the issues to be addressed, let us again consider the scenario depicted in Figure 7-11. Let's assume that the VoIP network has a softswitch architecture. Therefore, the VoIP network comprises a number of *media gateways* (MGs), SGs, and *media gateway controllers* (MGCs)/call agents. The purpose of the SGs is to connect directly with signaling points or STPs of the external SS7 network. The function of the MGs and MGCs is as described in Chapter 6, "Media Gateway Control and the Softswitch Architecture." Let's assume for now that the call agents communicate with each other using SIP and that they communicate with MGs using MEGACO.

Given that the MGC is the call control entity within the network, it follows that ISUP signaling received at an SG must be passed to the MGC so that the MGC can control call establishment and call release appropriately. If we assume a straightforward call from a trunking gateway at one side to a trunking gateway at the other side, then the scenario would be similar to Figure 7-12.

The carriage of ISUP signaling information from an SG to an MGC or from an MGC to an SG is indicated in the Figure 7-12 with the letters "IP." For example, an IAM carried in the IP network is denoted "IP IAM." Note that the content of the message does not change since it is assumed that the MGC contains a standard ISUP application. The only difference is that the ISUP messages are carried to and from the MGC over IP rather than over standard MTP. Thus, all that the SG does is to take the ISUP message from

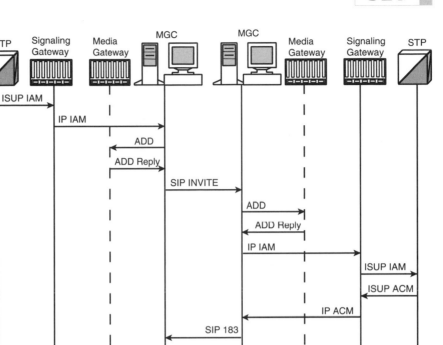

Figure 7-12
SIP/MEGACO/ISUP
interworking

the SS7 network and get it to the appropriate entity in the IP network. This task might sound simple—just a simple translation from point code to IP address, right? Unfortunately, things are rarely that simple.

Certainly, point code to IP address translation is one of the issues to be addressed, but we also face the issue of which transport (layer 4) protocol to use for carrying the messages. Furthermore, using a standard SS7 application such as ISUP in an IP environment is also an issue. How can we deploy such an application that expects certain services from lower layers such as MTP when those lower layers do not exist in the IP network?

For the transport protocol, the ISUP messages must be carried in the IP network with the same speed and reliability as in the SS7 network. In some implementations, proprietary mechanisms are used for getting information from an SG to a call agent. This approach requires, however, that the MGC and SG be sourced from the same vendor, thereby reducing the network operator's vendor selection choices. In some other implementations, the ISUP message is simply packaged as a UDP or TCP packet and sent from the SG to the call agent.

This approach can have problems, however. First, the strict performance requirements should be met with regard to message loss and message sequencing. These requirements lead us away from UDP, because it is inherently unreliable. We now have TCP as an alternative, but TCP is not suitable either because it cannot always meet the strict timing requirements of SS7. Although TCP can ensure the accurate in-sequence delivery of messages, it is not particularly fast. TCP also suffers from head-of-line blocking. Therefore, something else is needed.

In order to address these issues, the Sigtran working group has prepared a number of documents describing the issues and solutions. One such document is RFC 2719, "Framework Architecture for Signaling Transport," which describes an overall approach and methodology for signaling transport within IP networks.

Sigtran Architecture

The Sigtran architecture is defined in RFC 2719. The architecture uses the signaling components shown in Figure 7-13. Signaling over standard IP uses a common transport protocol that ensures reliable signaling delivery, and it uses an adaptation layer that supports specific primitives as required by a particular signaling application. In other words, the common signaling transport makes sure that messages are delivered error free and in sequence regardless of the vagaries of the underlying IP network. The adaptation layer provides an interface to the upper-layer protocols and applications such that those applications do not realize that the underlying transport is IP rather than traditional transport, such as MTP.

Although most of the adaptation layers fulfill SS7-related functions, the *Stream Control Transmission Protocol* (SCTP) (which corresponds to the common signaling transport of Figure 7-13) is a generic signaling transport

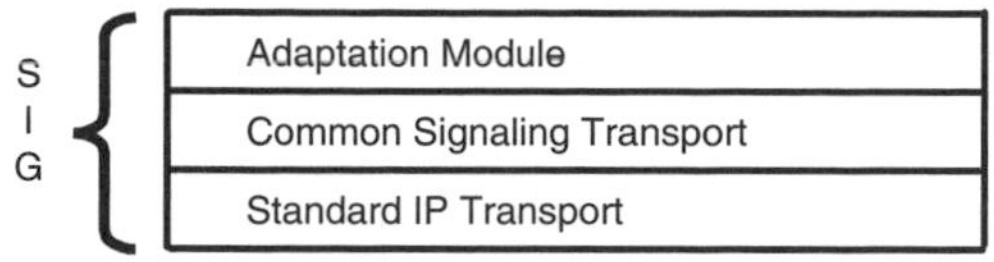

Figure 7-13
Sigtran signaling
transport
components

S
I
G
{

Adaptation Module
Common Signaling Transport
Standard IP Transport

protocol, which means that SCTP can also be used to reliably carry other types of signaling traffic. Therefore, Sigtran also includes adaptation layers that enable non-SS7 signaling to be carried on SCTP.

If we apply the Sigtran architecture to a situation where an SG is connected to the SS7 network on one side and to an MGC on the other, then the transport of ISUP signaling would appear as in Figure 7-14. Within the SG, the *Nodal Interworking Function* (NIF) is the function responsible for interworking between the SS7 network on one side and the IP network on the other.

Figure 7-15 shows the Sigtran architecture in more detail. SCTP corresponds to the common signaling transport shown in Figure 7-13. SCTP ensures the error-free, in-sequence delivery of user messages, enables fast delivery of messages, and avoids head-of-line blocking. Therefore, SCTP is a more suitable protocol than TCP. Moreover, SCTP supports network-level fault tolerance, something critical for carrier-grade network performance. SCTP is described in more detail a little later in this chapter.

Before delving into the details of SCTP, however, let's first take a quick look at the other components of the Sigtran architecture. These components are a number of adaptation layers that provide the interface between SCTP and upper-layer protocols. Several such components exist, most of which are SS7 specific, but two of which are related to other types of signaling. In the following sections, more focus is applied to those components that support ISUP interworking, because ISUP is the most commonly used SS7 application and is the protocol with which VoIP networks must most often interwork.

As mentioned, a number of adaptation modules operate on top of SCTP. These include the following:

- The *SS7 MTP2-User Adaptation Layer* (M2UA) provides adaptation between MTP3 and SCTP. M2UA provides an interface between MTP3 and SCTP such that standard MTP3 may be used in the IP network without the MTP3 application software realizing that messages are being transported over SCTP and IP instead of MTP2. For example, a standard MTP3 application implemented at the MGC could exchange MTP3 signaling network management messages with the external SS7

Figure 7-14
ISUP transport to
MGC

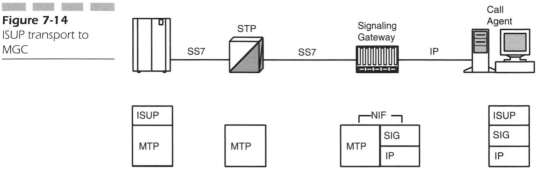

NIF = Nodal Interworking Function

Figure 7-15
Sigtran protocol stack

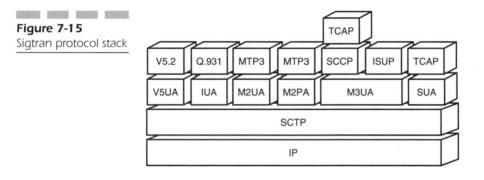

network. In the same manner that MTP2 provides services to MTP3 in the SS7 network, M2UA provides services to MTP3 in the IP network. M2UA has a registered port number of 2904.

The usage of M2UA is depicted in Figure 7-16. In this scenario, two SGs provide an interface to the outside SS7 network. Both are connected to an MGC. On the MGC side of the SGs, we have M2UA over SCTP over IP, whereas on the SS7 side, we have the standard SS7 MTP. At the MGC, we have a standard MTP3 operating over M2UA and IP. In a regular SS7 network, MTP3 utilizes the services of MTP2. In the scenario depicted in Figure 7-16, however, the MTP3 at the MGC utilizes the services of the MTP2 located at the SG without realizing that it is not local. The function of M2UA is to provide transparent access from the standard MTP3 at the call agent to the standard MTP2 at the SG. In the example of Figure 7-16, the MTP3 application at the MGC can receive MTP3 signaling network management messages such as *Transfer Allowed* (TFA) and *Transfer Prohibited* (TFP). MTP3 can

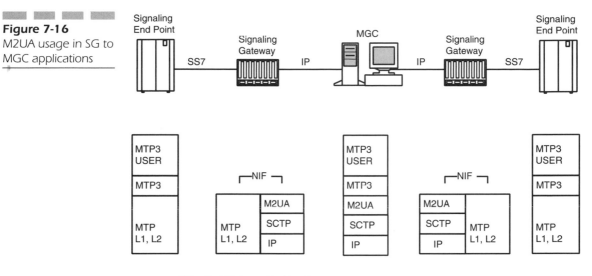

Figure 7-16

M2UA usage in SG to MGC applications

NIF = Nodal Interworking Function

use that information in determining how to route messages from a higher-level MTP3 user, such as ISUP.

■ The *SS7 MTP2 Peer-to-Peer Adaptation Layer* (M2PA) provides adaptation between MTP3 and SCTP. As is the case for M2UA, the MTP3 layer at an IP node (such as an MGC) communicates with M2PA as if it were regular MTP2. Consider Figure 7-17, which shows a potential usage of M2PA.

Even though M2UA and M2PA have some similarities, they are different in a number of respects. Although M2UA enables an IP-based node to utilize a standard MTP2 at a remote SG, M2PA is more akin to an IP-based version of MTP2. In other words, the SG-MGC connection shown in Figure 7-17 is effectively an SS7 link. That is not the case for the corresponding connection on Figure 7-16. One result of this difference is that an SG that utilizes M2UA is effectively a remote signaling terminal for the MGC shown in Figure 7-16. An SG that utilizes M2PA is a signaling node in its own right; it has its own signaling point code, and it is effectively an IP-based STP. These characteristics mean that the SG can also process higher-layer signaling functions, such as SCCP. Therefore, an SG that uses M2PA could perform *Global Title Translation* (GTT), for example. An M2UA-based SG cannot perform such functions.

Figure 7-17

M2PA usage in an SG
to MGC application

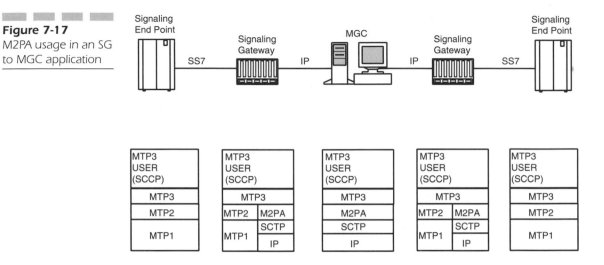

Both M2UA and M2PA can have valid roles to play when transporting
SS7 signaling over an IP-based transport. The choice of which protocol
to use will depend on how the network architect wants to design the
network. Specifically, the choice will depend on which functions are to
be performed at which nodes in the network. If, for example, an SG
needs to perform functions such as GTT, then M2PA will be the
appropriate choice. On the other hand, if the SG is simply meant as an
SS7 signaling terminal for an IP-based node (such as an MGC), then
M2UA will be sufficient. M2PA has a registered port number of 3565.

■ The *SS7 MTP3-User Adaptation Layer* (M3UA) provides an interface
between SCTP and those applications that typically use the services of
MTP3, such as ISUP and SCCP. M3UA and SCTP enable seamless
peer-to-peer communication between MTP3 user applications in the IP
network and identical applications in the SS7 network. The application
in the IP network does not realize that SCTP over IP transport is used
instead of the typical SS7. In the same manner that MTP3 provides
services to applications such as ISUP in the SS7 network, M3UA offers
equivalent services to applications in the IP network. M3UA has a
registered port number of 2905.

We must recognize that M3UA is simply an adaptation layer between
the upper protocols and SCTP. Therefore, although it provides the same
primitives to the upper layer as MTP3 offers in standard SS7, M3UA is

not an IP-flavored MTP3. For example, M3UA does not implement standard MTP3 signaling network management messages such as TFA, TFP, and so on. This point is important and will be emphasized shortly.

Consider an MGC that needs to run an application such as ISUP. The MGC can do so in several ways. For example, the MGC can run ISUP over MTP3 over M2UA (or M2PA) over SCTP, as shown in Figure 7-16 and Figure 7-17. Alternatively, the MGC can implement ISUP over M3UA over SCTP, as shown in Figure 7-18. The difference between the two approaches is a matter of where the MTP3 function really resides. In the scenario depicted in Figure 7-18, the real MTP3 exists at the SGs. M3UA simply enables the ISUP application at the MGC to remotely access the MTP3 function at the SG, without the ISUP application realizing that the MTP3 function is not local.

The MGC of Figure 7-18 might have its own point code, separate from that of the SG. In that case, the SG function likes an STP and appears as an STP to the outside SS7 network. The outside SS7 network views the MGC as a typical SS7 signaling endpoint to which access is achieved via one or more SG STPs.

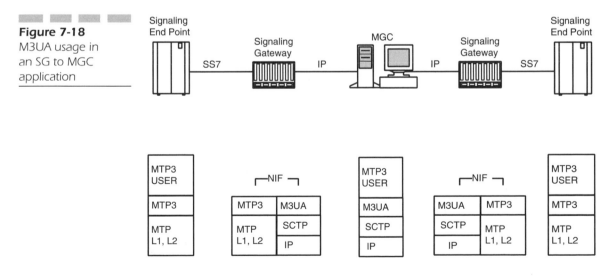

Figure 7-18
M3UA usage in an SG to MGC application

NIF = Nodal Interworking Function

Given that the M3UA provides services to ISUP and given the importance of ISUP interworking, the functioning of M3UA is described in more detail later in this chapter.

■ The *SS7 SCCP-User Adaptation Layer* (SUA) provides an interface between SCCP user applications and SCTP. Applications such as TCAP use the services of SUA in the same way that they use the services of SCCP in the SS7 network. In fact, those applications do not know that the underlying transport is different in any way. Similar to other adaptation layers, SUA merely provides a transparent conduit whereby an application such as TCAP at an IP node can use the services of SCCP at an SG. Hence, there can be transparent peer-to-peer communication between applications in the SS7 network and applications in the IP network. SUA has a registered port number of 14001.

■ The *ISDN Q.921-User Adaptation Layer* (IUA) also operates on top of SCTP. In ISDN, user signaling, such as Q.931, is carried by the Q.921 Data-link layer. The equivalent Sigtran specification in an IP network is IUA. Thus, Q.931 messages may be passed from the ISDN to the IP network, with identical Q.931 implementations in each network, neither of them recognizing any difference in the underlying transport. The registered port number for IUA is 9000.

■ The *V5.2-User Adaptation Layer* (V5UA) has a number of similarities with the IUA protocol and can be considered an extension of IUA. V5.2 defines an *European Telecommunications Standards Institute* (ETSI) interface for a connection between an access network and a local exchange. V5.2 supports network access for analog telephone lines, ISDN basic rate access, ISDN primary rate access, and other types of access that do not use out-of-band signaling. The main purpose of V5.2 is to enable an access concentrator system to be placed between the user device (such as an analog line) and the local exchange, as depicted in Figure 7-19. For example, a concentrator might be placed in a neighborhood, and thousands of subscriber lines might be attached to the concentrator. Up to 16 E1 links would connect the concentrator back to the local exchange. The links between the concentrator and local exchange operate according to the V5.2 protocol, which enables the multiplexing and concentration of user traffic across those links. This situation is depicted in Figure 7-19.

If the local exchange of Figure 7-19 were to be replaced by a softswitch configuration (MGC and MG), then an SG would be required between the access system and the MGC, as shown in Figure 7-20. The V5UA

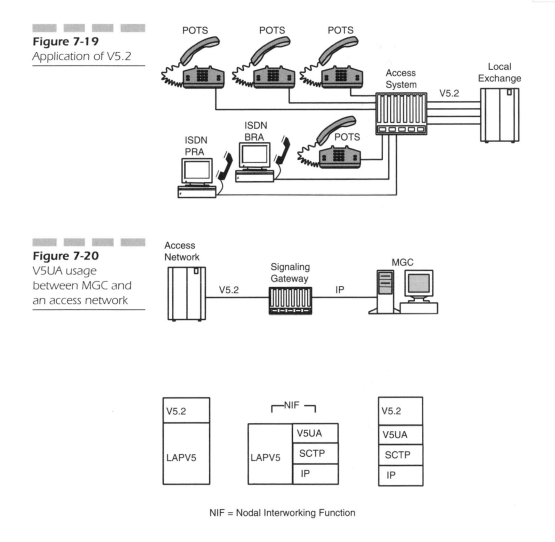

Figure 7-19
Application of V5.2

Figure 7-20
V5UA usage
between MGC and
an access network

NIF = Nodal Interworking Function

protocol enables the V5.2 applications at the MGC to utilize the native V5.2 functions on the access network side of the SG. V5UA has a registered port number of 5675.

SCTP

SCTP is specified in RFC 2960. The primary motivation behind the development of SCTP was that neither UDP nor TCP offered both the speed and reliability required of a transport protocol used to carry signaling. SCTP offers both speed and reliability, and it offers upper-layer applications the opportunity to take advantage of those characteristics. In the SCTP specification, such an application is known as an *Upper-layer Protocol* (ULP). A ULP can be any of the protocols directly above the SCTP layer, as illustrated in Figure 7-15, such as M2UA or M3UA.

Some SCTP Concepts   Before attempting to explain how SCTP provides this reliable transport to the various SCTP users, it is first worth becoming familiar with some SCTP terminology. Note that the following terms comprise just a selection of the most important terms necessary for an understanding of SCTP basics.

SCTP Endpoint   An SCTP endpoint is a logical sender or receiver of SCTP packets. In protocol terms, an SCTP endpoint is a combination of one or more IP addresses and a port number. SCTP enables an endpoint to have multiple IP addresses, meaning that the endpoint can be "multihomed" (spread over several physical platforms). This concept is important, because it enables for fault tolerance in the network, something critical to carrier-grade performance.

Even though a given SCTP endpoint can have multiple IP addresses, it can use only one port number. Thus, if an endpoint has several IP addresses, the same SCTP port number is applicable to each. The combination of an IP address and a port number is known as a *transport address*. Note that a given transport address may apply to only one SCTP endpoint, though that endpoint may have several transport addresses.

Association   SCTP works by establishing a relationship between SCTP endpoints. Such a relationship is known as an *association* and is defined by the SCTP endpoints involved and the current protocol state. Before applications at two endpoints can communicate, an association must be established. Once communication is complete, the association can be terminated. The association can also be terminated in error situations.

The upper-layer protocols such as ISUP, SCCP, or TCAP are not aware of such associations. After all, they are blind to the fact that the signaling is being carried by something other than standard MTP. Therefore, the task of instigating an SCTP association falls to the applicable adaptation layer.

Packets and Chunks   SCTP sits on top of IP. When SCTP wants to send a piece of information to the remote end, it sends what is known as an SCTP packet to IP, which routes the packet to the destination. The SCTP packet comprises a common header and a number of chunks, as depicted in

Figure 7-21
The SCTP packet
format

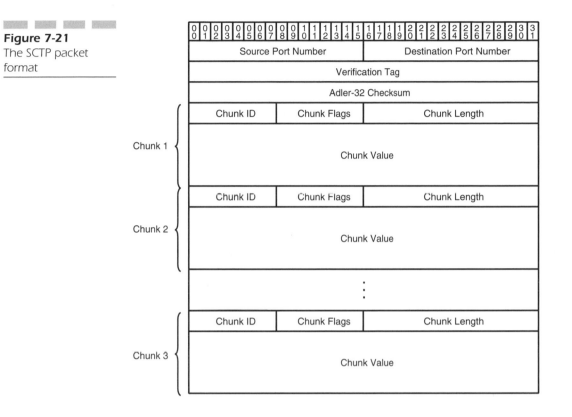

Figure 7-21. The common header includes the source and destination port numbers, which, when combined with the source and destination IP addresses, uniquely identify the endpoints. The header also includes a verification tag, which is used to validate the sender of the packet. The verification tag is described in further detail later in this chapter.

The common header also includes an Adler-32 checksum, which is a particular calculation based on the values of the octets in the packet. This checksum is used to ensure that the packet has been received without corruption and provides another level of protection over and above the IP header checksum.

A number of chunks follow the common header. Each chunk is comprised of a chunk header, plus some chunk-specific content. This content can be either SCTP control information or SCTP user information. In the case of SCTP user information from a ULP, the value of the Chunk ID is 0, indicating user payload data. Otherwise, the Chunk ID will have a value indicating a particular type of SCTP control information. The possible values for the chunk flags and chunk length depend upon the value of the Chunk ID.

Streams   A stream is a one-way logical channel between SCTP endpoints. One can think of a stream as a sequence of messages from one SCTP user to another. When an association is established between endpoints, part of the establishment of the association involves specifying how many streams are to be supported by the association. If we think of a given association as a one-way highway between endpoints, then the individual streams are analogous to the individual traffic lanes on that highway.

To understand the stream concept in a network scenario, imagine a call agent that uses ISUP and that communicates with the outside SS7 network via an SG. The physical signaling links from the SS7 network terminate at the SG and the actual speech circuits from the PSTN terminate at an MG controlled by the call agent, as shown in Figure 7-22.

If the call agent uses ISUP over MTP3 over M2UA, then the MTP3 function operating at the call agent will attempt to direct messages from ISUP to a particular signaling link since that is one of the functions of MTP3. No physical signaling link is used at the call agent, however; only an SCTP association is made between the M2UA layer at the call agent and the M2UA layer at the SG. In order to handle this situation, the SCTP association should include N streams, where N is the number of signaling links terminated at the SG. In this manner, the M2UA function at the call agent can map a particular signaling link, as specified by MTP3 to a particular stream. Equally, the SG can map the stream to a particular signaling link to the destination.

If the ISUP function at the MGC uses M3UA, then more flexibility can take place in how streams are allocated in the SCTP association between M3UA at the SG and M3UA at the MGC. Since the choice of signaling link is made by the MTP3 function at the SG, it is not necessary that the

Figure 7-22

Streams allocated according to signaling links or CIC values

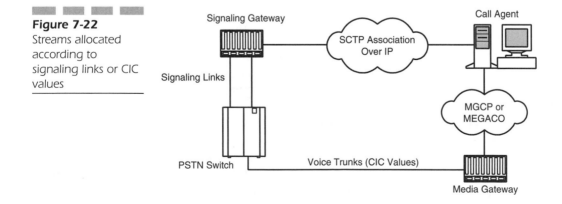

streams be allocated according to the signaling link, although they could be. Alternatively, the streams could be allocated according to the DPC/OPC combination or even according to the OPC/DPC/CIC range.

The use of these streams is an important tool to avoid head-of-line blocking and to ensure in-sequence delivery. In-sequence delivery is ensured because each message sent for a given stream contains a stream sequence number. The receiving entity can make sure that messages are transmitted to the user application in the order of stream sequence numbers. Secondly, each stream is processed separately so that message delivery for one stream is not held up while waiting for the next in-sequence message of a different stream.

Types of SCTP Chunks   Within an SCTP packet, individual chunks are identified by a chunk ID. Although a chunk ID has 256 possible values and hence 256 different chunk types exist, SCTP chunks can be grouped into four main categories: those that carry SCTP user data, those that carry SCTP control information, those that are reserved by the IETF, and those that carry IETF-defined extensions. The correlation between chunk ID values and chunk types is shown in Table 7-1.

Table 7-1

Chunk ID values and chunk types

ID value	Chunk type
00000000	Payload Data (DATA)
00000001	Initiation (INIT)
00000010	Initiation Acknowledgment (INIT ACK)
00000011	Selective Acknowledgment (SACK)
00000100	Heartbeat Request (HEARTBEAT)
00000101	Heartbeat Acknowledgment (HEARTBEAT ACK)
00000110	Abort (ABORT)
00000111	Shutdown Association (SHUTDOWN)
00001000	Shutdown Acknowledgment (SHUTDOWN ACK)
00001001	Operation Error (ERROR)
00001010	State Cookie (COOKIE ECHO)
00001011	Cookie Acknowledgment (COOKIE ACK)
00001100	Reserved for Explicit Congestion Notification Echo (ECNE)

Table 7-1 cont.

Chunk ID values
and chunk types

ID value	Chunk type
00001101	Reserved for Congestion Window Reduced (CWR)
00001110	Shutdown Complete (SHUTDOWN COMPLETE)
00001111 to 00111110	Reserved by IETF
00111111	IETF-defined Chunk Extension
01000000 to 01111110	Reserved by IETF
01111111	IETF-defined Chunk Extension
10000000 to 10111110	Reserved by IETF
10111111	IETF-defined Chunk Extension
11000000 to 11111110	Reserved by IETF
11111111	IETF-defined Chunk Extension

Chunk ID values are encoded such that the two highest-order bits specify the action that must be taken if the processing endpoint does not recognize the chunk type. The possible bits and their meanings are as follows:

- **00**   Stop processing this SCTP packet and discard it; do not process any further chunks within it.
- **01**   Stop processing this SCTP packet and discard it, do not process any further chunks within it, and report the unrecognized parameter in an Unrecognized Parameter Type (in either an ERROR or in an INIT ACK).
- **10**   Skip this chunk and continue processing.
- **11**   Skip this chunk and continue processing, but report an ERROR chunk using the Unrecognized Chunk Type cause of error.

SCTP Control Chunks   The INIT chunk is used to initiate an SCTP association between two endpoints. Unlike many other chunks, the INIT chunk must not share an SCTP packet with any other chunk. In other words, an SCTP packet containing an INIT chunk must contain no other chunks besides the INIT chunk. The INIT chunk is further described later in this chapter.

The INIT ACK chunk is used to acknowledge the initiation of an SCTP association. As is the case for the INIT chunk, the INIT ACK chunk must not share a packet with any other chunk.

The SACK chunk is used to acknowledge the receipt of DATA (payload) chunks and to inform the sender of any gaps in the received chunks. Not every received chunk merits a SACK chunk in response. Instead, the SACK chunk works by indicating where gaps occur. Let us suppose that a receiver has received chunks 1 through 5, plus chunks 8 and 9. The SACK message is used to indicate that chunks 6 and 7 are missing. These are the only chunks that need to be resent. This technique is more efficient that the retransmit mechanism of TCP.

The HEARTBEAT chunk is used to query the reachability of a particular endpoint. Let's assume that no chunks need to be sent from endpoint A to endpoint B during a particular period of time. In that case, endpoint A will send periodic HEARTBEAT messages to endpoint B, just to make sure that endpoint B is still alive. The HEARTBEAT contains sender-specific information. The receiver of the HEARTBEAT chunk should respond with a HEARTBEAT ACK chunk containing heartbeat information copied from the received HEARTBEAT chunk.

The ABORT chunk is sent by an endpoint to end an association abruptly. The ABORT chunk may contain cause information regarding the reason for aborting the association. The ABORT chunk may be multiplexed with other SCTP control chunks into one packet. In such cases, however, the ABORT chunk should be the last chunk in the packet. If it is not the last chunk, then subsequent chunks in the packet are ignored. DATA chunks should not be included in the same packet as an ABORT chunk.

The SHUTDOWN chunk is used for a graceful termination of an association. If a higher-layer application or management application wants to terminate an association, then the endpoint stops sending any new data to the far end. It will wait until all data sent to the far end has been acknowledged and will then send a SHUTDOWN to the far end in order to close the association. The SHUTDOWN will indicate the last data chunk received from the far end. The endpoint may, if necessary, retransmit user data to the far end before sending the SHUTDOWN chunk.

Upon receipt of a SHUTDOWN chunk, an endpoint will make sure that all the user data that it has sent has been acknowledged and, if necessary, retransmit data to the far end. Once it is sure that everything it sent in the past has been received, the endpoint shall send a SHUTDOWN ACK chunk.

Upon receipt of the SHUTDOWN ACK, the sender of the SHUTDOWN shall respond with a SHUTDOWN COMPLETE chunk and will remove all knowledge of the association. When the far end receives the SHUTDOWN COMPLETE chunk, it too will delete all knowledge of the association. At this point, the association is ended.

The ERROR chunk is sent to indicate that the endpoint has detected some error condition. The chunk will include an error cause to provide further information on the type of error. For example, an endpoint may have received a chunk for a nonexistent stream or may have received a chunk that is missing certain mandatory parameters. The receipt of an ERROR chunk does not, in itself, indicate a fatal condition. An ERROR chunk sent on its own may simply enable the receiver to correct the error condition. If the error condition is fatal, then the ERROR chunk can be sent in the same datagram as an ABORT chunk.

The COOKIE ECHO chunk is used only during the initialization of an association. When an endpoint receives an INIT chunk and responds with an INIT ACK chunk, it includes a cookie parameter within the INIT ACK chunk. This parameter contains information specific to the endpoint and to the endpoint's view of the association, a timestamp, and a cookie lifetime value (5 seconds is recommended). When the far end receives the INIT ACK, it copies the cookie information and returns it in a COOKIE ECHO chunk. The COOKIE ECHO chunk can be sent in a packet that also contains DATA chunks. In such a situation, however, the COOKIE ECHO chunk must be the first chunk in the packet.

The COOKIE ACK chunk is sent in response to a COOKIE ECHO chunk. Hence, the COOKIE ACK chunk is used only during the establishment of an association. Since the content of a COOKIE ECHO chunk is the same as what was sent in an INIT ACK, the sender of the INIT ACK can make sure that the initiator of the association has received the cookie information correctly. If the COOKIE ECHO chunk has been received without error within the lifetime specified for the cookie, the receiver of the COOKIE ECHO chunk sends a COOKIE ACK chunk. Otherwise, an ERROR chunk is sent. The COOKIE ACK chunk may be sent in the same datagram as DATA chunks, but it must be the first chunk in the datagram.

Payload Data (DATA) Chunk   The DATA chunk is used to carry information to and from the ULP. It has the format shown in Figure 7-23.

SCTP can possibly segment a given user message. This situation could occur if the path MTU is smaller than the size of the message to be sent. The *Beginning* (B) and *End* (E) bits are included because of such potential segmentation. The B bit indicates that this chunk contains the first segment of a user message, and the E bit indicates that the chunk contains the last segment of a user message. If a message is completely contained within one chunk, then both the B and E bits are set to 0. If the message contains more than 2 segments, then the first segment shall have the B bit set to 1

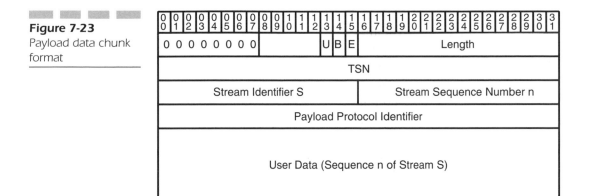

Figure 7-23
Payload data chunk format

and the E bit set to 0, while the last segment shall have the B bit set to 0 and the E bit set to 1. All segments in between shall have both the E and B bits set to 0.

The U bit indicates that this chunk belongs to an unordered data stream. In other words, the order of user messages in the stream is not significant and the value of the stream sequence number should be ignored. In such a case, SCTP passes the data to the appropriate upper layer without any concern for the order of the message. SCTP must still, however, ensure that segmented messages are reassembled before passing the data to the upper layer.

The *transmission sequence number* (TSN) is a 32-bit integer identifying this chunk in the context of the association. It is independent of any stream sequence number and is assigned by SCTP rather than by any user of SCTP. When an endpoint sends an INIT chunk, it includes a TSN value, which corresponds to the first DATA chunk that it plans to send. Thus, the first DATA chunk sent will contain the same value of TSN as in the INIT chunk. Thereafter, the TSN is incremented for each new DATA chunk that the endpoint sends in this association.

The stream identifier (S) is a 16-bit integer, identifying the stream to which the data belongs. The stream sequence number (n) is a 16-bit integer indicating the position of this message within the stream. For a given stream, the value of the stream sequence number begins at 0. It is incremented for each message sent in a given stream. Note that a segmented message shall have the same value of the stream sequence number in each segment.

The payload protocol identifier is passed from the user to SCTP at the sending end and passed from SCTP to the destination user at the receiving end. It is available to the users for passing further information about the chunk, but it is not examined or acted upon by SCTP.

Establishing an Association   The establishment of an association is typically instigated by an SCTP upper-layer protocol, which instructs SCTP to establish the association. The association can be established in advance of any data traffic being sent. For example, if a network administrator provisions an ISUP trunk group on an MGC and the MGC uses M3UA, then the act of defining the trunk group could cause M3UA to request SCTP to establish an association. Similarly, an association could be established when a new SS7 link is brought into service at an SG or when a new CIC range is brought into service.

The process is depicted in Figure 7-24. It begins by sending an SCTP packet containing an INIT chunk. The INIT chunk has the format shown in Figure 7-25.

The Chunk Flags field should be set to 0. The Initiate Tag is a 32-bit integer and is a random number. It must not have the value 0. The end that receives this INIT chunk (the far end) will store the value of the Initiate tag. For every SCTP data packet that the far end sends in this association, it will use the value of the Initiate tag as the Verification tag in the SCTP common header. The *Advertised Receiver Credit Window* (a_rwnd) represents the dedicated buffer space that the sender has allocated for this association. Although the buffer space available to the association may exceed this value during the lifetime of the association, it must never fall below this value. The *Number of Outbound Streams* (OS) and the *Number of Inbound Streams* (MIS) indicate the maximum number of streams that the initiator is willing to send and receive respectively for this association. The

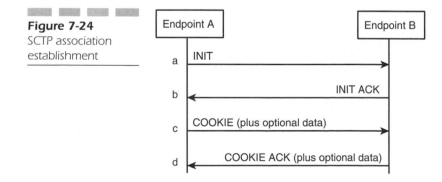

Figure 7-24
SCTP association
establishment

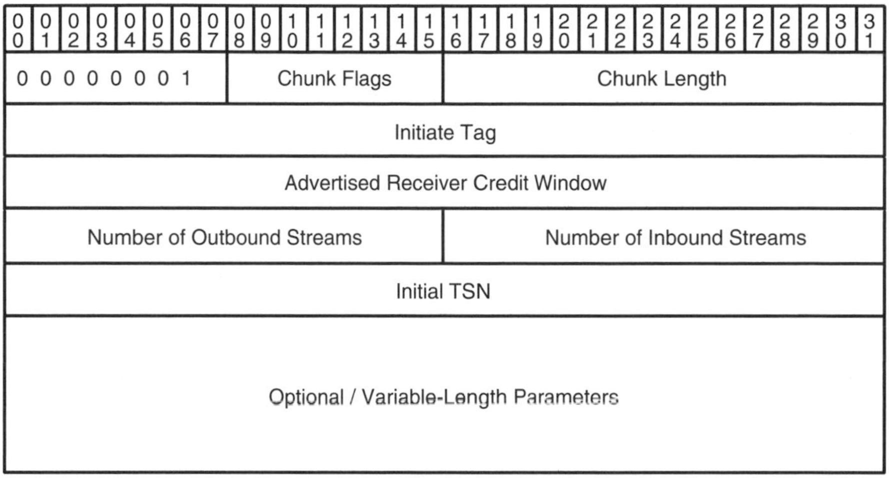

Figure 7-25
The INIT chunk
format

0 0 0 0 0 0 0 1	Chunk Flags	Chunk Length
Initiate Tag		
Advertised Receiver Credit Window		
Number of Outbound Streams		Number of Inbound Streams
Initial TSN		
Optional / Variable-Length Parameters		

Figure 7-25
The INIT chunk format

value 0 is not allowed for either of these parameters. The Initial TSN is the TSN value that the sender expects to use when sending the first DATA chunk in the association.

The INIT chunk can also contain a number of optional parameters and/or variable-length parameters. The INIT chunk can, for example, contain one or more IPv4 or IPv6 addresses, or it can contain a host name that can be resolved to one or more IP addresses.

When the other endpoint receives the INIT chunk, it responds with an INIT ACK chunk, which has a format identical to the INIT chunk, with the exception that the optional/variable component must include a cookie. Importantly, the Verification tag in the SCTP common header must contain the same value as the received Initiate tag in the INIT chunk.

The INIT ACK chunk can also contain a number of IPv4 or IPv6 addresses and possibly a host name that can be resolved to one or more IP addresses. Given that the INIT ACK chunk also contains numbers of inbound and outbound streams, both ends of the association know the maximum number of streams that the other can send and receive. Each endpoint must respect the other's requirements and should not send more than what the other endpoint can handle.

Upon receipt of the INIT ACK chunk, the initiator of the association sends a COOKIE ECHO chunk, the content of which is copied from the cookie parameter of the received INIT ACK. In order to quickly start the message exchange for which the association is being created, one or more additional DATA chunks may be included in the packet. The COOKIE ECHO chunk must be the first chunk in the packet, however.

Finally, the receiver of the COOKIE chunk sends a COOKIE ACK chunk. This is just a header containing the CHUNK ID value of 00001011, the chunk flags all set to 0, and a length indicator of 4. In other words, it's just a simple acknowledgement with no additional information. The packet containing the COOKIE ACK chunk can also contain a number of DATA chunks, provided that the COOKIE ACK chunk is the first chunk in the packet.

Transferring Data   The reliable transfer of user data is achieved by the use of two SCTP chunks. The first is the DATA chunk described earlier and depicted in Figure 7-23. The second is the SACK chunk, which is an SCTP control chunk and is shown in Figure 7-26.

The easiest explanation of the SACK chunk is by example. Let's assume that an endpoint has transmitted data chunks 1 through 11. Let's also assume that chunks with TSNs 1 through 4 and those with TSNs 7, 8, 10, and 11 have been received. Hence, chunks 5, 6, and 9 are missing. Let's also assume that chunks with TSN 8 and TSN 11 have been received twice. Therefore, the received data appears as depicted in Figure 7-27.

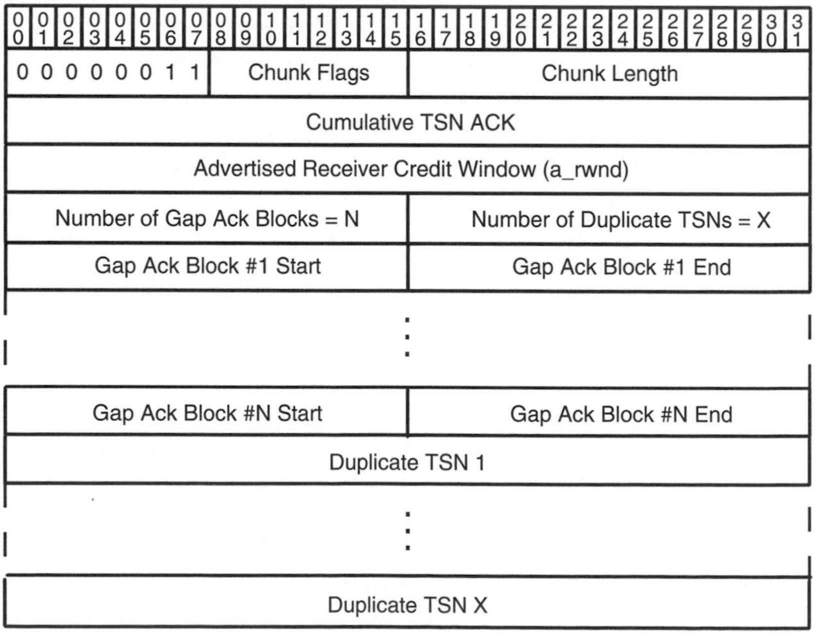

Figure 7-26

The SACK chunk format

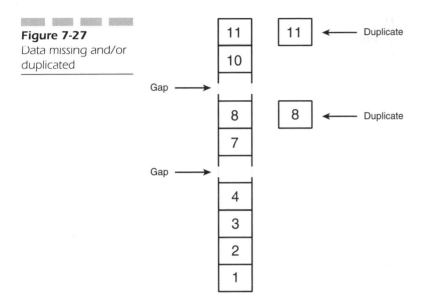

Figure 7-27
Data missing and/or
duplicated

In this scenario, the Cumulative TSN ACK field would contain the value 4, as this field indicates the highest TSN value received without any gaps. The number of Gap Ack Blocks (N) indicates the number of fragments received after the unbroken sequence. In our example, N has the value 2 to indicate the block 7 to 8 and the block 10 to 11. The number of duplicate TSNs (X) indicates the number of TSNs that have been received more than once.

The Gap Ack Block number 1 start field indicates the offset of the first segment from the unbroken sequence. This is the difference between the TSN value in the Cumulative TSN ACK field and the lowest TSN value of the first segment. In our example, this is the difference between 4 and the value 7 (i.e., 3). The Gap Ack Block number 1 end field indicates an offset from the Cumulative TSN ACK and the highest TSN in the first block after the Cumulative TSN. This field has the value 4 in our example (the difference between 8 and 4). This process is repeated for each fragment. Thus, for the next fragment, the Gap Ack Block start and end values are 6 and 7 respectively. Finally, we have a listing of the duplicate TSNs. In this case, the values 8 and 11 would be listed.

The *Advertised Receiver Window Credit* (a_rwnd) indicates the updated buffer space of the sender of this Selective ACK. This indicates how much receive buffer space is free at the time the SACK is sent. It enables the remote end to manage the amount of data sent, such that buffer overflow and consequent data loss can be avoided. SCTP specifies procedures based on the a_rwnd and the SACK chunk in general to ensure that congestion is avoided.

SCTP Robustness   Robustness is a key characteristic of any carrier-grade network. Robustness means that the network should implement procedures whereby failures or undesired occurrences are minimized. Robustness also means the capability to handle a certain amount of failure in the network without a significant reduction in quality. Furthermore, the network should provide a graceful rather than a drastic degradation in the event of failures or overload. SCTP addresses these issues in a number of ways.

As mentioned previously, SCTP implements congestion control mechanisms to ensure that one endpoint does not flood another with messages. Furthermore, SCTP incorporates Path MTU discovery so that messages are not sent if they are too long to be handled by the intervening transport network.

Recall that the SCTP common header contains source and destination port numbers. Recall also that the INIT and INIT ACK chunks may optionally include one or more IP addresses or a host name that can be resolved to one or more addresses. The inclusion of these parameters enables a given endpoint to be effectively multihomed (to have multiple IP addresses). If the INIT or INIT ACK chunk does not contain any IP addresses, then the IP address shall default to the IP address from which the packet is sent. Greater robustness is offered by having multiple IP addresses, however.

If a given endpoint supplies several IP addresses, then the other end shall choose one of those addresses as the primary destination address. If the primary address fails, then one of the other addresses is chosen as the destination address for subsequent messages. Furthermore, if one or more DATA chunks are transmitted to a given address and not acknowledged within the retransmission timer, then the sender should send the retransmission to one of the other addresses.

SCTP ensures that an endpoint is aware of the reachability of another endpoint through the use of SACK chunks if DATA chunks have been sent and through HEARTBEAT chunks if an association is idle. Therefore, the detection of path or endpoint failure is assured and appropriate action can be taken.

M3UA Operation

The following describes how M3UA over SCTP can be used in a VoIP network to provide seamless interworking with the external SS7 network for ISUP. First, let's become familiar with a number of terms applicable to M3UA. In fact, many of these terms apply to all of the adaptation layers:

- An *application server* (AS) is a logical entity handling signaling for a particular scope. For example, an AS could be a logical entity within a softswitch that handles ISUP signaling for a particular SS7 DPC/OPC/CIC range. Equally, an AS could be a logical entity within a database application that handles signaling for a particular SCCP DPC/OPC/*Subsystem Number* (SSN) combination. An AS contains a set of *application server processes* (ASPs), which are those actual processes operating within the AS. In fact, the AS can be considered as a list of ASPs, some of which are active and some standby.

- An *application server process* (ASP) is a process instance of an AS, acting as an active or standby process. An ASP could be, for example, that process within an MGC that is currently handling ISUP signaling. The ASP has an SCTP endpoint. Therefore, the ASP can be multihomed (spread across multiple IP addresses). In a robust network, there should be at least one active ASP and at least one standby ASP for a given application. A single ASP can serve multiple ASs. For example, an MGC could have several ASs, each corresponding to a unique OPC/DPC combination. That same MGC might have only one ASP handling all ISUP signaling (or two ASPs for redundancy or load sharing).

- A *routing key* is a set of SS7 parameters such as an SLS, DPC, OPC, or CIC range that identifies that signaling for a given AS. For example, if a given AS is to handle ISUP signaling for a particular combination of OPC/DPC/CIC range, then that OPC/DPC/CIC range is the routing key. Within an SG, a particular routing key will point to a particular AS. In other words, a one-to-one relationship exists between a routing key and an AS.

- *Network appearance* is a mechanism for separating signaling traffic between an SG and an ASP, where all the traffic uses the same underlying SCTP association. Imagine, for example, an SG that is an international SG. This SG will have at least one national point code and one international point code. The SG uses a national variant of MTP for communication within the national network and it uses the

ITU-T variant of MTP for communication within the international network. For communication between an SG and a call agent/MGC, the messages being exchanged must be placed in the context of the correct network appearance so that the appropriate link set can be used on the non-IP side of the SG.

Signaling Network Architecture   In order to ensure carrier-grade service, we must ensure that no single point of failure exists on the network between the SG and ASP. Consequently, SGs should be set up at least in pairs in a manner similar to the arrangement of STPs in the SS7 network. Furthermore, ASPs should be set up in a redundant or load-sharing configuration, spread out over different physical platforms (hosts). Such a robust configuration is shown in Figure 7-28. Note that the requirements for such a robust configuration are not something unique to the usage of M3UA; they apply to any system that needs to be fault tolerant (a requirement for carrier-grade operation).

Each ASP needs to be associated with a point code. The allocation of point codes to ASPs is completely flexible, however. For example, all ASPs connected to a given SG could share the same point code as the SG. In such a case, the combination of SG and ASPs appears to the SS7 network as a single signaling endpoint. Alternatively, all ASPs connected to a given SG could share the same point code, but a different point code from the SG. In such a case, the SG would appear to the SS7 network as an STP and the

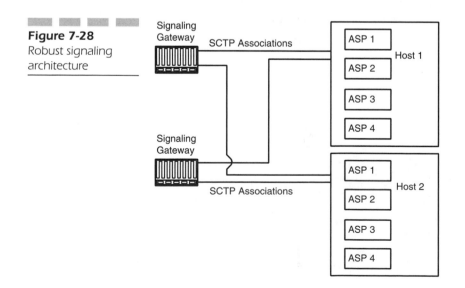

Figure 7-28
Robust signaling
architecture

combined ASPs as a single signaling endpoint located behind that STP. Another alternative would be for every ASP to have its own point code or for groups of ASPs to share a point code separate from the point code of the SG. In that case, the SG appears as an STP and each ASP or group of ASPs appears as a unique signaling endpoint.

The options available are dependent only upon the translation and mapping capabilities of the SG. If a given ASP or group of ASPs can communicate with the SS7 network via more than one SG, then the ASP or group of ASPs must have a point code different from those of the two SGs. In such a scenario, the SGs operate as STPs, as shown in Figure 7-29.

M3UA uses a client server model, where one end of the SCTP association acts as the client and initiates the association, while the other end acts as the server. Typically, the ASP will act as the client, with the M3UA application at an SG acting as a server. Product implementations should, however, enable a node to act either as a client or a server.

Services Provided by M3UA As mentioned, M3UA provides an interface between an application such as ISUP and the underlying transport. It enables the application to utilize the services of the real MTP3, which is located at an SG. M3UA accomplishes this task in a transparent manner, so that the application does not realize that the instance of MTP3 being used is at the SG rather than local.

In order to provide services to the upper layer, M3UA must offer the same primitives to the upper layer as are offered by MTP3. These are as follows:

- The *MTP-Transfer request* is sent from the upper layer to M3UA to request that a message be transferred to a particular destination.

- The *MTP-Transfer indication* is used by M3UA to pass an incoming message to the upper layer.

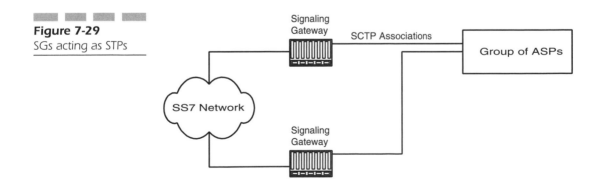

Figure 7-29
SGs acting as STPs

■ The *MTP-Pause indication* is sent by M3UA to the upper layer to indicate that signaling to a particular destination should be suspended. This primitive is used, for example, when the destination is not reachable.

■ The *MTP-Resume indication* is sent by M3UA to the upper layer to indicate that signaling to a particular destination can resume.

■ The *MTP-Status indication* is sent by M3UA to the upper layer to inform the upper layer of some change in the SS7 network such as network congestion or a destination user part becoming unavailable.

Transferring application messages from the application to the SS7 network is a relatively straightforward process. Let's take an example of an ISUP application at an MGC wanting to send a message to the SS7 network. The application issues an MTP-Transfer request to M3UA, and M3UA effectively sends this to SCTP as a DATA chunk to be transmitted on a particular SCTP association and in a particular stream. SCTP ensures that it reaches the SG, where the content of the DATA chunk is passed up to M3UA, which passes it to the interworking function of the SG. The interworking function passes it to MTP3 on the SS7 side of the SG, which takes care of routing the message correctly to the SS7 network.

In the opposite direction, the process is similar. A message arrives at the SG and is passed from MTP3 up to the interworking function and from the interworking function to M3UA. Based on the SS7 parameters of the message, such as the DPC/OPC/CIC range, the appropriate AS and the appropriate ASP are chosen. If more than one ASP is active, then one of the active ASPs is chosen (depending on the load-sharing algorithm). M3UA packages the message for transmission by SCTP as a DATA chunk on the appropriate SCTP association and within the appropriate stream. At the destination, the DATA chunk is passed to M3UA, which passes the information to the application using the MTP-Transfer indication primitive.

M3UA Messages   M3UA includes a number of messages between peer M3UA entities. These have the generic format shown in Figure 7-30. This format contains a header followed by the message content. The header is common across all adaptation layers. For M3UA, the protocol version is 0000 0001. The message class is as follows:

■ **0**   Management messages

■ **1**   Transfer messages

■ **2**   *SS7 Signaling Network Management* (SSNM) messages

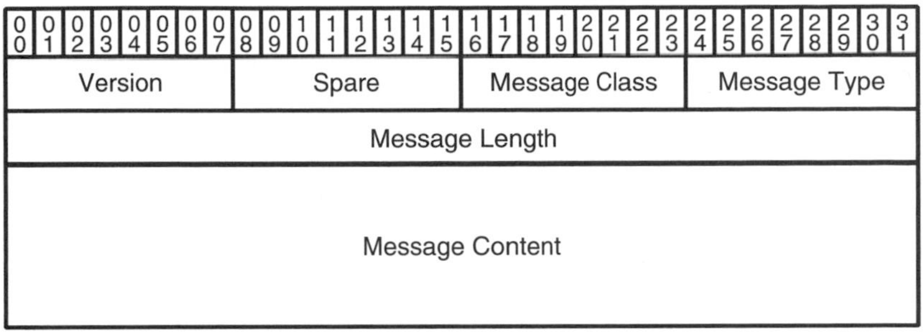

Figure 7-30
Common message
format

| 0 0 0 0 0 0 0 0 0 0 1 1 1 1 1 1 1 1 1 1 2 2 2 2 2 2 2 2 2 2 3 3 |
| 0 1 2 3 4 5 6 7 8 9 0 1 2 3 4 5 6 7 8 9 0 1 2 3 4 5 6 7 8 9 0 1 |

Version	Spare	Message Class	Message Type

Message Length

Message Content

- **3** *ASP State Maintenance* (ASPSM) messages
- **4** *ASP Traffic Maintenance* (ASPTM) messages
- **5** Reserved for IUA-related Q.921/Q.931 boundary primitive transport messages
- **6** Reserved for M2UA-related user adaptation messages
- **7** Reserved for SUA-related connectionless messages
- **8** Reserved for SUA-related connection-oriented messages
- **9** *Routing Key Management* (RKM) messages
- **10** Reserved for M2UA interface identifier management messages
- **11** Reserved for M2PA messages
- **12 to 127** Reserved by the IETF
- **128 to 255** Reserved for IETF-defined message class extensions

When M3UA passes user information between the SG and MGC, as described previously, it does so by sending M3UA DATA messages. These DATA messages are packaged as SCTP DATA chunks. When sending a DATA message, the message class field has the value 1 and the message type has the value 1 (see Table 7-2).

In addition to packaging user messages in the form of M3UA DATA messages, M3UA includes other messages between M3UA peer entities so that the entities may be able to communicate information regarding the SS7 network in general. For example, if a remote destination becomes unavailable, then the SG will become aware of this fact through SS7 signaling network management messages. The ISUP application at an MGC must also be made aware of the event so that it does not try to send messages that cannot reach their destination. M3UA at the MGC can indicate such an event through the use of the MTP-Pause indication primitive. But M3UA is not MTP3, and SS7 signaling network management messages are not passed directly from the SG to the MGC. So how does the M3UA at the MGC know that the destination is unavailable? The answer is that the information is passed in other M3UA messages besides the DATA message. Table 7-2 lists the various types of M3UA messages.

Table 7-2

M3UA messages

Message name	Message class	Message class value	Message type value
Error (ERR)	MGMT	00	00
Notify (NTFY)	MGMT	00	01
Data	Transfer	01	01
Destination Unavailable (DUNA)	SSNM	02	01
Destination Available (DAVA)	SSNM	02	02
Destination State Audit (DAUD)	SSNM	02	03
SS7 Network Congestion State (SCON)	SSNM	02	04
Destination User Part Unavailable (DUPU)	SSNM	02	05
Destination Restricted (DRST)	SSNM	02	06
ASP Up (ASPUP)	ASPSM	03	01
ASP Down (ASPDN)	ASPSM	03	02
Heartbeat (BEAT)	ASPSM	03	03
ASP Up Acknowledgment (ASPUP ACK)	ASPSM	03	04
ASP Down Acknowledgment (ASPDN ACK)	ASPSM	03	05
Heartbeat Acknowledgment (BEAT ACK)	ASPSM	03	06

Message name	Message class	Message class value	Message type value
ASP Active (ASPAC)	ASPTM	04	01
ASP Inactive (ASPIA)	ASPTM	04	02
ASP Active Acknowledgment (ASPAC ACK)	ASPTM	04	01
ASP Inactive Acknowledgment (ASPIA ACK)	ASPTM	04	02
Registration Request (REG REQ)	RKM	09	01
Registration Response (REG RSP)	RKM	09	02
Deregistration Request (DEREG REQ)	RKM	09	03
Deregistration Response (DEREG RSP)	RKM	09	04

Management Messages The *Error* (ERR) message is used when a message is received unexpectedly or with invalid contents. The ERR message includes an error code and optional additional information regarding the error condition.

The *Notify* (NTFY) message is used between M3UA peers to communicate the occurrence of certain M3UA-related events. For example, the NTFY message would be used to communicate a status change in an ASP.

Signaling Network Management Messages M3UA includes the following messages for signaling network management:

■ **Destination Unavailable (DUNA)** This message is sent from the SG to all concerned ASPs to indicate that a destination within the SS7 network is not available. The message indicates the affected destination (DPC) and, optionally, the network appearance and routing context (which identifies the routing key). The DUNA might include several destinations, so that one message can be used to indicate the unavailability of several signaling points. The message allocates 24 bits for each unavailable point code and also includes an 8-bit mask to indicate the unavailability of multiple contiguous point code values. The mask is effectively a wild card indicator and specifies how many digits in the point code value are wild-carded. For example, a network mask of 8 when used with an ANSI point code would indicate that the last 8 bits of the point code are wild-carded. In other words, all point codes in the cluster are unavailable.

The network appearance is used when the SG is logically partitioned across multiple networks and it serves to logically separate the signaling between SG and ASP according to each of the logical portions of the SG. The network appearance is also used to indicate the format of the DPC (14-bit or 24-bit). Therefore, the same DUNA message can be used in ANSI, ITU-T, or other national-specific signaling environments.

DUNA is generated at the SG when it determines from MTP3 network management messages that a destination is unavailable. The DUNA message is transmitted to the ASP, where M3UA uses it to create the primitive MTP-Pause indication, which it issues to the upper-layer protocol (such as ISUP).

■ **Destination Available (DAVA)**   This message is sent from the SG to all concerned ASPs when a destination becomes reachable. At the ASP, it is mapped to the primitive MTP-Resume indication. It contains the same information as the DUNA with respect to the destinations in question and the network appearance.

■ **Destination State Audit (DAUD)**   This message is sent from an ASP to an SG when the ASP wants to query the state of SS7 routes to one or more destinations. With the exception of the message-type value, the DAUD has the same format as the DUNA message.

■ **SS7 Network Congestion (SCON)**   This message is sent from the SG to all concerned ASPs to indicate that the route to one or more SS7 destinations is congested. At an ASP, it is mapped to the primitive MTP-Status indication, which is issued to the upper-layer protocol. The SCON message can also be sent from an ASP to the SG to indicate congestion at the ASP end of the interface. In that case, the SCON message includes a point code for the congested ASP, plus an optional congestion-level indicator.

■ **Destination User Part Unavailable (DUPU)**   This message is sent from the SG to all concerned ASPs to indicate that a given user part at a given destination is not available. It indicates the DPC in question and the user part in question, such as ISUP, SCCP, or TUP. At the ASP, the message is mapped to the primitive MTP-Status indication. The MTP-Status indication may be used to indicate congestion in the network or the unavailability of a given destination user part. Although in MTP3 these different events are indicated with the same primitive by using different cause codes, M3UA communicates these events between M3UA peers through the use of two different messages.

▪ **Destination Restricted (DRST)**   This message can be sent from an SG to concerned ASPs to indicate that one or more SS7 destinations are restricted from the perspective of the SG. Since this message has significance only from the perspective of the SG, it might be possible for signaling traffic to be routed to a destination via a different SG. The M3UA layer at the ASP should take such an alternative path if it is available.

ASP State Management Messages   M3UA includes the following messages for ASP state management:

▪ **ASP Up (ASPUP) and ASP Up Acknowledgment (ASPUP ACK)** The ASPUP message is sent from an M3UA layer to its peer to indicate that the M3UA layer is ready to receive ASP management and ASP traffic-related messages for all routing keys that the ASP is configured to handle. Prior to the establishment of an SCTP association between an *SG Process* (SGP) and an ASP, the ASP is considered to be in state ASP-DOWN. Once an SCTP association is established, then the affected ASP will send an ASPUP message. This message will cause the SG to consider the ASP to be in state ASP-INACTIVE, which means that the ASP is "awake" but not yet ready to handle traffic. The SG responds to the ASPUP, with the ASPUP ACK message. If the ASP does not receive that response within a timeout period, then the ASP can resend the ASP UP message.

▪ **ASP Down (ASPDN) and ASP Down Acknowledgment (ASPDN ACK)**   An ASP will send an ASPDN message to an SG when it wants to be removed from service in all ASs. Sending an ASPDN message will convey that no further DATA, SSNM, or ASPTM messages should be sent to the ASP. The ASPDN message can be generated automatically or as a result of some maintenance action. When an SG receives the ASPDN message, it considers the ASP to be in state ASP-DOWN and responds with an ASPDN ACK message.

- **Heartbeat messages (BEAT and BEAT ACK)**   The BEAT message is an optional message used by an M3UA entity to ensure that its peer entity is available. The BEAT message contains a Heartbeat Data parameter that is set by the sending node. The parameter could include, for example, a sequence number or timestamp. A node that receives a BEAT message will respond with a BEAT ACK message, which contains the same parameter values as the BEAT message. In other words, the BEAT ACK message is effectively an echo of the BEAT message.

  The heartbeat mechanisms of M3UA are optional. SCTP has its own heartbeat procedure and M3UA does not need to use a heartbeat technique of its own when M3UA operates over SCTP. M3UA can operate over other types of transports, in which case the BEAT and BEAT ACK messages might be required.

ASP Traffic Management Messages   M3UA includes the following messages for ASP traffic management:

- **ASP Active (ASPAC) and ASP Active Acknowledgment (ASPACN ACK)**   An ASP will send an ASPAC message when it is ready to begin processing traffic. When an SCTP association is established, the ASP will typically send the ASPAC immediately after receiving an ASPUP ACK from an SG. The ASPAC optionally contains a routing context, which identifies the routing keys for which the ASP is to be considered active. The inclusion of the routing context can be necessary if the ASP can serve multiple ASs. Recall that a one-to-one relationship exists between an AS and a routing key and the routing key can correspond to combinations of parameters such as OPC/DPC/CIC range. Recall also that an ASP is a process (and could correspond to a physical processor), and that an ASP could be shared among several ASs. For example, an ASP could be configured to support traffic processing for several combinations of DPC/OPC/CIC range. If the ASP is to actively process traffic for a limited set of those combinations, then the routing context will indicate the applicable combinations. If the routing context is omitted, then the ASP is prepared to handle traffic for all combinations for which it is configured. Upon receipt of an ASPAC message, the SG will respond with ASPAC ACK.

- **ASP Inactive (ASPIA) and ASP Inactive Acknowledgment (ASPIA ACK)**   An ASP will send an ASPIA message when it no longer wants to process traffic for some or all the routing keys applicable to it. This action can be initiated automatically or as a result of some maintenance action. As is the case for the ASPAC message, the ASPIA can include a routing context parameter to indicate those routing keys for which the ASP is to be considered inactive. If the routing context parameter is omitted, then the ASP is to be considered inactive for all applicable routing keys.

Upon receipt of the ASPIA message, the SG marks the ASP as inactive and stops sending any traffic towards it. If the ASP was part of a group of ASPs in load-sharing mode, then the traffic will be distributed to the other ASPs in the group. The SG responds to the ASPIA with ASPIA ACK.

Figure 7-31 and Figure 7-32 show examples of ASP state management and ASP traffic management messages. In Figure 7-32, we also see the use of the NOTIFY message to inform an ASP of the change in status of another ASP in the AS.

Routing Key Management Messages   M3UA includes the following messages for routing key management:

■ **Registration Request (REG REQ) and Registration Response (REG RSP)**   In the previous discussions, we have addressed the fact that a given ASP can be associated with one or more routing keys (such as a DPC/OPC/CIC range). This association must be recognized at both the ASP and the SG, and it will occur as a result of some management or provisioning action. One way to establish the association is to provision the necessary data at both the ASP and the SG. If such provisioning is done manually, then one runs the risk that human error can intervene and lead to a data mismatch between the

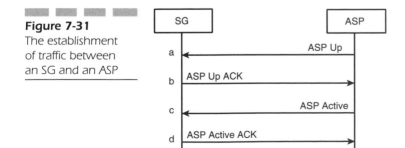

Figure 7-31

The establishment of traffic between an SG and an ASP

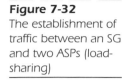

Figure 7-32
The establishment of
traffic between an SG
and two ASPs (load-
sharing)

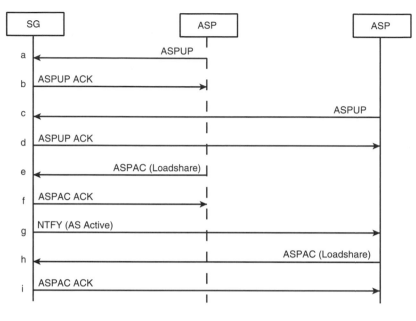

ASP and the SG. An alternative approach is to provision the necessary data only at an ASP and for the ASP to then communicate the information to the SG, which will ensure information consistency between ASP and SG. Routing key management messages are the means by which such consistency is maintained.

An ASP uses the REG REQ message to register with an SG. The message indicates to the SG that the ASP wants to participate in the traffic handling for the AS identified by a specific routing key. In fact, since a given ASP can serve multiple ASs (corresponding to multiple routing keys), the REG REQ message can include multiple routing keys, each of which specifies a combination such as DPC; DPC and OPC; the DPC, OPC, and CIC range; DPC and the network appearance; and so on. In the case of a successful registration, the SG will respond with a successful REG RSP message that will include a number of registration result parameters corresponding to the number of routing keys included in the REG REQ message and a routing context value associated with a successfully registered routing key. The inclusion of the routing context value means that further communication regarding a given routing key can be referenced by the routing context value rather than having to send the complete description of the routing key (such as DPC/OPC/network appearance/CIC range) every time.

A registration can be unsuccessful in several cases. For example, the SG might not consider the routing key to be valid. Alternatively, the SG might not support the dynamic registration of routing keys. Dynamic registration means that an SG will self-provision a new routing key if it receives a valid, new, and unique routing key in an REG REQ message. In other words, dynamic registration enables routing key data to be provisioned only at the ASP and to be propagated to one or more SGs.

Regardless of the result of registration at the SG, the SG will respond to an REG REQ message with at least one REG RSP message, which indicates the success or failure of the registration. The REG RSP can include multiple registration result parameters. Thus, if a single REG REQ message specifies multiple routing keys, a single REG RSP message can indicate the success or failure of registration for each of the routing keys. In the case of failure, the registration result indicates the reason for failure.

■ **Deregistration Request (DEREG REQ) and Deregistration Response (DEREG RSP)** The DEREG REQ and DEREG RSP messages are simply the opposite of the corresponding registration messages. An ASP sends a DEREG REQ to an SG when it wants to deregister a given routing key. The message contains a list of routing contexts that the ASP wants to deregister. The SG will respond with a DEREG RSP indicating the success or failure of each routing context.

The deregistration of a routing key by an ASP does not necessarily mean that the routing key will be deleted at the SG. Multiple ASPs may support a given routing key in a load-sharing or active/standby arrangement. If, however, a deregistration occurs for the last ASP handling a given routing key, then the SG can choose to also delete the routing key itself.

M2UA Operation

The previous discussion focuses on a situation where the call agent/MGC implements a function such as ISUP, but it does not implement any of the lower layers of the SS7 stack. In particular, MTP3 is not implemented at the MGC. The direct consequence of this approach is that the call agent does not have the same view of the SS7 network as an entity that implements MTP3. In particular, it does not send or receive signaling network management messages.

If it were necessary (or desirable) for the call agent to be more involved in signaling network management, then one option is to implement MTP3 at the MGC and use M2UA over SCTP to access the MTP2 functions at an SG. Although this gives the MGC more visibility of the SS7 network, another consequence is that the MGC can be considered more tightly coupled to the SG. In fact, the SG becomes a remote signaling terminal that is logically part of the MGC from an SS7 perspective.

The MTP3 function at the MGC is a normal MTP3 implementation that includes routing and distribution capabilities and that needs to know about the details of all signaling links that it might use. The main purpose of the SG is to take care of the MTP2 functions, which means that the SG acts as a signaling terminal for the MTP3 application at the MGC.

M2UA uses concepts similar to those used by M3UA. These include the concept of an AS and an ASP. It also includes similar messages, such as ASPUP, ASPDN, ASPAC, ASPIA, NTFY, BEAT, ERR, and the applicable acknowledgements. These messages are used in exactly the same way as used in M3UA, with the difference that the ASPs in M2UA perform different functions to those in M3UA. In M3UA, the ASP can be related to something like a DPC/OPC/CIC range or some other set of characteristics; in M2UA, the ASP is an instance of MTP3 at a node such as an MGC.

The messages sent between M2UA peers use the same format of message used by M3UA. The difference is the Message Type field of the common header. Those messages that are M2UA-specific have different message class values (values 6 and 10) to those messages that are used for other adaptation layers. The management, ASP state management, and ASP traffic management messages are common across all adaptation layers. The following section describes the M2UA-specific messages.

MTP2 User Adaptation (MAUP) Messages (Message Class 6)   Unlike M3UA where the M3UA message immediately follows the common header, M2UA utilizes an M2UA header that applies to all *MTP2 User Adaptation* (MAUP) messages. In other words, a given MAUP message begins with the common header, followed by the M2UA header, followed by the message itself. The M2UA header contains an interface identifier. This identifier identifies the physical interface at the SG on which a given signaling message is sent or received. In practical terms, the interface identifier indicates a specific signaling link. The following are the MAUP messages:

- *DATA* (message type value 01) is used to carry an MTP2-user *protocol data unit* (PDU). The DATA message can optionally contain a Correlation ID. When an M2UA peer receives a DATA message, it can respond with a DATA ACKNOWLEDGE message (message type value 15) to indicate successful receipt of the message. That message contains the same value of Correlation ID. This process can be useful in the event of SCTP peer congestion or SCTP association failure.

■ *ESTABLISH REQUEST* (message type value 02) is used to establish a link at the SG or to indicate that a link has been established. In setting up signaling links, the establishment of those links at the SG may take place in advance of management action on the MGC to establish links. In such a case, when the MGC sends an establish request to the SG, the SG may simply return the ESTABLISH CONFIRM (message type value 03).

■ *RELEASE REQUEST* (message type value 04) is used by the MGC to request that the SG take a particular signaling link out of service. The request will include the reason why the link should be released. Once the link has been taken out of service, the SG responds with a RELEASE CONFIRM (message type value 05). It is also possible for the SG to autonomously take a link out of service, which would occur, for example, in the case of a hardware failure at the SG. In such a case, the SG would send a RELEASE INDICATION (message type value 06) to the MGC.

■ *STATE REQUEST* (message type value 07) is sent from an MGC to the SG to cause the SG to perform some action on a particular signaling link. The actions include tasks such as requesting normal link alignment or flushing transmit and retransmit buffers. Upon completion of the action at the SG, the SG responds with a STATE CONFIRM (message type value 08). It is also possible for the SG to autonomously send a STATE INDICATION (message type value 09) to the MGC to indicate the condition on a link. The message might be sent upon the occurrence of a remote processor outage or a change in the physical state of the link itself.

■ *CONGESTION INDICATION* (message type value 14) can be sent from an SG to an ASP (at an MGC, for example) to indicate the congestion status and discard status of an SS7 link. When the MSU buffer fill increases above an onset threshold, decreases below an abatement threshold, or crosses a discard threshold in either direction, the SG will send a CONGESTION INDICATION message when it supports SS7 MTP2 variants that support multiple congestion levels.

The SG will send the message only when a change in either the discard level or the congestion level can actually be reported, meaning that the value is different from that of the previously sent CONGESTION INDICATION message. In addition, the SG will limit the frequency of CONGESTION INDICATION messages.

A number of messages are used during link changeover. In such an event, it is necessary for the ASP to retrieve certain information from the SG, particularly information that is stored at the SG for retransmission on a link. DATA RETRIEVAL REQUEST (message type value 10) is used during link changeover procedures. This message is sent by the ASP to request the BSN or to retrieve PDUs from the transmit and retransmit queues. The SG responds with the DATA RETRIEVAL CONFIRM (message type value 11). It is also possible for the SG to just send a PDU to the ASP from the transmit or retransmit queue, in which case the SG will use a DATA RETRIEVAL INDICATION (message type value 12). Multiple RETRIEVAL INDICATION messages may be sent from the SG to the ASP in the case of link changeover. When the final PDU from the retransmit queue is being sent, the SG will use the DATA RETRIEVAL COMPLETE INDICATION (Message type value 13) instead. This message should be sent only once in a given changeover event.

Interface Identifier Management (IIM) Messages (Message Class 10)   Much like M3UA, M2UA uses ASP state management messages and ASP traffic management messages. The difference is that an ASP in M2UA will handle the processing of traffic for one or more signaling links, while an ASP in M3UA will have traffic for one or more routing keys (such as OPC/DPC/CIC range). In other words, when an M2UA ASP becomes active, then M2UA at an SG will send the ASP traffic related to a specific signaling link. In order to do so, the SG must know which links a given ASP is to handle. The SG can know this information through data provisioning at the SG or through registration of an ASP with an SG. In other words, an ASP can register with an SG to inform the SG of which links are to be associated with a given ASP.

This process is similar to the routing key management process of M3UA. In fact, the same message names are used: REG REQ, REG RSP, DEREG REQ, and DEREG RSP. The REG REQ message is sent from an ASP to an SG to associate one or more link keys with the ASP. The link key includes a *Signaling Data Terminal Identifier* (SDTI) and a *Signaling Data Link Identifier* (SDLI). The REG REQ message also includes a *Local-LK-Identifier*. This parameter is used to correlate a REG REQ message and its response. An SG will respond to an REG REQ with an REG RSP, which contains the same value of Local-LK-Identifier and a registration result to indicate the success or failure of the registration.

An ASP sends the DEREG message to deregister a particular interface, that is, to indicate that the ASP no longer wants to handle traffic for that interface. The reply to the DEREG REQ is the DEREG RSP.

Although the *Interface Identifier Management* (IIM) messages of M2UA have the same names and serve similar purposes to the equivalent messages of M3UA, they are not the same. The registration and deregistration messages of M2UA have a different message class value (10) compared to the M3UA routing key management messages (message class value 9). Unlike the M2UA MAUP messages, the IIM messages do not include the M2UA common header. The IIM messages do, of course, include the SCTP common header.

M2PA Operation

As described previously, M2PA effectively enables IP-based SS7 links between an SG and another node (such as an MGC), whereas M2UA simply enables the MGC to have a remote SS7 terminal, with IP used on the interface between the MGC and SG. M2PA also enables IP-based SS7 signaling between two IP signaling points without traversing the traditional SS7 network.

As mentioned earlier in this chapter, three main types of SS7 messages are used in a traditional SS7 network: MSUs, which carry actual user payload; LSSUs, which convey link status information; and FISUs, which are sent when no other data is waiting to be sent. Since M2PA effectively enables IP-based SS7 links, it offers similar types of messaging, with the exception of the FISU for which no M2PA equivalent exists. The reason why M2PA has no FISU equivalent is because the IP network is a shared resource. Constantly sending such messages in an IP network would be wasteful of shared resources. Moreover, the SCTP BEAT message ensures the correct operation of the IP-based links, and the M2PA messages for data transfer include acknowledgment functions. Consequently, two M2PA messages are available: the User Data message and the Link Status message.

Like other adaptation layers, M2PA establishes SCTP associations between M2PA peers. M2PA-initiated SCTP associations use two streams in each direction. One stream is used for the sending of User Data messages, while the other is used for sending Link Status messages.

The User Data message of M2PA is simply an SS7 MSU with a few of the standard SS7 fields excluded. The message contains the *Length Indicator*

(LI), *Service Information Octet* (SIO), and *Signaling Information Field* (SIF). The Flag, *Backward Sequence Number* (BSN), *Backward Indicator Bit* (BIB), *Forward Sequence Number* (FSN), *Forward Indicator Bit* (FIB), and *Check bits* (CK) are not included. The LI is included in the User Data message because some national SS7 variants use the two spare bits of the LI as a priority indicator.

The Link Status message is sent between M2PA peers to indicate the current state of the link and is analogous to the SS7 LSSU. In much the same way as the SS7 LSSU contains a "status" field, the M2PA Link Status message contains a state parameter. The values for the state parameter are as follows:

- **1**   Alignment
- **2**   Proving Normal
- **3**   Proving Emergency
- **4**   Ready
- **5**   Processor Outage
- **6**   Processor Outage Ended
- **7**   Busy
- **8**   Busy Ended
- **9**   Out of Service
- **10**   In Service

M2PA Link Alignment   Link alignment is the process by which a link is brought into service in a controlled manner, so that one end does not try to send user traffic before the other end is ready. A common situation for link alignment is when a new link is being established.

Once an SCTP association is established, M2PA sends the Link Status Out of Service message to its peer. The actual link activation process is started by MTP3, which sends a Start request to M2PA. This request causes M2PA to send a Link Status Alignment message to its peer and start a timer (typically 10 seconds). A Link Status Alignment message is expected in return, and if it is not received, then M2PA informs MTP3 that the link is out of service. Assuming, however, that a Link Status Alignment message is received from its M2PA peer before the timeout, M2PA begins the link-proving process.

To begin the link-proving process, each M2PA usually sends a Proving Normal Link Status message to its peer at a regular rate. M2PA can, however, use the Proving Emergency Link Status message if requested by

MTP3. The proving duration is typically 10 seconds, but it can be as little as 400 milliseconds in the case of emergency proving.

Once the proving period has ended, M2PA sends Link Status Ready messages to its peer at regular intervals. Upon receipt of a Link Status Ready message in return, or a User Data message, the link is considered to be in service and M2PA sends an In Service message to its peer. The link is now ready for traffic.

In addition to the Link Alignment messages just described, other messages are used to convey fault conditions. The Processor Outage message indicates a failure at a layer above M2PA (such as MTP3) and the Processor Outage Ended message indicates that the failure condition has ceased. The Busy message conveys an indication of receiver congestion to the far end, so that the far end can stop sending messages. The Busy Ended message indicates that the congestion condition has ceased and enables user traffic to flow again.

Interworking SS7 and VoIP Architectures

The primary thrust of this chapter has been to describe how a VoIP network can interwork with an external SS7 network. Finally, we consider any differences that the architecture of the VoIP network might cause—specifically whether the network follows the softswitch or the H.323 model.

Interworking Softswitch and SS7

The concepts behind Sigtran are aimed at the reliable transfer of signaling information. Many of the adaptation layers are SS7 specific and lend to interworking between SS7 and the softswitch architecture. One reason is that the softswitch architecture includes a number of MGs placed close to the sources and sinks of media in the PSTN (at the edges of the IP network). In the distributed softswitch architecture, however, the MGCs that control the MGs may be far removed from the gateways themselves. Thus, having SGs at the edge of the IP network makes sense.

The interworking is typically achieved by the use of at least two SGs that terminate SS7 links and pass SS7 messages to the actual applications at an MGC. The architecture is shown in Figure 7-33 and the protocols used include SCTP, M3UA, and M2PA or M2UA, as previously discussed.

Figure 7-33
Interworking SS7 and
softswitch
architecture

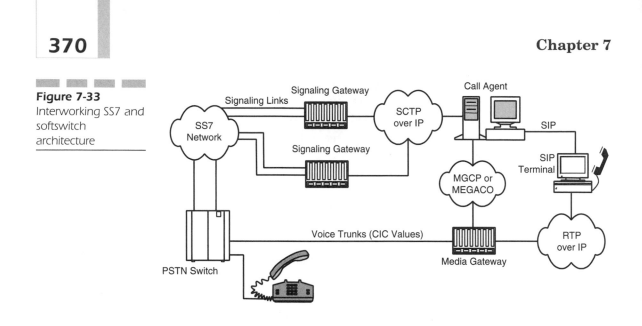

Figure 7-33 shows the use of SCTP only for transporting SS7-related signaling. The underlying transport for SIP and MGCP/MEGACO is not shown. Although these could be TCP or UDP (UDP only for MGCP), one could easily imagine a situation where SCTP is used. In fact, an Internet draft already exists that addresses the use of SCTP as a transport for SIP.

ISUP Encapsulation in SIP   SCTP and the associated adaptation layers provide the capability to convey SS7 application protocol information from an endpoint in the PSTN to an MGC in the center of the VoIP network. The MGC then has access to the signaling information needed to route calls in the VoIP network. SS7 application protocols themselves might or might not be used internally in the VoIP network. Often, protocols such as ISUP are mapped to equivalent VoIP protocols, such as SIP. In fact, SIP-ISUP mapping is the most common, and we have already discussed that topic in Chapter 5, "The Session Initiation Protocol (SIP)." Such mapping works very well if one end of the call is a VoIP node and the other is a PSTN node. A problem may occur, however, when the softswitch network provides a transit function between a circuit-switched network at the originating end and a circuit-switched network at the terminating end.

Because protocols are designed in different ways, one rarely finds a direct match between the messages and parameters of one protocol and those of another. In addition, interworking has to address the differences between state machines, timers, and so on. The result, although workable, is often less than perfect. The reason is because protocol A might be slightly better in one aspect, while protocol B might be better in another. Consequently, interworking situations often lead to a lowest common denominator result where the functionality offered is limited to that supported by both protocols. For example, SIP header fields enable the inclusion of information that simply does not exist in ISUP. One example of this is the Retry-after header field that just does not map to an equivalent in ISUP.

Equally, ISUP provides information that does not easily map to SIP headers (if at all). Often, such information might need to be mapped to the closest SIP equivalent or perhaps be discarded. This situation is not so bad if a call is placed from a PSTN user to a SIP user, because the SIP user agent server would not be able to interpret such information in any case. Many networks exist today that provide long-distance service using VoIP, however. These networks connect to the PSTN at numerous points. Long-distance calls from one PSTN user to another can transit via the VoIP network, entering the VoIP network at one point and leaving again at another. If ISUP is converted to SIP at the point of ingress and SIP is converted back to ISUP at the point of egress, the result may be that the ISUP messages leaving the network may be different from those that entered the network, as shown in Figure 7-34.

In order to counteract this problem, SIP enables the message body to encapsulate an ISUP message. Of course, SIP messages and responses are used between MGCs and, where appropriate, those messages must contain an SDP message body in order for the MGCs to describe the media to be sent between the MGs they control. Some of the messages can also contain an ISUP message in binary form within the message body. Therefore, the message body can have multiple parts: an SDP session description used by the MGCs and MGs, and an encapsulated ISUP message carried transparently across the network.

Consider the scenario depicted in Figure 7-35. The incoming IAM is mapped to a SIP INVITE that contains an SDP session description as well as the original ISUP message. At the egress point, the SDP information is

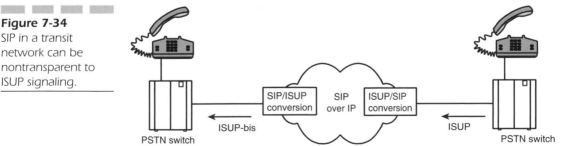

Figure 7-34
SIP in a transit network can be nontransparent to ISUP signaling.

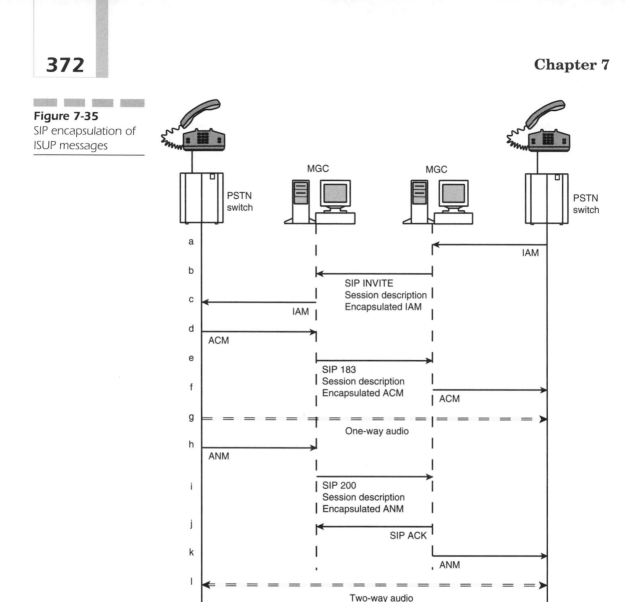

Figure 7-35
SIP encapsulation of
ISUP messages

used to set up the RTP session between the gateways, and the ISUP information is extracted to send to the PSTN. When the call is through-connected, the received ACM is mapped to the SIP 183 (session progress) response, which contains a temporary SDP media description, and the message also contains an encapsulated ISUP ACM to pass to the originating PSTN node. A similar process occurs when the call is answered and an ANM message is returned from the called end. Note that the figure does not include SIP messages for the reliable delivery of provisional responses. We can assume, however, that reliable delivery would be applied (the use of the

Provisional Response ACK [PRACK] method). Even though the 183 response is considered provisional, it is important that it be delivered reliably.

In order to encapsulate an ISUP message with a SIP message body, the Content-Type: header field of a SIP message should not indicate application/sdp. Instead it should have the following format:

```
Content-type: multipart/mixed; boundary = SDP-ISUP-boundary
```

The Content-type indicates that the message body contains different sets of information in different formats. The boundary field indicates the boundary between one description and the other within the message body. Its usage is described in RFC 2046, "Multipart Internet Mail Extensions (MIME) Part Two: Media Types."

Within the message body, the beginning of each different payload is indicated by "--boundary," where boundary is as specified in the Content-Type: header. Therefore, if a SIP message were to contain a Content-type field as shown previously, each payload within the message body would be indicated by the string "--SDP-ISUP-boundary." After the last payload is another line with the format "-- boundary--." In our example, the last line would be "--SDP-ISUP-boundary--."

An SIP INVITE using ISUP encapsulation might look something like the following:

```
INVITE SIP: 9725551234@telco2.net SIP/2.0
From: SIP: 2145556789@telco1.net; tag=abcd12345
Call-ID: 123456789@telco1.net
Content-Length: xxx
Content-Type: multipart/mixed; boundary = SDP-ISUP-boundary
MIME-Version: 1.0

-- SDP-ISUP-boundary
Content-Type: application/sdp

v=0
o=collins 22334455 123 IN IP4 444.333.222.111
s=
c= IN IP4 444.333.222.110
t=0 0
m=audio 5555 RTP/AVP 0

-- SDP-ISUP-boundary
Content-Type: application/ISUP; version = ANSI
Content-Encoding: binary
1A 00 01 00 60 00 0A 03 06 0D 03 80 90 A2 07
03 10 18 27 85 31 48 0A 07 03 11 12 74 66 69
53 EA 01 00 00
-- SDP-ISUP-boundary--
```

Interworking H.323 and SS7

Interworking between an H.323 architecture and SS7 could work in a similar manner to interworking between softswitch and SS7. The approach would be to use SGs that terminate SS7 signaling. The SS7 application information would be carried using Sigtran from the SS7 gateways to one or more H.323 gateways.

Equally, however, the gateway itself could simply terminate SS7 links directly. In fact, such an approach has merit. In the H.323 architecture, the gateway itself contains a great deal of the application logic required for call setup, and the gateway is also the entity that performs the media conversion. The fact that the gateway holds the application logic means that it must send and receive call control signaling information. The fact that the gateway performs the media conversion means that it should be placed at the edge of the IP network. Since the gateway is at the edge of the network, it is likely easier and more efficient for the gateway itself to terminate SS7 links from the PSTN directly. The architecture would be as shown in Figure 7-36.

In fact, interworking SS7 with H.323 is in some ways an easier proposition than interworking SS7 and softswitch (at least for ISUP). The reason is because a close relationship exists between ISUP and Q.931. After all,

Figure 7-36

Interworking SS7 and
H.323 architecture

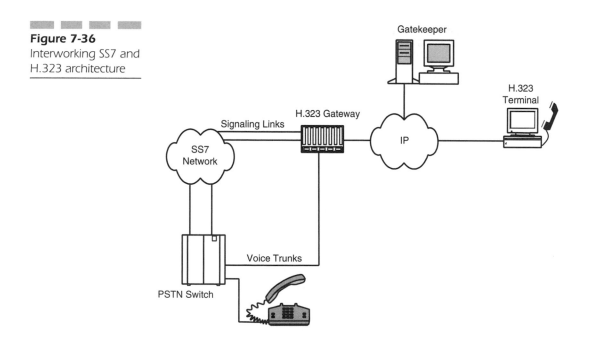

ISUP is the ISDN User Part, the interswitch network protocol corresponding to the ISDN access protocol Q.931. Consequently, mapping is easy between Q.931 messages and ISUP messages. For example, the setup message in Q.931 maps to IAM in ISUP, and ACM in ISUP maps to the Connect message in Q.931. Therefore, if an H.323 network were to receive a call from the PSTN using ISUP, the signaling would be similar to that shown in Figure 7-37.

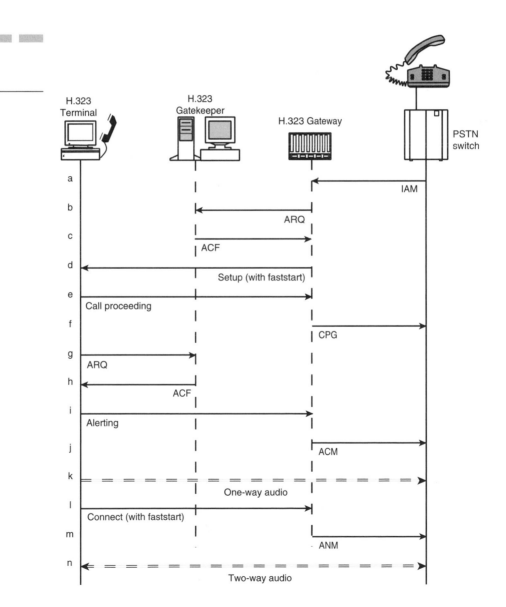

Figure 7-37
ISUP/H.323
interworking

The example assumes the use of the H.323 Fast Connect procedure. If the called H.323 terminal did not support that procedure, then the Connect message would not include a faststart element. In that case, the arrival of the Connect message at the gateway would not immediately lead to an ANM back to the PSTN. Rather, the logical channel procedures of H.245 would first be used to establish the media path between the gateway and the called terminal.

Quality of
Service (QoS)

To many, *quality of service* (QoS) is one of the biggest, if not *the* biggest, issue facing *Voice over IP* (VoIP). QoS (or the lack thereof) is one reason why many of the VoIP solutions on the market today that provide voice over the Internet are free services. The philosophy behind such offerings is that customers can hardly complain about poor quality when they are being provided with a free service. Such a philosophy is fine if the business model is based upon advertising or some other revenue stream rather than paying subscribers. Such VoIP providers may choose to continue providing free service by sending voice over the Internet and not worrying about QoS. For many others, however, the massive revenue generated by the *Public Switched Telephone Network* (PSTN) operators every year is very attractive. Competitors to traditional local and long-distance providers would like to win some of that business, and they see VoIP as a means to do so. The big hurdle is QoS.

Not only do new carriers see VoIP as a means to compete with traditional carriers, but both existing and new carriers would like to have just a single multifunctional network to carry both voice and data. The support of both voice and data traffic on a single network means lower cost. Again, however, the challenge is in ensuring a high-quality service and making sure that one type of traffic does not consume excessive network resources to the detriment of other types of traffic.

This chapter focuses on a number of technical solutions for providing QoS in a VoIP network. The various solutions are described separately, followed by a description of how they may be combined to complement each other. First, however, it is worth considering what QoS means and why it is such an issue for VoIP networks.

The Need for QoS

QoS is a collective measure of the level of service delivered to a customer. It can be considered as the level of assurance from a particular application that the network can meet its service requirements. From a technical perspective, QoS can be characterized by several performance criteria such as availability (low downtime), throughput, connection setup time, the percentage of successful transmissions, and the speed of fault detection and correction. In an *Internet Protocol* (IP) network, QoS can be measured in terms of bandwidth, packet loss, delay, and jitter. In order to provide a high QoS, the IP network needs to provide assurances that, for a given session or a set of sessions, the measurement of these characteristics will fall within certain bounds.

Unfortunately or fortunately (depending on your viewpoint), IP is by nature a best-effort service and offers no guarantees. For this reason, the *Transmission Control Protocol* (TCP) was developed as a layer above IP to ensure error-free, in-sequence delivery of information. Unfortunately, TCP provides this service at the expense of some delay, which is not a big problem for typical IP applications such as e-mail, web browsing, and file transfer. The picture changes significantly, however, when real-time applications such as voice are added to the mix. These applications are time sensitive to the extent that delays and packet loss can very easily make a service unusable. When it comes to voice, delay and jitter are of particular significance. Consequently, TCP cannot be used, which is why the *User Datagram Protocol* (UDP) is used instead (as described in Chapter 2, "Transporting Voice by Using IP").

Using UDP for transporting voice is fine, provided that packet loss is relatively low and the network has relatively little congestion. Traffic in an IP network can be bursty and unpredictable, however, which can lead to a situation where one application consumes network resources for a period of time (however brief), forcing packets from other applications to wait and/or be discarded. Again, TCP can take care of this for non-real-time applications, but for other applications that use UDP, the application itself needs to take care of retransmission.

Unfortunately, however, retransmission is not an option for voice because it introduces delay. If just a few packets are lost, some voice-coding schemes can handle the interruption by replaying previous packets or by extrapolating from previous packets. In the case of significant packet loss, however, voice-coding schemes will be unable to cope, which means that a speaker will have to repeat what he or she just said. Such a situation is hardly desirable, and certainly not one that a paying customer will tolerate. Therefore, we need to ensure that significant delay or packet loss is avoided.

Why should we be concerned about all these issues? The short answer is money. The long answer is that high-quality service is imperative if an operator is to attract and retain paying subscribers. If a provider is to charge for voice service, no matter how low the price, then the quality of that voice service must at least match that provided by traditional circuit-switched network operators. When it comes to quality, however, circuit switching has a distinct advantage. Because a dedicated two-way transmission pipe is established between the parties in a call and because no buffering of packets or packet loss takes place, the quality is excellent. High quality is to be expected, because circuit switching was designed specifically for voice from the outset.

The problem with circuit switching is that it is ill suited to other forms of communication. IP, on the other hand, is well suited to non-real-time communication and can also be made suitable for real-time communication, provided that QoS solutions are in place. Fortunately, solutions do exist. The primary thrust of this chapter is to describe these technical solutions.

End-to-End QoS

Before delving into the technical solutions for QoS, it should be pointed out that QoS is not just a technical matter on one operator's network. In general, a customer does not know or care about how a provider gets a call from one place to another, provided that the call gets there and meets the customer's quality expectations. This means that QoS must be end to end and requires the support of all networks in the chain. A call might originate on one provider's network and terminate on another provider's network, perhaps traversing other networks in between. Each of the networks must cooperate to ensure that the quality provided matches what is expected. This requirement raises the issue of *Service Level Agreements* (SLAs) between different operators. SLAs are agreements in which operators make commitments to each other regarding the type and QoS to be offered to each other and the penalties involved if such commitments are not met. We will discuss SLAs again later in the section "The Need for SLAs."

Meanwhile, however, we must again remember that VoIP and voice over the Internet are not necessarily the same thing. We must recognize that the Internet is a huge collection of different networks and is not managed by any one entity. For those parts that are actively managed, they are not necessarily managed in the same way or according to the same set of criteria. Although SLAs may be possible between certain VoIP carriers, too many networks are involved on the Internet to think that it will be possible to place a call from anywhere to anywhere else over the Internet and always experience superior quality. Many operators readily agree that a particular QoS can be provided in managed networks, but shake their heads at the suggestion that QoS is a realistic expectation for voice over the Internet. For this reason, this book makes a significant distinction between VoIP and voice over the Internet.

This does not mean that QoS is only achievable when a call stays within a single network. Rather, SLAs at least offer the potential of ensuring quality across multiple networks. Until all *Internet service providers* (ISPs) offer the same quality commitments and implement equivalent technical solutions, however, anywhere-to-anywhere voice over the Internet is likely to remain a best-effort, low-quality service.

It's Not Just the Network

Not only is QoS not confined to a single operator's network, but it is not confined to network issues in general. In other words, a quality service means a lot more than just good voice quality. In the competitive environment of telecommunications, low prices and advanced features may attract customers, but the best way to retain customers is to provide a consistently high service across all facets of the business. Given that a potentially high cost may be associated with acquiring a customer in the first place, the last thing a provider wants is to see a customer leave and go to a competitor. Therefore, QoS is provided not only by technical solutions in the network, but it is also provided through superior customer service, rapid service provisioning, 100 percent accurate billing, clear and concise product descriptions, and so on. Although this is a technical book that focuses on network solutions, we must remember that those network solutions provide only one piece of the puzzle when it comes to overall service quality.

Overview of QoS Solutions

A number of solutions have been developed to address the QoS issue. These solutions approach the problem from various angles. One approach is to ensure that resources for a given session are available and reserved for that session prior to establishing the session. Conceptually, this approach has certain similarities to circuit switching, where the bandwidth needed for a call is reserved for that call before the called party's phone rings.

Another approach is to categorize traffic into different classes or priorities with higher-priority values assigned to real-time applications, such as voice, and lower-priority values assigned to non-real-time traffic, such as e-mail. In some ways, this approach is easier to implement, but it requires that no one application can totally shut another out. Just because there is voice on a network that supports voice and data applications does not mean that the data never gets a reasonable share of resources. Therefore, the need exists to implement fair resource allocation techniques.

Finally, perhaps the most common technique today is overprovisioning the network, that is, provisioning greater bandwidth than the traffic demand would require. After all, a lack of bandwidth is often the reason for packet delay or loss. Bandwidth costs money, however, and needs to be utilized efficiently.

More Bandwidth

Allocating more bandwidth might sound like a very simplistic and expensive approach to QoS—simple because it does not require major system development, but expensive because it means significant overbuild. This overbuild would need to exist so that network resources would be available in times of traffic bursts. Unfortunately, the additional bandwidth would remain unused for most of the time. That method would be a very inefficient way of solving the QoS problem, but it does have some merit and should not be dismissed completely.

We live in an age where huge advances continue to be made in squeezing bandwidth from facilities that appeared to have reached their limit. Not that long ago, a 9,600-baud modem was considered almost a utopian situation for a modem connection. Now, 56 Kbps modems are standard for dial-up access, and *Digital Subscriber Line* (DSL) technology means that homes can have several megabits per second in each direction.[1] And this is just in the access part of the network. In the core of the network, we now have *Dense Wave Division Multiplexing* (DWDM) offering the capability to support tens of gigabits per second and hundreds of gigabits per second in the future on a single optical fiber. Consequently, as these new transmission techniques have become available, the cost of bandwidth has been decreasing and will likely continue to decrease.

On the other hand, however, Moore's Law says that computing power doubles roughly every 18 months, which means that applications are demanding bandwidth at an ever-increasing rate. In addition, the number of Internet users continues to increase. The result is that the demand for bandwidth is increasing rapidly. Although we currently have a glut of transmission capacity (a good deal of excess dark fiber currently exists), that situation will not last forever. Historically, bandwidth availability and bandwidth demand have tended to move almost in lock-step. This is simply a supply and demand issue. If bandwidth is scarce, then applications that

---

[1]DSL has many variants. The most common is *Asymmetric DSL* (ADSL), which can theoretically provide up to 640 Kbps uplink and up to 9 Mbps downlink. Whether a given customer experiences these rates is a different issue with commercial as well as technical considerations.

require high bandwidth are either not developed because they would be impossible or too expensive to use, or they are deployed only in a limited scope. Once more bandwidth becomes available, however, a major barrier to the deployment of such applications is removed and application developers can proceed with implementing new ideas. They can do so without worrying about a bandwidth demand that might have been unconscionable a short time earlier.

Therefore, excess bandwidth is often needed, if only to support additional traffic as demand continues to grow. Although additional bandwidth is a necessity for a network that is required to support voice in addition to the data traffic that may have been carried all along, excess bandwidth is not by itself a complete solution to the QoS issue. Mechanisms need to be in place to manage the available bandwidth so that it is used most effectively to support the services provided.

QoS Protocols and Architectures

The following is a high-level overview of the QoS solutions developed within the *Internet Engineering Task Force* (IETF). The details of these solutions are provided in subsequent sections of this chapter.

Resource Reservation Resource reservation techniques for IP networks are specified in RFC 2205 on the *Resource Reservation Protocol* (RSVP), which is part of the IETF integrated services suite. RSVP is a protocol that enables resources to be reserved for a given session (or sessions) prior to any attempt to exchange media between the participants. Of the solutions available, RSVP is the most complex, but it is also the solution that comes closest to circuit emulation within the IP network. RSVP provides strong QoS guarantees, significant granularity of resource allocation, and significant feedback to applications and users.

RSVP currently offers two levels of service. The first level is *guaranteed*, which comes as close as possible to circuit emulation. The second level is *controlled load*, which is equivalent to the service that would be provided in a best-effort network under no-load conditions.

Basically, RSVP works as depicted in Figure 8-1. A sender first issues a PATH message to the far end via a number of routers. The PATH message contains a *traffic specification* (TSpec), which provides details of the data that the sender expects to send in terms of bandwidth requirement and packet size. Each RSVP-enabled router along the way establishes a *path state* that includes the previous source address of the PATH message (that

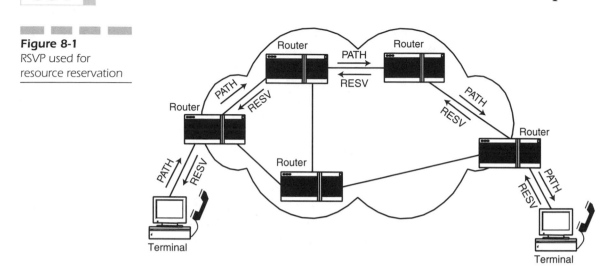

is, the next hop back towards the sender). The receiver of the PATH message responds with a *reservation request* (RESV) message that includes a flowspec. The flowspec includes a TSpec and information about the type of reservation service requested, such as controlled-load or guaranteed service.

The RESV message travels back to the sender along the same route that the PATH message took (but in reverse). At each router, the requested resources are allocated, assuming that they are available and that the receiver has the authority to make the request. Finally, the RESV message reaches the sender with a confirmation that resources have been reserved.

One interesting point about RSVP is that reservations are made by the receiver, not by the sender of data. This approach accommodates multicast transport, where there may be large numbers of receivers and only one sender.

Note that RSVP is a control protocol that does not carry user data. The user data (voice) is transported later using the *Real-Time Transport Protocol* (RTP). This occurs only after the reservation procedures have been performed. The reservations that RSVP makes are soft, which means that they need to be refreshed on a regular basis by the receiver(s).

Differentiated Service (DiffServ)   *Differentiated Service* (DiffServ) is a relatively simple means for prioritizing different types of traffic. The DiffServ protocol is described in RFC 2475, "An Architecture for Differentiated Services." Basically, DiffServ makes use of the IPv4 *Type of Service* (ToS) field and the equivalent IPv6 Traffic Class field. The portion of the ToS/Traffic Class field used by DiffServ is known as the DS field. The field is used in specific ways to mark a given stream as requiring a particular type of forwarding. The type of forwarding to be applied is known as *per-hop behavior* (PHB), of which DiffServ defines two types: *Expedited Forwarding* (EF) and *Assured Forwarding* (AF).

RFC 3246 defines EF as a service in which a given traffic stream is assigned a minimum departure rate from a given node, one that is greater than the arrival rate at the same node provided that the arrival rate does not exceed a pre-agreed maximum. This technique ensures that queuing delays are removed. Since queuing delays are a major cause of end-to-end delay and the main cause of jitter, this technique ensures that delay and jitter are minimized. In fact, EF can provide a service that is equivalent to a virtual leased line.

RFC 2597 defines AF as a service in which packets from a given source are forwarded with a high probability provided that the traffic from that source does not exceed some pre-agreed maximum. AF defines four classes, with each class allocated a certain amount of resources (buffer space and bandwidth) within a router. Within each class, a given packet can have one of three drop rates. At a given router, if there is congestion within the resources allocated to a given AF class, then the packets with the highest drop rate values will be discarded first so that packets with a lower drop rate value receive some protection. In order to work well, it is necessary that the incoming traffic does not have packets with a high percentage of low drop rates. After all, the purpose is to ensure that the highest-priority packets get through in the case of congestion and that cannot happen if all the packets have the highest priority.

Label Switching   Label switching is something that has seen significant interest in the Internet community, and significant effort has been placed behind the definition of a protocol called *Multiprotocol Label Switching* (MPLS). Like DiffServ, MPLS marks traffic at the entrance to the network. The primary function of the marking is not to allocate priority within a router, however, but to determine the next hop in the path from the source to destination.

MPLS involves the attachment of a short label to a packet in front of the IP header. This procedure is effectively similar to inserting a new layer between the IP layer and the underlying link layer of the *Open System Interconnection* (OSI) model. The label contains all the information that a router needs to forward a packet. The value of a label can be used to look up the next hop in the path and forward the packet to the next router. The difference between this approach and standard IP routing is that the match is exact and is not a case of looking for the longest match (the match with the longest subnet mask). This process enables faster routing decisions within routers.

The MPLS label identifies something called a *Forwarding Equivalence class* (FEC). All packets that belong to a given FEC are treated equally for the purposes of forwarding. All packets in a given stream of data, such as a voice call, will have the same FEC and receive the same forwarding treatment. Therefore, we can ensure that the forwarding treatment applied to a given stream can be set up such that all packets from A to B follow exactly the same path. If that stream has a particular bandwidth requirement, then that bandwidth can be allocated at the start of the session. Thus, we can ensure that a given stream has the bandwidth that it needs and the packets that make up the stream arrive in the same order as transmitted. Hence, a higher QoS is provided. Not only can we ensure that a given session can be given the required bandwidth and that all packets in a session are forwarded the same way, but we can apply the same technique to a traffic aggregate. For example, we can apply the same FEC to all voice packets between two *media gateways* (MGs) that are geographically separate.

Consider a VoIP network that provides long-distance voice service. Such a network would have numerous gateways in different locations. Provided that we have an understanding of the traffic demand between a given pair of media gateways, we can allocate a specific path (with specific resources) through the IP network to handle the traffic between those gateways. Also, if the traffic between the gateways does not exceed expectations, then the traffic will always have access to the necessary network resources to ensure that QoS requirements are met. This approach (whereby we allocate specific paths and network resources to specific types of traffic) is known as *traffic engineering*. In fact, MPLS is effectively a traffic-engineering tool and it ensures QoS by enabling us to ensure that the network is correctly engineered to meet our QoS objectives.

QoS Policies

Given that mechanisms exist to apply higher QoS to certain streams compared to others, the issue arises as to which QoS levels should be applied to which types of traffic. This raises the issue of QoS policies. Although protocols such as RSVP, DiffServ, and MPLS provide the mechanisms to differentiate traffic and allocate resources to specific types of traffic, QoS policies specify how those mechanisms are used.

One important implication of QoS is the fact that certain users or types of traffic will receive better service than others. In many cases, the QoS for a given type of traffic can have monetary implications. For example, it is common to pay for a long-distance phone call (on a per-call basis). One cannot be expected to pay for a phone call if the call does not receive the necessary network resources to ensure high quality. Other services might not have such monetary implications and might be offered on a best-effort basis. Such a situation creates an incentive for certain applications (or individuals) to steal better service without paying for it. Therefore, we need authentication functions to identify users and ensure that a given user is who he or she claims to be. Furthermore, a given user may be entitled to a given level of QoS under certain circumstances but not under other circumstances.

Finally, the Internet has already seen *denial of service* (DoS) attacks, in which certain systems (or persons) flood the network with traffic, causing congestion and service failure for everyone else. Therefore, certain rules must be established to specify which circumstances lead to which actions, and there must be methods for ensuring that those rules are applied. This can be considered as a type of policing function.

Appropriately, the IETF has developed a protocol known as the *Common Open Policy Service protocol* (COPS), which RFC 2748 specifies. COPS is a client-server protocol, as depicted in Figure 8-2. A *Policy Enforcement Point* (PEP), such as a router that needs to enforce certain rules, queries a *Policy Decision Point* (PDP) that makes the actual policy decisions. The PEP can be considered the policeman and the PDP can be considered the judge. Given these analogies, the acronym COPS is particularly fitting.

The Resource Reservation Protocol (RSVP)

The previous sections of this chapter gave a brief overview of QoS issues and solutions. Next we delve deeper into the technical solutions and start with RSVP.

Figure 8-2
COPS
request/response

In order to use RSVP, routers and hosts must implement certain functions. These functions are shown in Figure 8-3. Prior to data transfer when a QoS reservation is being established, the admission control function determines whether sufficient resources exist to satisfy the requested QoS, and the policy control function determines whether the user has the authority to make the requested reservation. The policy control function could, for example, use COPS to communicate with a PDP to make the policy determination. During data transfer, the packet classifier determines the QoS to be applied to a given packet, and on the outgoing interface, the packet scheduler or other link-layer function determines when a given packet is to be forwarded. Together, the packet classifier, packet scheduler, and admission control are known as traffic control.

RSVP Syntax

Before describing the messages and functions of RSVP in greater detail, we must first mention a few points about the syntax of RSVP messages. The syntax of the information contained in RSVP messages is specified in RFC 2215, "General Characterization Parameters for Integrated Service Network Elements." Basically, RSVP is used to implement a number of services and these services are identified by particular service numbers. For example, the guaranteed service has the service number 2 and the controlled-load service has the service number 5. A number of parameters are associated with each service, and each parameter has a parameter number.

Both services and parameters are specified in messages with the format *Type-Length-Value* (TLV). Thus, a type field identifies each service or parameter, followed by a length field, followed by the data itself.

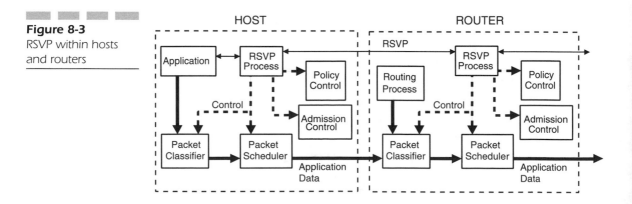

Figure 8-3
RSVP within hosts
and routers

In some cases, data needs to be sent that is not associated with any particular service or that is common to a number of different services. For this purpose, service number 1 is defined. If a message is to be sent and contains both general and service-specific information, then the message will specify service number 1 containing the general information parameters, followed by service information (such as for service number 2) containing service-specific information.

Establishing Reservations

As previously described, RSVP reserves resources along a path from the receiver of user data back to the sender of that data. The reason is because RSVP is designed with multicast in mind. With multicast, there may be a limited number of senders and multiple receivers. Figure 8-4 depicts a scenario where only one sender and many receivers are used. From sender to receiver, the data path branches. In the reverse direction, the data path merges. Let's assume that Receivers 1 and 2 have different QoS objectives.

Figure 8-4
RSVP used with multicast

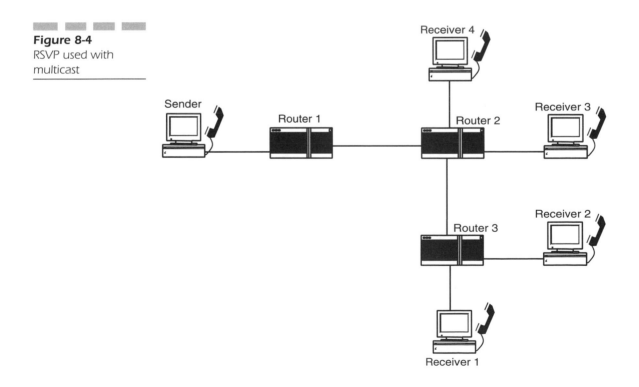

In that case, the resources needed between Routers 2 and 3 do not need to be the sum of the requirements of Receivers 1 and 2, just the maximum. Equally, the resources required between Routers 1 and 2 do not need to be the sum of the requirements of all four receivers, just the maximum.

RSVP deals with unidirectional data transfer only. In a typical two-party VoIP conversation, resource reservation must be carried out in both directions, assuming that both parties intend to speak.

The sender of data sends a PATH message to the intended receiver of the data and includes within that message a TSpec, which describes the characteristics of the data to be sent. The receiver responds with an RESV message, which contains a flowspec indicating the type of resource reservation required. In the case of multiple receivers, the flowspec might be modified as it is passed back towards the sender as a result of different receivers having different QoS requirements. This is called *merging flowspecs*.

TSpec   A TSpec is sent from a sender in a PATH message and is also included in the information sent from a receiver to a sender in an RESV message. It takes the form of a token bucket specification plus a *peak rate* (p), a *maximum packet size* (M), and a *minimum policed unit* (m). The token bucket specification defines a *bucket size* (b) and a *token rate* (r), as depicted in Figure 8-5.

This conceptual bucket has a capacity of b bytes and is filled with tokens at a rate of r bytes per second. Upon arrival of a packet, the packet will be

Figure 8-5
Token bucket

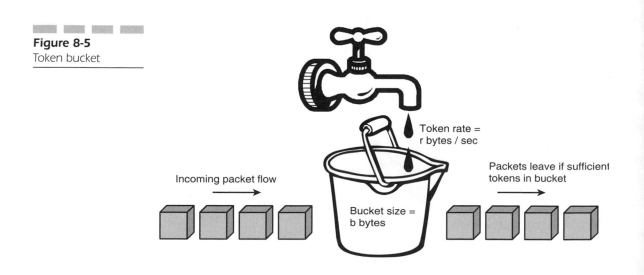

Token rate =
r bytes / sec

Incoming packet flow

Packets leave if sufficient
tokens in bucket

Bucket size =
b bytes

transmitted onwards only if the number of tokens in the bucket is at least as large as the packet. Otherwise, the packet might be dropped or queued until the bucket accumulates enough tokens. If the incoming flow is less than the taken bucket rate r, then the bucket will eventually fill up, at which point the generation of tokens is suspended until such time as a packet arrives to consume some tokens. Provided that the average rate flowing toward the bucket is less than or equal to r, the token bucket should not constrain the flow. Furthermore, the rate of flow toward the bucket may exceed r for a period of time, but it cannot do so indefinitely without running out of tokens and being subject to delay or packet loss as a result. Note that a token bucket is not a physical container of packets, but a numerical counter of tokens to schedule the releasing time of packets.

A token bucket approach is more flexible than the classic leaky bucket scheduling mechanism. A leaky bucket technique maintains a constant outward flow regardless of the incoming flow rate. Although a token bucket technique can constrain flow, it does enable variations in the incoming flow rate and permits the outbound flow rate to vary in accordance with the incoming flow rate.

The p, which should be greater than r, is the maximum data rate that the sender will use. The sender should not use that rate for any extended period of time, since to do so would be incompatible with the token bucket specification. The M is the maximum size of IP data packet that the sender will send. If any packets exceed that size, then they will be considered to be outside the limits of the TSpec and will not be treated with the same QoS as packets that are within the specification.

The m is measured in bytes. All packets less than m bytes in length are considered to be m bytes long for two reasons. The first is to enable nodes in the network to estimate the amount of resources that will be needed to process each packet. The second is to bound the bandwidth overhead consumed by link-level headers. The maximum bandwidth overhead of link-layer headers will be the ratio of link-layer header size to m. Putting all these components together, Figure 8-6 shows the format of a sender TSpec. The Token Bucket TSpec is defined in RFC 2215 as having parameter number 127.

Flowspec   The flowspec, sent in an RESV message from a receiver, contains a TSpec of its own, plus an indication of the QoS control service being requested. Currently, two services can be requested: the guaranteed service and controlled-load service. The flowspec for controlled-load service is depicted in Figure 8-7 and the flowspec for guaranteed service is depicted in Figure 8-8.

Figure 8-6
Sender TSpec

3 1	3 0	2 9	2 8	2 7	2 6	2 5	2 4	2 3	2 2	2 1	2 0	1 9	1 8	1 7	1 6	1 5	1 4	1 3	1 2	1 1	1 0	0 9	0 8	0 7	0 6	0 5	0 4	0 3	0 2	0 1	0 0
V = 0			Reserved													Overall length = 7 (excluding header)															
Service header = 1						0		Reserved								Length of service 1 data = 6															
127								Flags (0)								Parameter length = 5 (excluding header)															
Token Bucket Rate (r) (32-bit IEEE floating point number)																															
Token Bucket Size (b) (32-bit IEEE floating point number)																															
Peak Data Rate (p) (32-bit IEEE floating point number)																															
Minimum Policed Unit (m) (32-bit integer)																															
Maximum Packet Size (M) (32-bit integer)																															

V = message format version

Figure 8-7
The flowspec for
controlled-load
service

3 1	3 0	2 9	2 8	2 7	2 6	2 5	2 4	2 3	2 2	2 1	2 0	1 9	1 8	1 7	1 6	1 5	1 4	1 3	1 2	1 1	1 0	0 9	0 8	0 7	0 6	0 5	0 4	0 3	0 2	0 1	0 0
V = 0			Reserved													Overall length = 7 (excluding header)															
Service header = 5						0		Reserved								Length of service 5 data = 6															
127								Flags (0)								Parameter length = 5 (excluding header)															
Token Bucket Rate (r) (32-bit IEEE floating point number)																															
Token Bucket Size (b) (32-bit IEEE floating point number)																															
Peak Data Rate (p) (32-bit IEEE floating point number)																															
Minimum Policed Unit (m) (32-bit integer)																															
Maximum Packet Size (M) (32-bit integer)																															

V = message format version

The flowspec for controlled-load service is simply a TSpec. The flowspec for guaranteed service is a TSpec plus a *Rate* (R) term and a *Slack* (S) term. The R term indicates the bandwidth required and should be equal to or greater than r. If it is greater, then extra bandwidth is being requested, which will reduce queuing delays. The S term is a number of microseconds. It indicates the difference between the desired delay and the delay that would be achieved if the rate R is used. A network element may use the Slack to reduce the amount of resources reserved for this flow if necessary.

Figure 8-8
The flowspec for
guaranteed service

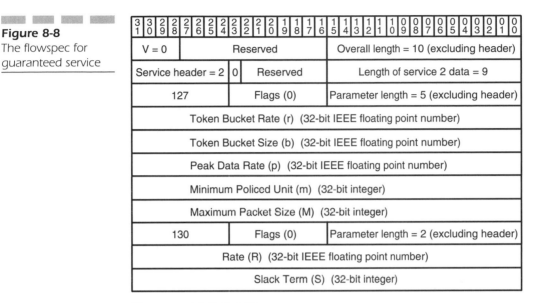

3 3 2 2	2 2 2 2	2 2 2 2	1 1 1 1	1 1 1 1	1 1 1 0	0 0 0 0	0 0 0 0
1 0 9 8	7 6 5 4	3 2 1 0	9 8 7 6	5 4 3 2	1 0 9 8	7 6 5 4	3 2 1 0

V = 0	Reserved	Overall length = 10 (excluding header)
Service header = 2	0	Reserved	Length of service 2 data = 9
127	Flags (0)	Parameter length = 5 (excluding header)
Token Bucket Rate (r) (32-bit IEEE floating point number)		
Token Bucket Size (b) (32-bit IEEE floating point number)		
Peak Data Rate (p) (32-bit IEEE floating point number)		
Minimum Policed Unit (m) (32-bit integer)		
Maximum Packet Size (M) (32-bit integer)		
130	Flags (0)	Parameter length = 2 (excluding header)
Rate (R) (32-bit IEEE floating point number)		
Slack Term (S) (32-bit integer)		

V = message format version

Filter Spec An RSVP session is identified only by a destination IP address and a protocol ID. The session identification can optionally contain a destination port number. Thus, an RSVP session identification does not include any information about the sender of data. Given that the destination port is optional, and the session is not identified by any sender-related information, a problem might occur in determining the specific data flow for which a given reservation is requested. For example, a given receiver might be receiving two streams of data from two different senders (as could be the case in a three-party VoIP call). A need exists to indicate what QoS is being requested for the flow from each sender, but the RSVP session identification itself does not include sender information. This is where the filter spec comes into play.

A filter spec defines the set of data packets (that is, the flow) to which a particular QoS is to be applied, and it is an input to the packet classifier function. The format of the filter spec depends on whether IPv4 or IPv6 is being used. At a minimum, the filter spec contains a sender IP address and optionally a sender port number. The IP address is used to identify the sender of the data to which the requested QoS is to be applied. The source port number is used for situations such as a video conference where both a video and an audio stream are set up from a given sender to a given receiver with a different QoS requirement for each stream.

Given that the filter spec can contain a port number, each router in the path must be able to examine the details of the header to determine the port number, which raises a number of considerations. First, IP datagrams in the flow must not be fragmented, which in turn means that we must use path *maximum transmission unit* (MTU) discovery. Second, IP security procedures might encrypt the header, which means that RSVP must include security functions to decrypt the header. Third, the IPv6 header is of variable length, which might cause a significantly greater processing effort at routers.

The filter spec is included in PATH messages from the sender to the receiver and in RESV messages from the receiver to the sender. When sent in a PATH message, it is known as a sender template.

ADSpec As previously described, a sender first sends a PATH message downstream towards one or more receivers. The TSpec in the PATH message indicates the type of data to be sent and is passed from the sender to the receiver without modification. The receivers respond with an RESV message to indicate the QoS requested and the RESV is propagated upstream towards the sender.

Having a receiver request a QoS that intervening routers cannot meet is pointless, however. For example, the RESV message might specify a bandwidth requirement that exceeds the bandwidth of one of the links in the path from sender to receiver. Therefore, a need exists for a receiver to be informed about the network between the sender and receiver, so that the receiver does not request something that the network cannot provide. To meet this need, RSVP includes the capability for the sender and routers to indicate their QoS capabilities within a PATH message from the sender to receiver. This is done through the use of an ADSpec, where the AD indicates advertising.

A sender constructs an initial ADSpec, indicating what it can support, and it includes the ADSpec in the PATH message. At each router along the path from the sender to receiver, the ADSpec is updated and passed on. By the time the PATH message gets to the receiver, the ADSpec provides a pretty good indication of what the receiver can reasonably request in terms of QoS from the routers and from the sender.

One interesting point about the ADSpec is that it can also indicate that one or more routers in the path from the sender to receiver does not support RSVP. Such a situation is perfectly possible. In the event that an RSVP-capable router sends a PATH message on a link to a router that does not support RSVP, or if an RSVP-capable router receives a PATH message on a link from a router that does not support RSVP, then the RSVP-capable router sets a bit in the ADSpec to indicate that RSVP is not supported end to end.

The general format of the ADSpec is depicted in Figure 8-9. The format includes general information (service number 1), optionally followed by either or both guaranteed service information and controlled-load service information.

The general information specifies whether a *nonintegrated-service-capable* (non-IS) hop is involved in the path (bit marked X in Figure 8-9). For the purposes of RSVP, non-IS means lacking RSVP support. If this bit is set, then the information in the rest of the ADSpec is no longer relevant, since QoS has to be end to end and an obvious break has taken place in the QoS chain from the sender to receiver.

The ADSpec also indicates the number of hops between IS-capable nodes, the path bandwidth, the path MTU, and the minimum path latency. The minimum path latency is the delay that would be experienced along the path if no additional queuing delays took place.

Following the general information in the ADSpec, a guaranteed service data fragment and/or a controlled-load data fragment may contain information related to those two services. These data fragments are inserted if the application might use either or both services for the session in question.

Figure 8-9

The format of the ADSpec

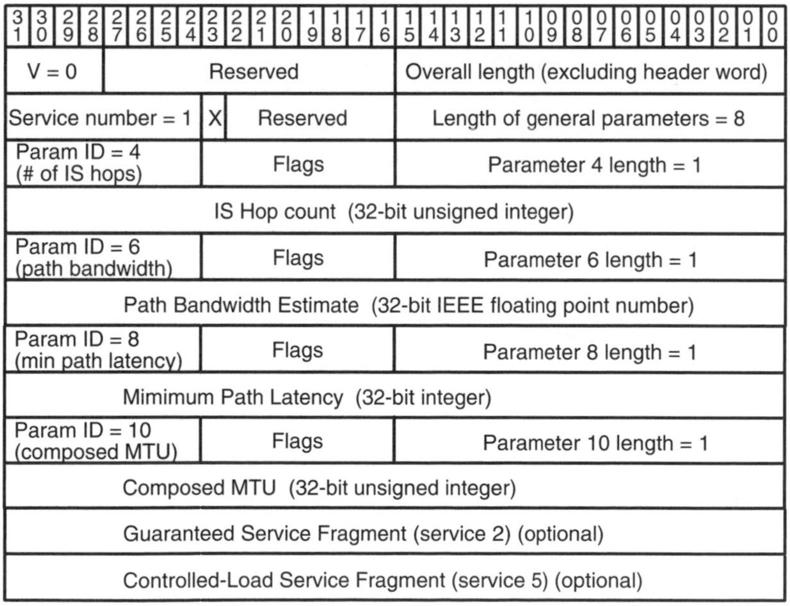

| 3 3 2 2 2 2 2 2 2 2 2 2 1 1 1 1 1 1 1 1 1 1 0 0 0 0 0 0 0 0 0 0 |
| 1 0 9 8 7 6 5 4 3 2 1 0 9 8 7 6 5 4 3 2 1 0 9 8 7 6 5 4 3 2 1 0 |

V = 0	Reserved	Overall length (excluding header word)
Service number = 1	X	Reserved	Length of general parameters = 8
Param ID = 4 (# of IS hops)	Flags	Parameter 4 length = 1
IS Hop count (32-bit unsigned integer)		
Param ID = 6 (path bandwidth)	Flags	Parameter 6 length = 1
Path Bandwidth Estimate (32-bit IEEE floating point number)		
Param ID = 8 (min path latency)	Flags	Parameter 8 length = 1
Mimimum Path Latency (32-bit integer)		
Param ID = 10 (composed MTU)	Flags	Parameter 10 length = 1
Composed MTU (32-bit unsigned integer)		
Guaranteed Service Fragment (service 2) (optional)		
Controlled-Load Service Fragment (service 5) (optional)		

V = message format version

RSVP Messages We have already discussed a number of RSVP messages, such as PATH and RESV, and have also looked at some of the information carried in these messages. We now look at the format of these message and others in a little more detail.

Every RSVP message contains a common header, as shown in Figure 8-10. The various messages are Path (1), Resv (2), PathErr (3), ResvErr (4), PathTear (5), ResvTear (6), and ResvConf (7). The checksum is used to ensure that the message has been received without error. The Send_TTL is the IP TTL (time to live) value of the message. With normal IP forwarding, RSVP can use this field to determine if a non-RSVP hop has been involved by comparing the value in the RSVP header with the IP TTL value, because a non-RSVP node will update the IP TTL value but not the RSVP Send_TTL.

Following the common header, the message contains a number of objects, such as a sender TSpec, an ADSpec, and so on. Each object has a header and contents, as shown in Figure 8-11. The Class-num and C-type identify the object. The Class-num identifies the object and the C-type enables a different version of the object to exist. For example, the format of a filter spec varies depending on whether IPv4 or IPv6 is being used. In that case, the Class-num indicates that the object is a filter spec and the C-type indicates whether it is the IPv4 or IPv6 variant.

The following are examples of the object types defined in RSVP. The list is not exhaustive, and the reader should consult RFC 2205 for a complete listing and associated definitions:

- **SESSION Class** This has a Class-num value of 1. It has a C-type value of 1 for IPv4 and a C-type value of 2 for IPv6. This object

Figure 8-10
The RSVP common header

Byte 0		Byte 1	Byte 2	Byte 3
Vers = 1	Flags (0)	Message type	RSVP checksum	
Send_TTL		Reserved	RSVP length	

Figure 8-11
The RSVP object format

Byte 0	Byte 1	Byte 2	Byte 3
Length (bytes)		Class-num	C-type
Object contents			

includes the IP destination address, the protocol ID for the type of data being sent, and optionally the destination port.

- **FLOWSPEC Class** This has a Class-num of 9 and a C-type value of 2 (the C-type value of 1 is obsolete). The flowspec is described earlier in this chapter.

- **SENDER_TEMPLATE Class** This has a Class-num value of 11. For IPv4, the C-type value is 1 and for IPv6, the C-type value is 2. The sender template is sent in a Path message and is, in fact, a filter spec sent by the sender.

- **RSVP_HOP Class** This object contains the IP address of the interface through which the last RSVP-capable node passed this message. The object is used in the Path message and is saved at each node along the route. The purpose is to ensure that an RESV message from the receiver to the sender follows the same path back through the network as was taken by the PATH message. This object has a Class-num value of 3, a C-type value of 1 for IPv4, and a C-type value of 2 for IPv6.

- **TIME_VALUES Class** This is a timeout period in milliseconds and indicates how long the message is to be considered valid. This object has a Class-num value of 5 and a C-type value of 1.

- **ERROR_SPEC Class** This is an object that is included in RSVP error messages. The object includes the IP address of the node at which the error was detected, plus an error code identifying the type of error, plus some additional error information. This class has a Class-num value of 6. The C-type value is 1 for IPv4 and 2 for IPv6.

- **STYLE Class** We can select different reservation styles in the case of multiple receivers and/or multiple senders. A fixed-filter style means that a receiver uniquely identifies the individual sender and the corresponding data stream from that sender to which a given reservation is to apply. A wildcard-filter style indicates that the reservation request applies to all data streams from all senders in a given session. The shared-explicit style lists specific senders, indicating that a given request applies to data streams from those senders. The applicable Class-num value for the STYLE class is 8 and the C-type value is 1.

Example Reservations Figures 8-12 and 8-13 show two examples of successful reservations. Figure 8-12 shows a simple case of one sender and one receiver. In this case, a PATH message is propagated through the

Figure 8-12
*Successful reservation
for a single sender
and a single receiver*

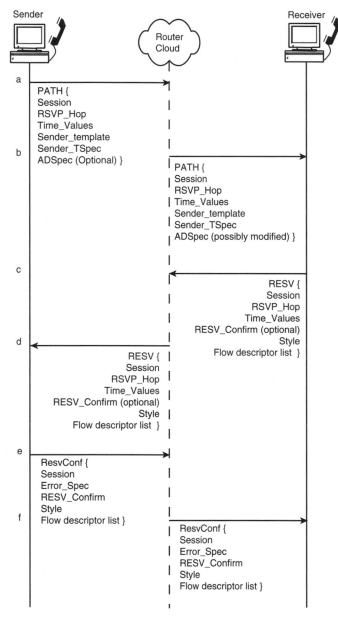

Sender

Router
Cloud

Receiver

a

PATH {
Session
RSVP_Hop
Time_Values
Sender_template
Sender_TSpec
ADSpec (Optional) }

b

PATH {
Session
RSVP_Hop
Time_Values
Sender_template
Sender_TSpec
ADSpec (possibly modified) }

c

RESV {
Session
RSVP_Hop
Time_Values
RESV_Confirm (optional)
Style
Flow descriptor list }

d

RESV {
Session
RSVP_Hop
Time_Values
RESV_Confirm (optional)
Style
Flow descriptor list }

e

ResvConf {
Session
Error_Spec
RESV_Confirm
Style
Flow descriptor list }

f

ResvConf {
Session
Error_Spec
RESV_Confirm
Style
Flow descriptor list }

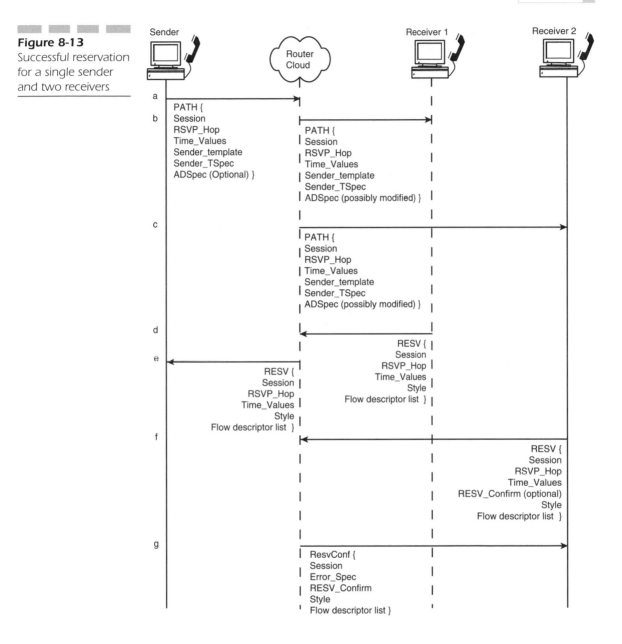

Figure 8-13

Successful reservation for a single sender and two receivers

network from the sender to receiver. The network nodes do not modify the sender template and TSpec, but they can modify the ADSpec.

The receiver responds with an RESV message, which can optionally contain a request that the reservation be confirmed. The RESV message indicates the style of reservation to be applied and a list of flow descriptors. Each such descriptor is a flowspec and/or filter spec, depending on the reservation style in question.

In Figure 8-12, the receiver has requested confirmation of the success of the reservation. Consequently, a ResvConf message is returned. The Error_Spec contains an error code of 0 in the case of a successful reservation. The Resv_Confirm object is copied from the RESV message. The style and flow descriptor indicate the particular reservations that are being confirmed. The information in these elements might not be identical to the corresponding information on the RESV message, because in the case of multiple senders, the reservation might not be successful for data streams from all the senders.

In the multicast example of Figure 8-13, Receiver 1 sends an RESV message just before Receiver 2. In this example, the QoS requirement of Receiver 1 is stronger than that of Receiver 2. In that case, there is no need to modify the reservation already sent back towards the receiver, because the reservation from Receiver 1 is good enough to support the requirements of both receivers. In the example, Receiver 2 requested a confirmation, but Receiver 1 did not. Note that the confirmation is sent from the router at which the reservations merge.

This procedure is used when a router receives an RESV message with a request for confirmation and a reservation is already in place that meets or exceeds the requirements of the reservation to be confirmed. In certain cases, this situation can lead to a false confirmation, particularly if the existing reservation request fails at some point closer to the sender. Therefore, a reservation confirmation should only be considered an indication of high probability that a reservation has succeeded.

Reservation Errors

Of course, a given resource reservation could fail, in which case an error message is returned. Two error message exist: PathErr and ResvErr. PathErr messages are simply passed back to the sender whose PATH message led to the error.

ResvErr messages are sent to a receiver when a given reservation request fails. With merged flowspecs, it is important that a reservation request that fails for a given reason does not deny service for a different request that would have succeeded on its own. For example, imagine two receivers who both send reservation requests and those requests meet at a single router that resides in the path from the sender to receivers. If one receiver's request fails at the router due to an excessive bandwidth request, the other receiver's request should not fail if it has requested a smaller bandwidth that could be supported.

Guaranteed Service

RFC 2212 defines guaranteed service. This service involves two parts: first, ensuring that there will not be any packet loss, and second, ensuring minimal delay. Ensuring that there will not be any loss is a function of the token bucket depth (b) and token rate (r) specified in the TSpec. At a given router, provided that buffer space of value b is allocated to a given flow and that a bandwidth of r or greater is assigned, then no loss should be experienced.

Delay is a function of two components. The first is a fixed delay due to processing within the individual nodes and is only a function of the path taken. The second component of delay is the queuing delay within the various nodes. Let us assume that a given node supports a token bucket model (b, r) and offers a given stream a bandwidth of R for a particular flow. If R is less than r, then the token bucket algorithm will not introduce any queuing delay, because the rate of tokens supplied to the token bucket will exceed the rate of the incoming data flow. If R is greater than or equal to r, then the token bucket can introduce a queuing delay. The delay in such a case is bounded by b/R.

In reality, other queuing delay components will also play a part: C and D. C is a rate-dependent error term and represents the delay that a flow might experience as a function of the rate of the flow. For example, C/R could represent the delay involved in breaking up a datagram into the payload of *Asynchronous Transfer Mode* (ATM) cells. D is a fixed delay term within the node itself, not dependent on the rate. C is measured in bytes and D is measured in microseconds. A node that provides guaranteed service must ensure that the total delay is less than $b/R \leq C/R \leq D$.

Recall that the ADSpec can contain a guaranteed service object. This object has the form shown in Figure 8-14. One can see that the object includes cumulative values for C(tot) and D(tot). In addition, values of C and D since the last traffic-reshaping point in the path are also included. Traffic reshaping involves buffering at a node in the network to correct deviations from the original TSpec in the received data. This buffering is set

Figure 8-14

ADSpec guaranteed
service object

3 1	3 0	2 9	2 8	2 7	2 6	2 5	2 4	2 3	2 2	2 1	2 0	1 9	1 8	1 7	1 6	1 5	1 4	1 3	1 2	1 1	1 0	0 9	0 8	0 7	0 6	0 5	0 4	0 3	0 2	0 1	0 0
Service number = 2	X	Reserved	Length of service 2 data
Param ID = 133 (composed Ctot)	Flags	Parameter 133 length = 1
End-to-end composed value for C (Ctot) (32-bit integer)
Param ID = 134 (composed Dtot)	Flags	Parameter 134 length = 1
End-to-end composed value for D (Dtot) (32-bit integer)
Param ID = 135 (composed Csum)	Flags	Parameter 135 length = 1
Since last reshaping point composed C (Csum) (32-bit integer)
Param ID = 136 (composed Dsum)	Flags	Parameter 136 length = 1
Since last reshaping point composed C (Csum) (32-bit integer)
Service-specific data (if present)

up so that the nonconforming data can be brought back into conformance. Note that such nonconformance might not necessarily be the fault of the source application. Nonconformance could also be the fault of queuing effects in routers.

When a receiver receives an ADSpec with the guaranteed service data fragment, it can make a decision as to the value of R that should be requested in the flowspec within the RESV message. Because the path that the data flow will take determines the value of Dtot, the receiver can do nothing to minimize the impact of the term. The delay impact of Ctot is reduced as R increases, however. Therefore, the receiver can request a value of R such that the total delay will always be within limits that are acceptable. Of course, the value of R cannot be so high that it would exceed the supported bandwidth along the path from sender to receiver. That value is included in the ADSpec, so that the receiver knows the maximum value of R that could possibly be supported.

Controlled-Load Service

Controlled-load service is a close approximation to the QoS that an application would receive if the data were being transmitted over a network that is very lightly loaded. This service means that a very high percentage of packets will be delivered successfully and that the delay experienced by a very high percentage of packets will not exceed the minimum delay experienced by any successfully delivered packet. RFC 2211 describes controlled-load service.

The service specification does not identify specific characteristics of network elements that need to be minimized. Rather, a node that implements controlled-load service should ensure that it offers a given flow the necessary bandwidth and buffer space to support the TSpec indicated in a given request. This is the reason why the flowspec for controlled-load service is just a TSpec with no additional information. Equally, the controlled-load data fragment of an ADSpec offers no additional information beyond that contained in the general part of the ADSpec.

One important point about the controlled-load service is the QoS that would be experienced by packets that fall outside of the TSpec. There is no guarantee that such packets would get a QoS anything near the QoS of packets that fall within the TSpec. In fact, such packets are likely to experience significant delay or loss.

Removing Reservations and the Use of Soft State

Reservations can be removed in two ways: explicitly by a sender or receiver, or as a result of a timeout.

Two messages exist for the explicit teardown of the reservation state within network nodes: PathTear and ResvTear. PathTear is used by a sender and travels towards all receivers, deleting path state and reservation state in all nodes along the path. ResvTear is used by a receiver, and it effectively does the same thing in reverse. A given receiver does not have the authority to delete the reservation state for other receivers, however. Therefore, in the case of merged reservations, a ResvTear message will propagate upstream in the network only to the extent that deletion of a reservation state applies only to the receiver in question.

RSVP uses a soft state approach to managing reservations. In other words, reservations need to be refreshed on a regular basis or they will time out. Both PATH and RESV messages contain a refresh period (R) within the TIME_VALUES object. To keep a reservation active, the receiver should send a new RESV message after R seconds. Within each node is a timer, L, that is some function of R. If the node does not receive a refreshing message within L seconds, then the reservation state is deleted. L should be set to a value that is significantly larger than R so that one or more RESV messages could be lost without deleting the reservation state.

This setting is important because RSVP does not include reliability functions for ensuring that RSVP messages have guaranteed delivery and it is possible that refreshing messages could be lost. In fact, the timeout of reservations exists for the same reason. A PathTear or ResvTear message at the end of a session might be lost. If the reservation does not time out, then we could have a situation where network resources are being reserved when they are no longer needed, while new reservations fail due to a lack of available resources.

DiffServ

RSVP is a comprehensive QoS mechanism. Of the QoS solutions available, it comes the closest to circuit emulation, but it does so at a significant cost. RSVP requires that RSVP-enabled routers examine IP headers in some detail, and it requires that routers maintain state for all existing reservations. If RSVP is used on a session-by-session basis (such as on a per-VoIP-call basis), then there might be millions of concurrent reservations. Such a situation can mean significant overhead. Consequently, RSVP has difficulty scaling to very large systems.[2] In many cases, we can provide a good QoS to numerous applications without the complexity and overhead of RSVP. DiffServ is one means for achieving this goal.

If we think of QoS purely in terms of bandwidth, the only way to increase the QoS offered to a given data stream is to increase the bandwidth. Increasing bandwidth can be achieved by physically adding bandwidth or by offering one application extra bandwidth at the expense of another application. RSVP works by reserving a set bandwidth for a particular data stream. Once all the bandwidth is consumed, further reservation attempts fail. The approach with DiffServ is that it is acceptable to offer one application greater QoS at the expense of another application if the first application really needs it for acceptable performance and the second application will not really notice a big difference. For example, we can offer a greater QoS to interactive traffic (such as voice) at the expense of non-real-time traffic (such as e-mail), because it matters greatly if voice is subject to delay.

---

[2]This statement is true when RSVP is used to reserve resources for each individual session. As we shall see later in this chapter, however, traffic-engineering extensions for RSVP (RSVP-TE) mean that RSVP can be used for traffic-engineering functions. Since RSVP-TE is not used on a per-session basis, it does not have the same scaling limitations as basic RSVP.

It does not greatly matter if an e-mail takes minutes instead of seconds to be delivered. DiffServ simply offers the capability to prioritize certain types of traffic and it offers certain types of handling in the network depending on the priority of a given traffic stream.

The DiffServ Architecture

Recall that the IPv4 header has a ToS field, and the IPv6 header has a corresponding Traffic Class field. DiffServ renames this field the DS field. As shown in Figure 8-15, DiffServ uses six bits of the field to encode the *DS codepoint* (DSCP) and does not currently use the remaining bits of the field. In fact, those two bits are used for explicit congestion notification, as specified in RFC 3168. Within the network, the values of the DSCP in a given IP packet are used to handle the packet according to a certain PHB. Therefore, all that is required within the core of the network is the capability to examine the DSCP and act accordingly.

The DiffServ architecture is defined in RFC 2475, "An Architecture for Differentiated Services," with clarifications added by RFC 3260. In the DiffServ architecture, the packets of a given stream must be marked with the appropriate DSCP value so that routers in the network can provide the correct PHB. To avoid situations where every packet from every stream is marked to get the highest priority, functions at the edge of the network must ensure that only qualified packets are marked with a particular DSCP value. These functions include metering to measure the packet rate, policing to ensure that traffic meets a particular agreed profile, traffic shaping and dropping if necessary, and the actual marking itself. Collectively, these functions are known as traffic conditioning. Of course, it is also necessary to classify packets to identify the stream to which they belong or other characteristics that lead to the selection of a particular DSCP value. These functions work in a combined manner, as depicted in Figure 8-16.

One advantage of the DiffServ architecture is that the classification, policing, and traffic-shaping functions are pushed to the edge of the network, while the requirements on routers within the network are kept to a minimum. This means that the bulk of the effort and investment involved

Figure 8-15
DiffServ usage of IP
ToS/Traffic Class octet

Figure 8-16
Classifying and
conditioning traffic

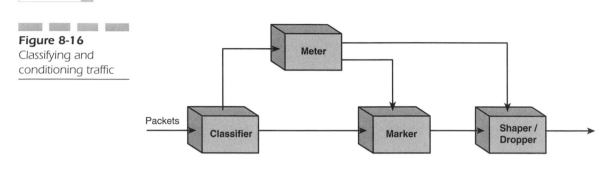

in using DiffServ is placed at or near the host that runs an application that wants to use DiffServ. The amount of change required in the core of the network is minimal and does not change with the number of user applications that want to use DiffServ. Consequently, DiffServ scales extremely well.

A disadvantage of DiffServ is that it works on a per-hop basis. In other words, each router in the network provides priority to a certain packet according to that packet's DSCP value, but DiffServ does not offer a mechanism that reserves resources from end to end. Thus, although DiffServ can enable better QoS for certain applications compared to others, it does not provide a QoS guarantee for any application, particularly in the case of network congestion. Therefore, we must strive to ensure that the network does not become congested in the first place.

The Need for SLAs

The most critical component of successful DiffServ operation is at the edge of the network. It is vital that application packets be correctly classified and marked, and applications must be prevented from abusing the system. Therefore, a given network domain and packet source must agree on the rules related to packet classification and traffic conditioning, and these functions must be implemented at or near the source of the packets. The functions can be implemented at the source itself (such as within a customer's host[s]), or they could be implemented in an edge router provided by the network operator (such as an ISP or an *Internet telephony service provider* [ITSP]).

In either case, an SLA must be in place between the customer and network operator. Such an agreement should include a definition of the traffic profile to be sent from the source to the network, the classification and marking rules, and the behaviors to be applied within the network for specific DSCP values. The traffic profile might include a token bucket specification, as described previously. The classification rules could be based on combinations of source address, destination address, source port, destination port, protocol ID, time of day, or other criteria.

The customer and provider should agree on the classification criteria and, just as important, the SLA should specify what is to happen if traffic is received that is outside the traffic profile. In such a situation, actions could include the dropping of packets, marking nonconformant packets with a different DSCP value than that used for conforming packets, traffic shaping, or additional charges to the customer.

SLAs should not only be used between network operators and their customers, but they should also be placed between different network operators. In the context of this discussion, a network operator can be considered to be a particular DiffServ-capable domain with a common set of policies and PHB definitions. If traffic needs to be passed from one operator to another, the two operators should agree on the rules related to the service that particular streams receive and in particular the meaning to be applied to particular DSCP values. Only in this way can we achieve end-to-end QoS.

Per-Hop Behavior (PHB)

PHB is the treatment that a DiffServ router applies to a packet with a given DSCP value. Note that a router deals with multiple flows from many sources to many destinations. Many of the flows may have packets marked with a DSCP value that indicates a certain PHB. That set of flows from one node to the next that share the same DSCP codepoint is known as an *aggregate*. From a DiffServ perspective, a router operates on packets that belong to specific aggregates. When a router is configured to support a given PHB (such as dedicating bandwidth), the configuration is with respect to aggregates rather than specific flows from a specific source to a specific destination.

Currently, two PHBs are defined: EF and AF. These are described in RFC 3246 and RFC 2597, respectively.

Expedited Forwarding (EF)   The objective with EF PHB is to provide a service that is low loss, low delay, and low jitter, such that the service approximates a virtual leased line. The basic approach is to minimize the loss and delay experienced in the network by minimizing queuing delays. This job can be done by ensuring that, at each node, the rate of departure of packets from the node is a well-defined minimum and that the arrival rate at the node is always less than the defined departure rate. To meet this objective, traffic-conditioning functions at the edge of the network are necessary to ensure that the incoming rate is always below the configured outgoing rate. The recommended DSCP value for the EF PHB is 101110.

The EF PHB can be implemented in network nodes in a number of different ways. For example, the EF PHB could be implemented by a priority queue. Such a mechanism could enable the unlimited preemption of other traffic such that EF traffic always gets first access to outgoing bandwidth. This approach could, however, lead to unacceptably low performance for non-EF traffic. Therefore, it is important that the implementation of EF in a given node is designed such that EF traffic does not inflict enormous damage to other traffic. One way to do this would be to police the rate of incoming EF traffic through a token bucket limiter. Another way to implement the EF PHB would be through the use of a weighted round-robin output scheduler, where the share of the output bandwidth allocated to EF traffic is equal to some configured rate. That output rate should be greater than or equal to the rate expected for incoming EF traffic.

Assured Forwarding (AF)   AF PHB is defined in RFC 2597. Its objective is to provide a service that ensures that high-priority packets are forwarded with a greater degree of reliability than lower-priority packets.

Normally, the traffic into a DiffServ network from a particular source should conform to a particular traffic profile. In particular, the rate of traffic should not exceed some pre-agreed maximum. In the event that it does, the source of the traffic runs the risk that the data will be lumped in with normal best-effort IP traffic and will be subject to the same delay and loss possibilities. In fact, in a DiffServ network, certain resources will be allocated to certain behavior aggregates, which means that a smaller share is allocated to standard best-effort traffic. Thus, receiving best-effort service in a DiffServ network could be worse than receiving best-effort service in a non-DiffServ network. A given subscriber to a DiffServ network might want the latitude to occasionally exceed the requirements of a given traffic profile without being too harshly punished. AF PHB offers this possibility.

AF PHB is a means for a provider to offer different levels of forwarding assurances for packets received from a customer. AF PHB enables packets to be marked with different AF classes and, within each class, to be marked with different drop precedence values. Within a router, resources are allocated according to the different AF classes. If the resources allocated to a given class become congested, then packets must be dropped. The packets to be dropped are those with higher drop precedence values. From a customer's point of view, this offers the possibility of saying something like, "Please don't drop my packets, but if you have to, then drop these particular ones before any others, as these are less important."

AF has four classes and three drop precedence levels in each class. Therefore, a given packet has a class (i) and a drop precedence (j), and the associated DSCP is AFij. The recommended values of the codepoints are shown in Table 8-1. We can see that the three most significant bits indicate the class and the three least significant bits indicate the drop precedence within the class.

Within a DiffServ network, the AF implementation must detect and respond to long-term congestion by dropping packets and respond to short-term congestion by queuing. These requirements imply a function that monitors short-term congestion and, based on short-term congestion levels, derives a smoothed long-term congestion level. When the smoothed congestion level is below a particular threshold, no packets should be dropped. If the smoothed congestion level is between a first and second threshold level, then packets with the high precedence level should be dropped. As the congestion level rises, more and more of the high-drop-precedence packets should be dropped until such time as a second congestion threshold is reached, at which point 100 percent of the high-drop-precedence packets are dropped. If the congestion continues to rise, then packets of the medium precedence level should also start being dropped.

The implementation must treat all packets within a given class and precedence level equally. Thus, if 50 percent of the packets with a given class and precedence value are to be dropped, then that 50 percent should be spread evenly across all packets for that class and precedence. Put another way, a given flow should experience the same drop rate as every other flow with the same class and precedence level. Note that different AF classes are treated independently and are given independent resources. Thus, when packets are to be dropped, they are dropped for a given class and drop precedence level. Packets of one class and precedence level might experience a 50 percent drop rate, while the packets of a different class with the same precedence level are not dropped at all. Regardless of the amount

Table 8-1

Recommended AF codepoint values

Drop Precedence	Class 1	Class 2	Class 3	Class 4
Low	001010	010010	011010	100010
Medium	001100	010100	011100	100100
High	001110	010110	011110	100110

of packets that need to be dropped, a DiffServ node must not reorder AF packets within a given AF class regardless of their precedence level.

Multiprotocol Label Switching (MPLS)

We can see that RSVP and DiffServ take very different approaches to solving the QoS problem. RSVP, in its basic form, tackles the issue by reserving resources in advance of a given session. DiffServ, on the other hand, takes the approach of sharing network resources according to priority. Since RSVP was initially designed to operate on a session-by-session basis (and millions of concurrent sessions might take place), it has difficulty scaling to large systems. DiffServ can scale well, but it operates on a hop-by-hop basis and does not ensure that end-to-end resources are available for a given session. An ideal solution is one that offers the best of both worlds: the capability to ensure end-to-end resource availability for a large number of sessions. MPLS is one such solution.

MPLS is not primarily a QoS solution, although it may be used to support QoS requirements. Rather, MPLS is best viewed as a new switching architecture. Standard IP switching requires that every router must analyze the IP header and make a determination of the next hop, based on the content of that header. The primary driver in determining the next hop is the destination address in the IP header. The destination address is compared with entries in a routing table, and the longest match between the IP header and the addresses in the routing label determines the next hop. The approach with MPLS is to attach a label to the packet. The content of the label is specified according to an FEC, determined at the point of ingress to the network. The packet and label are passed to the next node, where the label is examined and the FEC is determined. The FEC is then used as a simple lookup in a table that specifies the next hop and a new label to use. The new label is attached and the packet is forwarded.

So what is the big difference between this and standard routing based on the IP header? To begin with, the FEC is determined at the point of ingress to the network, where information might be available that cannot be indicated in the IP header. For example, the FEC could be chosen according to a combination of the destination address, QoS requirements, the ingress router (or interface), or any of a multitude of criteria. The FEC can implicitly indicate such information, and routing decisions in the network can automatically take that information into account. In particular, a given FEC can force a packet to take a particular route through the network without having to cram a list of specific routers into the IP header. This fact can be important for ensuring QoS, where the bandwidth available on a given path and the path length both have a direct impact to the quality that is achieved. Imagine a network that supports different types of traffic. Between any two points in the network, we might want one type of traffic to take a particular path while another type of traffic takes a different path. For example, VoIP traffic from point A to point B should take a very direct path (to minimize latency), while other traffic is forced along a less direct route.

In some ways, MPLS is similar to ATM. MPLS enables a specific type of traffic to take a specific path from ingress to egress, and that path can be established in advance. The path in question can be engineered with appropriate resources, and the point of ingress can include policing functions to ensure that the traffic demand does not exceed the allocated resources. This approach is similar to the connection admission functions invoked at the establishment of an ATM *Virtual Channel Connection* (VCC).

In fact, MPLS and ATM are competing technologies. Although most people in telecommunications recognize that MPLS is likely to be the long-term choice for supporting multiple services on a single network, many question its maturity. ATM, on the other hand, is stable, mature, and widely deployed. Despite these advantages, however, ATM has a number of disadvantages, including its complexity and the famous (or infamous) ATM cell tax. As MPLS sees more deployment, we can expect MPLS to supplant ATM in the long term.

Meanwhile, however, a middle ground is apparent. The fact that MPLS places a label in front of an IP packet does not necessarily imply a new layer between layers 2 and 3. Obviously, the fact that a new label is attached to the packet can mean that a new layer exists, if only for the purpose of interpreting the contents of the label. The label itself can be carried in an existing layer 2 protocol, however, provided an appropriate field exists in which to carry it. For example, the label could be carried in a Frame Relay *Data Link Connection Identifier* (DLCI) field. Similarly, the label could be carried in ATM *Virtual Path Identifier* (VPI) or *Virtual Channel Identifier* (VCI) fields. If one has an existing ATM infrastructure and wants to migrate to MPLS, the mapping of MPLS labels to ATM VPI/VCI fields is one way to begin the migration without risking a quantum leap from one technology to another.

The MPLS Architecture

The MPLS architecture is specified in RFC 3031, "Multiprotocol Label Switching Architecture." As already described, MPLS involves the determination at the point of ingress to the network of an FEC value to apply to a packet. That FEC value is then mapped to a particular label value and the packet is forwarded with the label. At the next router, the label is evaluated and the corresponding FEC is determined. A lookup is then performed to determine the next hop and the new label to apply. The new label is attached and the packet is forwarded to the next node.

This process suggests that the value of the label can change as the packet moves through the network, which it can. What does not change is the FEC.

Basically, the process works as shown in Figure 8-17. An incoming packet is assigned a particular FEC based on some criteria. In the figure, the packet has an FEC of F. Note that the FEC is not a value carried in the packet. Rather, the FEC can be considered as a combination of factors: the port number, originating address, destination address, ToS bits, and so on.

At the ingress router (R1), the FEC of F means that the packet must be sent to router R2 and that it should have a label value of L1. When the packet and label arrive at R2, R2 knows that the label value L1 means an FEC of F for packets received from R1. R2 then proceeds to look up the next

Figure 8-17

Label value and FEC relationships

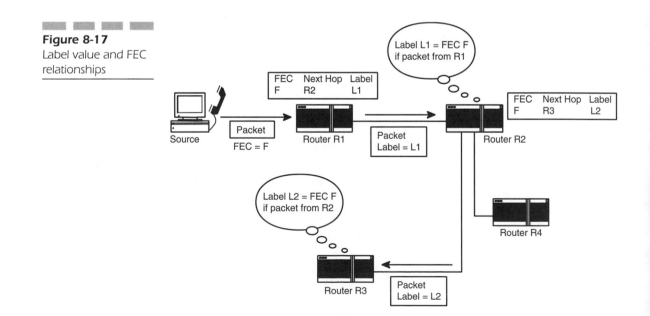

hop and label value. It determines that the packet should be forwarded to R3 with a label value of L2. At R3, the packet and label are received. R3 knows in advance that the label value L2 means an FEC of F for packets from R2 and it uses this information for its routing decision, and so on.

An important point to note is the fact that the ingress router (R1 in our example) has the job of assigning the first label. That initial assignment determines subsequent handling within the network. In other words, the first router determines the FEC. The assigned FEC can be based on information that only the first router knows, such as the ingress interface. The ingress router is known as a *label edge router* (LER).

We can see that the relationship between FEC and the label value is a local affair between two adjacent *label-switching routers* (LSRs). If a given router is upstream from the point of view of the data flow, then it must have an understanding with the next router downstream as to the binding between a particular label value and FEC.

For example, in Figure 8-17, the label value L1 is the outgoing label at R1 for packets with FEC F going to R2. At R2, the incoming label value of L1 is bound to FEC F for packets incoming from R1. Also at R2, the label value L2 is the outgoing label for packets with FEC F going to R3, and so on. Thus, the binding between the label value and FEC is limited to the two ends of an interface between adjacent LSRs for a particular direction of data flow. Also, if R2 were to receive packets with the label value L1 on a link from a different LSR, such as R4, then a different FEC might apply. The fact that the meaning of a given label depends on the incoming interface means a greater routing flexibility than with standard routing tables (where the outgoing path from a router depends on the destination IP address but not on the incoming interface).

Label-Switched Paths (LSPs) and Label Distribution   The foregoing discussion shows that the path through a network is determined solely by the FEC that applies at the point of ingress. Such a path is a known as a *label-switched path* (LSP), and it may be established in a number of ways. First, we can easily imagine a scenario where a network engineer determines that a certain FEC shall apply to a certain type of traffic from a certain source to a certain destination. That engineer could then determine the most appropriate path through the network and configure each of the applicable router interfaces with the necessary FEC/label bindings. Alternatively, we can use a protocol that enables the routers to automatically share FEC/label bindings. Such a protocol is known as a *label distribution protocol*.

The primary function of label distribution protocols is the establishment and maintenance of LSPs, which includes the establishment of FEC/label bindings at each LSR in the LSP. In the MPLS architecture, the downstream LSR decides on the particular binding. The downstream LSR uses a label distribution protocol to communicate the binding to the upstream LSR. A downstream LSR can directly distribute a label/FEC binding to an upstream LSR, which is known as *unsolicited downstream*. Additionally, we have *downstream on demand*, whereby an upstream LSR requests a binding from a downstream LSR.

Several label distribution protocols are available. In some cases, existing protocols such as RSVP have been extended to provide support for label distribution. In addition, brand new protocols have been developed, such as the *Label Distribution Protocol* (LDP) and *Constraint-Based LDP* (CR-LDP), which is an extension of LDP. If a given label distribution protocol is used between two LSRs to exchange label/FEC binding information, then the two LSRs are known as *label distribution peers*. Although the MPLS architecture does not mandate that any particular label distribution protocol be used, we discuss three such protocols later in this chapter: LDP, CR-LDP, and RSVP-TE.

FEC and Label Formats

In order for a downstream LSR to communicate a label/FEC binding to an upstream LSR, the label distribution protocol messages must include fields that specify FEC and label values. For example, a downstream LSR can use the LDP Label Mapping message to communicate a particular label/FEC binding to an upstream LSR. The message includes specific FEC values and label values. Clearly, FECs and labels must have specific formats. Those formats depend on the label distribution protocol being used.

In LDP, an FEC is comprised of a list of FEC elements, as shown in Figure 8-18, with each FEC element either a host address or address prefix. The format of a label to be used between any two adjacent LSRs, however, depends on the layer 2 protocol used on the link between them. For protocols (such as the *Point-to-Point Protocol* [PPP]) that do not already support a field that can carry a label value, the label takes the generic format of a four-octet field placed in front of the IP packet. The label contains four fields: a 20-bit label field, which carries the actual label value; a 3-bit *experimental* (Exp) field, which is for experimental use; an 8-bit TTL field; and a 1-bit stack (S) field, which indicates whether this label is the bottom of the label stack. We will discuss the concept of a label stack shortly.

When such a label format is communicated using a label distribution protocol (during LSP establishment), the generic format of Figure 8-19 is

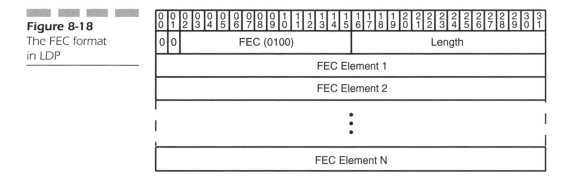

Figure 8-18
The FEC format in LDP

used for layer 2 protocols such as PPP. For example, in the Label Mapping message of LDP, a parameter of the form shown in Figure 8-19 would be passed to indicate that a generic-format, four-octet label of a particular value is to be associated with a particular FEC. The first four-octet word indicates the format of the label and the second word includes the label itself. Once the LSP is established, packets from the upstream LSR to the downstream LSR that correspond to the FEC in question would be prepended by the corresponding four-octet label (that is, the second word of Figure 8-19).

Both ATM and Frame Relay contain fields that could be used to carry an MPLS label. For example, the VPI and VCI fields of the ATM header could carry an MPLS label. Equally, the Frame Relay DLCI field could carry an MPLS label. In the case of ATM, the label is also four octets long, but it takes the form shown in Figure 8-20. In a Label Mapping message, the format and value of the label would be communicated with the label format of Figure 8-20. The first word of Figure 8-20 indicates the format of the label, and the second word conveys the label value itself. The V-bits indicate whether both the VPI and VCI are significant or just one of them. 00 indicates that both are significant, 01 indicates that only the VPI is significant, and 10 indicates that only the VCI is significant.

When Frame Relay is used at layer 2, the label is mapped to the DLCI, as shown in Figure 8-21. The Len field indicates the number of bits in the DLCI. 0 indicates that the DLCI is 10 bits long, and 2 indicates that the DLCI is 23 bits long. Values 1 and 3 are reserved.

The Label Stack   Much of the discussion so far has suggested that a given packet can have only one label. Although such a situation is valid, a given packet can have more than one label. In fact, MPLS describes a label stack, where there may be several labels, one after the other, in advance of

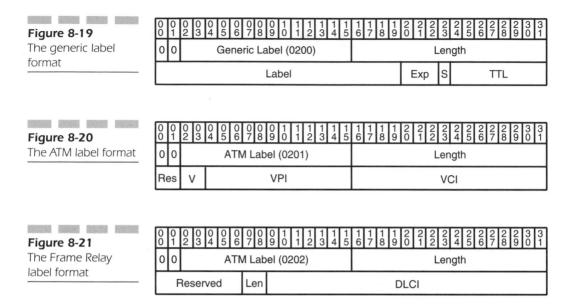

Figure 8-19
The generic label format

Figure 8-20
The ATM label format

Figure 8-21
The Frame Relay label format

the actual IP packet. In such a situation, an LSR that receives the labeled packet will base its actions on the first (top) label. The obvious question is, "Why might we need more than one label?" The answer is tunneling.

Imagine a VoIP network operator that provides international service between the United States and the United Kingdom That provider will have gateways and routers on both sides of the Atlantic, but might not own the network in between. Rather, the network operator could purchase service from another party, such as a large IP network operator. That IP network operator might offer service to many different customers, and the traffic from each of those customers would be carried over the same network, but would need to be kept logically separate within that common network.

Suppose that a given packet with FEC F needs to follow the LSP R1, R2, R3, and R4. Let's assume that R2 and R3 are not directly connected, but form the two ends of the tunnel R2, R2A, R2B, R2C, and R3. Therefore, the actual path taken is R1, R2, R2A, R2B, R2C, R3, and R4, as depicted in Figure 8-22. When a packet travels from R1 to R2, it has a single label. R2 determines that the packet P must enter the tunnel. Therefore, R2 attaches another label on top of the first, so that the label stack has a depth of 2. In fact, it may replace the first label with a label that is meaningful to R3 and then place a new label on top. The packet P has two labels as it passes through the tunnel. When it reaches R2C, the LSR recognizes that it is the penultimate LSR in the tunnel. (Both R2 and R3 are effectively part of the

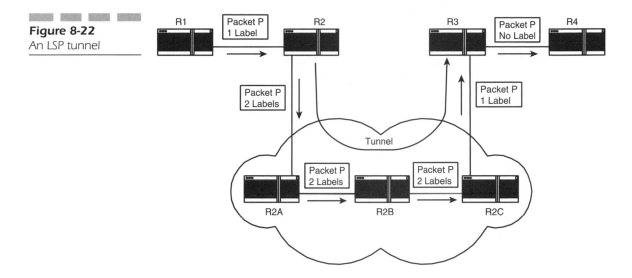

Figure 8-22
An LSP tunnel

tunnel.) R2C pops the stack and forwards the packet to R3 with only one label. R3 uses the label to route to R4. In fact, R3, being the penultimate LSR in the overall LSP, may also pop the stack and pass the packet to R4 unlabeled.

The scenario depicted in Figure 8-22 is effectively an LSP nested within another LSP. The MPLS architecture enables multiple levels of nesting, which can be useful in many situations. Consider again our transatlantic VoIP provider. That provider might want to send more than just VoIP traffic between the two countries. In addition to the actual revenue-generating VoIP traffic, we can easily imagine a situation where the *Information Technologies* (IT) department must send billing information, and the Network Operations department needs to send network alarm and provisioning data. Each department could have its own IP networks on either side of the Atlantic. The company as a whole could purchase service from an international IP network operator, with its total traffic carried in a tunnel through the international IP operator's network. Within that tunnel, each department's traffic would be carried within its own tunnel, as depicted in Figure 8-23.

Actions at LSRs

An LSR's actions depend on the value of the label it receives from an upstream LSR. In fact, the action taken by the LSR is specified by the *Next*

Figure 8-23
MPLS tunnels within
tunnels

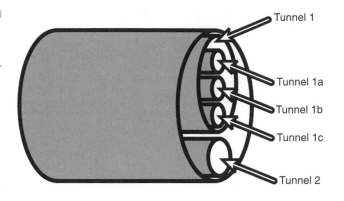

Hop Level Forwarding Entry (NHLFE), which indicates the next hop, the operation to perform on the label stack, and the encoding to be used for the stack on the outgoing link. The operation to perform on the stack may mean that the LSR should replace the label at the top of the stack with a new label. The operation might require the LSR to pop the label stack or replace the top label with a new label, and then add one or more additional labels on top of the first label.

The next hop for a given labeled packet might be the same LSR. In such a case, the LSR pops the top-level label of the stack and forwards the packet to itself. At this point, the packet might still have a label to be examined, or it might be a native IP packet without a label, in which case the packet is forwarded according to standard IP routing.

A given label could possibly map to more than one NHLFE. This situation might occur where load-sharing takes place across multiple paths. In such a case, the LSR chooses one of the NHLFEs to use according to some internal procedures.

In general, if a router happens to know that it is the penultimate LSR in a given path, then it should remove any labels and pass the packet to the final LSR without a label. This minimizes the amount of processing that the ultimate LSR needs to undertake. If the penultimate LSR passes a labeled packet to the final LSR, then the final LSR must examine the label, determine that the next hop is itself, pop the stack, and forward the packet to itself. The LSR must then reexamine the packet to determine what to do with it. If, on the other hand, the packet arrives without a label, the final LSR has one less step to execute. How a particular LSR determines that it is the penultimate LSR for a given path is a function of label distribution and the distribution protocol used.

MPLS Traffic Engineering

One of the most important applications of MPLS is in the area of traffic engineering, which can be summarized as the modeling, characterization, and control of traffic to meet specified performance objectives. As described in RFC 2702, the performance objectives in question may be traffic oriented or resource oriented. The former deals with QoS and includes aspects such as minimizing delay, jitter, and packet loss. The latter deals with optimum usage of network resources, particularly network bandwidth.

These two objectives are not necessarily mutually exclusive. For example, congestion avoidance is a major goal related to both network resource objectives and QoS objectives. From a resource point of view, we want to avoid situations where one part of the network is congested while another part of the network is underutilized, particularly if the underutilized resources could be used to carry some of the traffic that is experiencing congestion. Equally, from a QoS point of view, we want to allocate traffic streams to available resources to ensure that those streams do not experience congestion and the consequent packet loss and delay.

Congestion is primarily caused in two ways. The first is simply due to a lack of sufficient resources in the network to accommodate the offered load. The second is the steering of traffic towards resources that are already loaded, while other resources remain underutilized. The expansion of capacity or the application of congestion control techniques such as flow control can correct the first situation. The second situation can be addressed by good traffic engineering, that is, ensuring that traffic is directed through the network in a manner that is consistent with the needs of that traffic. For example, we might want to direct VoIP traffic along the shortest path in order to minimize latency. Traffic that is less delay sensitive could be sent along a less-direct path.

The current situation with IP routing and resource allocation is that the routing protocols are not well equipped to deal with traffic-engineering issues. For example, a protocol such as *Open Shortest Path First* (OSPF) can actually promote congestion because it tends to force traffic down the shortest route, even though other acceptable routes might be less loaded. In many network deployments, the solution to this problem has been to use traffic-engineering functions at layer 2 in the network. ATM, for example, enables virtual circuits to be easily rerouted in cases of congestion. MPLS also enables traffic-engineering functions.

Traffic Trunks RFC 2702 introduces the concept of a traffic trunk, which can be considered to be a set of flows that share specific attributes.

In practice, these attributes include the ingress and egress LSRs, the FEC, and possibly other characteristics such as the average rate, peak rate, priority, and policing attributes. For example, all VoIP packets from a given source to a given destination might be considered a single traffic trunk. Meanwhile, other traffic (e-mail, for example) between the same source and destination might be considered a different traffic trunk.

A traffic trunk is something that can be (and is) routed over a given LSP. In fact, it is possible to explicitly specify the LSP that a traffic trunk should use. This has the immediate advantage of being able to steer certain traffic away from the shortest path, which is likely to get congested before other paths. If we understand the available network resources and the expected demand for each type of traffic, then we can route each type of traffic through the network in a manner that best meets QoS requirements. In other words, we can ensure that those applications with strong QoS needs are routed along paths that have the resources to meet those QoS needs. Moreover, we can change the LSP that a given traffic trunk will use. This capability enables the network to adapt to changing load conditions whether through administrative intervention or through automated processes within the network. Thus, we can choose to send a certain type of traffic over a certain path, and we can change to a second path if the first path fails.

As described in RFC 2702, traffic engineering in an MPLS network has three main aspects:

- Mapping packets to FECs
- Mapping FECs to traffic trunks
- Mapping traffic trunks onto the physical network topology through LSPs

How individual packets are assigned to a given FEC and how FECs are assigned to traffic trunks are decisions that are made at the ingress to the network. These decisions can made according to any number of criteria (such as the physical ingress interface, the IP ToS value, or the source IP address). We simply need to ensure that the criteria are fully understood by both the MPLS network provider and the source of packets (either a customer or another network provider). This requirement again emphasizes the need for SLAs between providers and between providers and customers.

Of the three mappings that need to take place, the third is at the heart of ensuring that the network provides the quality that is needed for a given type of traffic. It involves constraint-based routing, whereby traffic is matched with network resources according to the characteristics of the traffic and the characteristics of available resources. For example, one characteristic of traffic is a bandwidth requirement and one characteristic of a path is the maximum bandwidth that it offers. Obviously, path selection should be limited to those paths that can offer at least the bandwidth needed by the traffic trunk.

Label Distribution Protocols and Constraint-Based Routing

We have already described the concept of label distribution protocols, which is the means by which LSRs share label/FEC bindings so that traffic belonging to a particular FEC can follow a particular LSP.

The Label Distribution Protocol (LDP) One of the most common such protocols is LDP. LDP includes a number of messages that are exchanged between LSRs for the purpose of establishing and maintaining signaling relationships between the LSRs, requesting and providing label/FEC bindings, and maintaining the overall health of an LSP. LDP conveys information through LDP *protocol data units* (PDUs), which comprise a common header, followed by an LDP message.

Figure 8-24 shows the format of the LDP common header. The LDP identifier indicates the label space of the LSR that is sending the message in question. The first four octets of the LDP identifier include a globally unique identifier for the LSR. The remaining two octets indicate the label space of the LSR. A given LSR might reuse the same label values across different interfaces, in which case the value of the label space identifier indicates the interface to which a label value applies. If an LSR uses just a single label space for the whole platform, then the label space value of 0 is used.

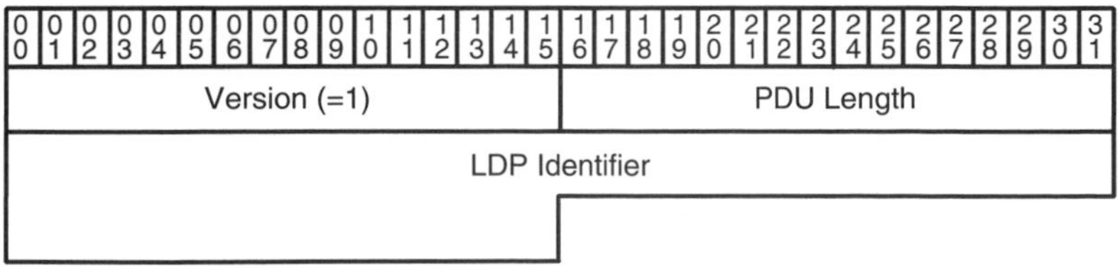

Figure 8-24
The LDP PDU
common header

| 0 0 0 0 0 0 0 0 0 0 1 1 1 1 1 1 1 1 1 1 2 2 2 2 2 2 2 2 2 2 3 3 |
| 0 1 2 3 4 5 6 7 8 9 0 1 2 3 4 5 6 7 8 9 0 1 2 3 4 5 6 7 8 9 0 1 |

Version (=1)	PDU Length
LDP Identifier	

LDP Messages　LDP offers four categories of messages:

- Discovery messages
- Session Management messages
- Advertisement messages
- Notification messages

Discovery messages are used to announce and maintain the presence of an LSR on the network. An LSR periodically sends Hello messages to other LSRs to indicate its presence on the network. The LSR can use basic discovery or extended discovery. Basic discovery means that LDP Hello messages are sent to the well-known multicast address for "all routers on this subnet," which means that the Hello messages are sent only to other routers that are directly connected at the link layer (layer 2). Extended discovery enables an LSR to send Hello messages to routers that are not directly connected at layer 2. Such Hello messages are known as targeted Hello messages and are sent to specific IP addresses that the LSR is aware of (perhaps through configuration data loaded in the LSR).

Upon receipt of a Hello message, an LSR can reply with a Hello message of its own. At this point, both LSRs are aware of the other's presence and know each other's IP addresses.

Session Management messages are used to establish, maintain, and terminate sessions between LDP peers. Once two LSRs have exchanged Hello messages, they proceed to exchange Initialization messages. An Initialization message includes a number of mandatory and optional parameters. Among the mandatory parameters is the receiver's LDP identifier. Since the LDP common header contains the sender's LDP identifier, the Initialization message effectively contains both LDP identifiers: the sender's and the receiver's. This combination uniquely identifies a relationship between LSRs that began with the exchange of Hello messages.

Among the optional parameters of the Initialization message are ATM and Frame Relay address ranges. These values indicate the ATM VPI/VCI or Frame Relay DLCI ranges that can be used if the link between the LSRs uses ATM or Frame Relay.

Once two LSRs have exchanged Initialization messages, an LDP session is established between them. Thereafter, the session is maintained through the exchange of LDP PDUs. If an LSR does not receive an LDP PDU after a "keep-alive" time period, then it will conclude that the connection between the LSRs has failed. In the event that no LDP PDUs need to be sent between LSRs, they can periodically send LDP keep-alive messages to ensure that the LDP session does not time out.

Advertisement messages are used to create, change, and delete label mappings for FECs. LDP includes two main messages related to label mapping: the Label Request message and the Label Mapping message. The Label Request message is used by an LSR that operates in downstream-on-demand mode and indicates that the upstream LSR wants to obtain a label/FEC binding. The Label Mapping message contains the actual label/FEC binding. In downstream-on-demand mode, the Label Mapping message is sent as a response to a Label Request message. In unsolicited downstream mode, the message can be sent without first receiving a Label Request message.

LSRs can operate according to either independent label distribution control or ordered label distribution control. With independent label distribution control, an LSR can advertise a label mapping to an LDP peer any time it desires. Thus, an LSR can send a Label Mapping message to an upstream LSR without first having received a Label Mapping message from a downstream LSR. With ordered label distribution control, an LSR can advertise a label mapping to an upstream LSR only if the LSR has an FEC/label binding for the next hop or if the LSR itself is the MPLS egress point.

If we combine downstream-on-demand with ordered label distribution control, then label distribution would work as follows. The ingress LSR would send a Label Request message to a downstream LSR and that message would propagate through the network to the egress LSR. The egress LSR would respond to its next upstream LSR with a Label Mapping message. That upstream LSR would store the FEC/label binding for its egress interface, create a new FEC/label binding for its ingress interface, and communicate that binding to the next upstream LSR with a Label Mapping message of its own. The process continues until the ingress LSR receives the Label Mapping message in response to its original request. Figure 8-25 depicts the signaling sequence for such a scenario.

Figure 8-26 shows the format of the LDP Label Request message and Figure 8-27 shows the format of the LDP Label Mapping message. The term TLV refers to the structure Type-Length-Value, whereby a given field includes an identifier to indicate the type of field in question (T), a length subfield (L) and the actual value of the field itself (V).

The Label Request message has two optional parameters: the Path Vector TLV and the Hop-count TLV. Both of these parameters help to detect loops. The Path Vector TLV is simply a list of LSRs. Each LSR that propagates a Label Request message adds its own LSR ID (the first four octets of the LDP identifier described previously) to the list of LSR IDs. Similarly, each LSR that forwards a Label Request message increments the

Figure 8-25
LSP establishment
with downstream-on-
demand and ordered
label distribution
control

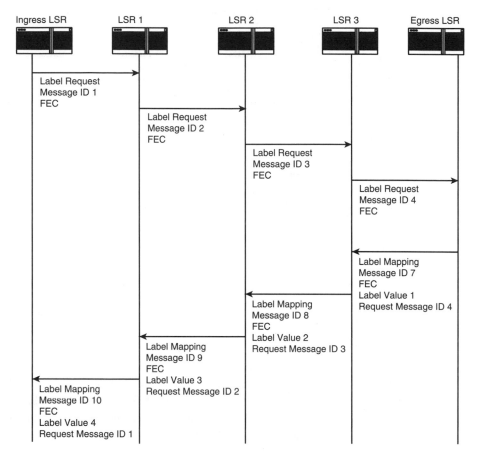

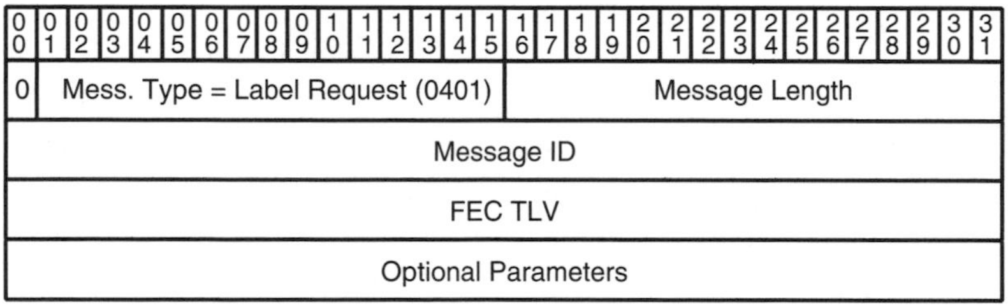

Figure 8-26
The LDP Label
Request message

Hop-count TLV by 1 (except when the Hop-count TLV was previously set to 0, which means unknown).

The Label Mapping message contains the same optional parameters as the Label Request message for the same reason—loop detection. In addition, the Label Mapping message can optionally carry a Label Request

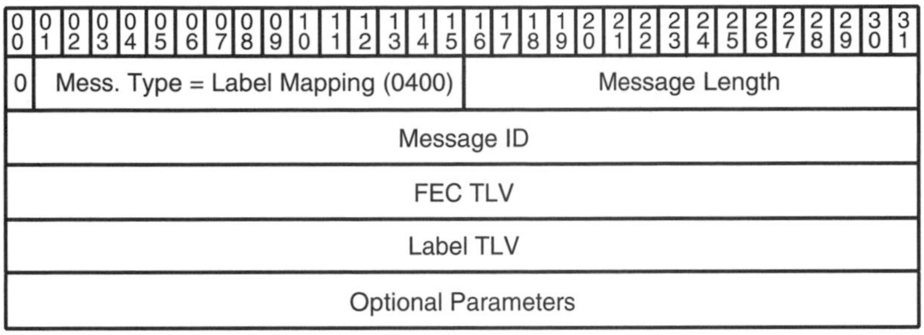

Figure 8-27
The LDP Label
Mapping message

0 0	0 0 0 0 0 0 0 0 0 1	1 1 1 1 1 1 1 1 1 2	2 2 2 2 2 2 2 2 2 3	3
0 1	2 3 4 5 6 7 8 9 0 1	2 3 4 5 6 7 8 9 0 1	2 3 4 5 6 7 8 9 0	1
0	Mess. Type = Label Mapping (0400)		Message Length	
Message ID				
FEC TLV				
Label TLV				
Optional Parameters				

message ID TLV. This parameter is simply the message ID of the Label Request message that prompted this Label Mapping message, which enables an upstream LSR to correlate a given Label Mapping message with the request that instigated the message. The example signal flow of Figure 8-25 shows the use of the Label Request message ID TLV.

In addition to the Label Request and Label Mapping messages, we also have the Label Abort Request message, the Label Withdraw message, the Label Release message, the Address message, and the Address Withdraw message.

An LSR sends a Label Abort Request message when it wants to cancel a Label Request and has not yet received a Label Mapping message in response to the original request. The Label Abort Request message includes the message ID of the original request so that the downstream LSR knows which request is being cancelled.

A downstream LSR sends a Label Withdraw message to an upstream LSR to indicate that the upstream LSR can no longer use a particular FEC/label binding. An upstream LSR sends a Label Release message to a downstream LSR to indicate that it no longer requires the label/FEC mapping that was previously provided by the downstream LSR. An upstream LSR must send a Label Release message in response to a Label Withdraw message.

An LSR uses an Address message to advertise its interface address(es) to LDP peers. The LSR performs such advertisements prior to the exchange of any Label Request or Label Mapping messages. Moreover, the LSR sends an Address message whenever a new interface is added to the LSR. An LSR uses the Address Withdraw message to withdraw previously advertised interface addresses. An LSR would send also the message if, for example, an interface is taken out of service. Both the Address and Address Withdraw messages contain an Address List TLV. This TLV comprises an address family field indicating the type of addresses in the list (IPv4 or IPv6), followed by a list of addresses corresponding to the interfaces being advertised or withdrawn.

Notification messages are used to provide advisory information and signal error information. The Notification message includes a Status TLV, which can carry status or error information. If, for example, a loop were detected during LSP establishment, then a Notification message would be returned with Status Code 0000000B (loop detected). A Notification message can optionally contain an LDP PDU (or part of one) or an LDP message (or part of one) if a particular PDU or message generated the error. For example, if a given LDP message caused an error at the receiving LSR, then that LSR would return a Notification message with an appropriate status code plus as much of the received message as necessary to indicate the problem in the message.

Constraint-Based LDP (CR-LDP)   The foregoing description of LDP shows how an LSP can be established end to end. The LSP is effectively a network path that is associated with a given FEC. As we have also described, the significance of a given FEC is established at the network edge, where information such as the bandwidth requirements, IP ToS setting, the ingress router, and the incoming interface can be known. Unfortunately, LDP does not enable that additional information to be passed in LDP messages as part of LSP establishment. LDP only passes the destination address or address prefix, which does not enable downstream LSRs to make LSP routing decisions that can take other information into account. For example, an LER might want to establish an LSP that has certain bandwidth requirements. We need a mechanism whereby such requirements can be conveyed during label advertisement and mapping so that the LSP establishment can include the allocation of the appropriate resources. CR-LDP is one way to do this.

CR-LDP is specified in RFC 3212 and is an extension of LDP, whereby an LSP can be established subject to certain constraints, such as traffic parameters, resource requirements, and other characteristics. In fact, CR-LDP can enable the establishment of LSPs according to explicit routing requirements such that the LSP establishment messages can specifically indicate which LSRs should be included in the LSP.

Explicit Routes (ERs)   A *Constraint-Based LSP* (CR-LSP) is an LSP that is established subject to a number of criteria. It is established based on information available at the edge of the network. An *explicit route* (ER) is one type of CR-LSP where some or all the nodes to be used are specified. In a strict ER, all nodes in the path are specified and the path is not allowed to include nodes that are not specified. In a loose ER, a number of nodes in the path are specified, but other nodes can also be allowed. CR-LDP enables ER information to be included in the LDP Label Request message. The overall idea is to define very specific paths for traffic that has specific characteristics.

Traffic Characteristics CR-LDP enables a number of traffic characteristics to be specified through the use of a traffic parameters TLV, as shown in Figure 8-28. The peak rate indicates the maximum rate at which traffic should be sent to the CR-LSP. It includes two components: the *peak data rate* (PDR), measured in bytes/second, and the *peak burst size* (PBS), measured in bytes. Each of these could be components of a token bucket specification where the bucket is filled at a rate equal to PDR and the bucket size equals PBS.

The committed rate indicates the rate that the network commits to be available to the CR-LSP. It has the components *committed data rate* (CDR) and *committed burst size* (CBS). These may also be used in a token bucket specification. The value of PDR should be greater than or equal to the value of CDR.

The *excess burst size* (EBS) can be used at the edge of the MPLS domain for the purposes of traffic conditioning and can be used to measure the extent by which traffic exceeds the committed rate. The EBS could also form part of a token bucket specification, where the token bucket size is EBS and the token rate is CDR.

The frequency indicates how often the CDR should be made available. Frequent (1) means that the available bandwidth should average at least

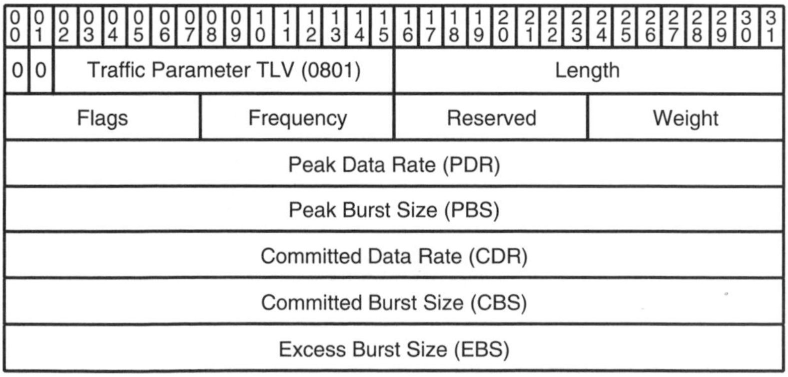

Figure 8-28
CR-LDP traffic parameter

0 0 0 0 0 0 0 0 0 0 1 1 1 1 1 1 1 1 1 1 2 2 2 2 2 2 2 2 2 2 3 3
0 1 2 3 4 5 6 7 8 9 0 1 2 3 4 5 6 7 8 9 0 1 2 3 4 5 6 7 8 9 0 1

0 0	Traffic Parameter TLV (0801)	Length		
Flags	Frequency	Reserved	Weight	
Peak Data Rate (PDR)				
Peak Burst Size (PBS)				
Committed Data Rate (CDR)				
Committed Burst Size (CBS)				
Excess Burst Size (EBS)				

the CDR over any short interval. Very frequent (2) means that the available rate should average at least the CDR over any packet interval at the CDR.

The weight determines the CR-LSPs share of any possible excess bandwidth above the committed rate and can be used within a given MPLS domain to determine which traffic streams get access to any available additional bandwidth.

The Flags field of the Traffic Parameters TLV is shown in Figure 8-29. The field includes a number of indicators related to each of the other fields in the Traffic Parameters TLV. Specifically, the flags indicate whether each of the traffic parameters requirements is negotiable or not as follows:

- F1 relates to PDR: 0 = non-negotiable, 1 = negotiable.
- F2 relates to PBS: 0 = non-negotiable, 1 = negotiable.
- F3 relates to CDR: 0 = non-negotiable, 1 = negotiable.
- F4 relates to CBS: 0 = non-negotiable, 1 = negotiable.
- F5 relates to EBS: 0 = non-negotiable, 1 = negotiable.
- F6 relates to weight: 0 = non-negotiable, 1 = negotiable.

If a given component of the Traffic Parameters TLV is negotiable, then an LSR might offer a lower value than is requested. For example, an LER might request a PDR of 10,000 bytes/second. If, however, the LER indicates that the requested PDR is negotiable, then the resultant LSP might support a lower value (say, 8,000 bytes/second).

Given that a request to establish an LSP might include an ER requirement as well as a Traffic Parameters TLV, one might find an inconsistency between the different components of the request. For example, an LER might request that a CR-LSP support a specific traffic profile and follow a specific path. We can easily imagine a situation where the path in question does not have the available resources to support all the requested traffic parameters. In such a case, the request to establish a CR-LSP could be rejected. If, however, certain traffic parameters are negotiable, the network might still be able to establish the CR-LSP along the requested path but with lesser resource commitments than originally requested.

Figure 8-29
CR-LDP traffic parameter flags

| Reserved | F6 | F5 | F4 | F3 | F2 | F1 |

Given that many of the traffic parameters of CR-LDP can form part of a token bucket specification, we can make sure that an LSP is established to meet criteria that will be policed at the network edge. In other words, the ingress router could implement a token-bucket-based traffic-policing technique for a given stream and could establish an LSP through the network that meets the requirements of the token bucket specification. Provided that the source of traffic complies with the token bucket specification, network resources will support the traffic.

Resource Classes   CR-LDP enables the use of resource classes to specify which links may be used in a given CR-LSP. The purpose is to help limit the set of possible links to those that meet certain resource requirements. For example, a resource class could indicate OC-48, meaning that all links of OC-48 capacity belong to that resource class. Similarly, *Gigabit Ethernet* (Gig-E) could be a resource class. Resource classes are sometimes known as colors, where resources of the same color belong to the same class. A given resource can belong to more than one resource class. Note that CR-LDP does not specify the specifics of resource classes; it simply provides a means for indicating a resource class in LDP messages. CR-LDP does so by allowing for a resource class TLV. It is only necessary that all LSRs within a given domain have the same understanding of resource classes and the contents of the resource class TLV.

Preemption   CR-LDP enables a CR-LSP to be allocated to a given traffic trunk, so that a CR-LSP is established based on certain resource requirements. If a CR-LSP cannot be established due to a lack of available resources, then CR-LDP enables one to reroute other traffic in order to make room. This is called *preemption*.

Preemption involves the assignment of two priority levels to a given CR-LSP. The first, called setupPriority, indicates the authority assigned to a given CR-LSP to preempt (bump) another. The second, called holdingPriority, indicates how much authority is required by another CR-LSP to bump the CR-LSP that currently has the resources. B can only bump A, if B's setupPriority is higher than A's holdingPriority. Actually, priority values are assigned in descending order. The value 0 is most important and the value 7 is least important. For a given CR-LSP, its holdingPriority should be at least as good as its setupPriority. Otherwise, having bumped a less important CR-LSP, it can be immediately bumped by a CR-LSP with the same priority as itself.

Modified LDP Messages for CR-LDP Using the new information defined in CR-LDP, the Label Request and Label Mapping messages are modified as shown in Figures 8-30 and 8-31.

The LSPID is a unique identifier for the CR-LSP in an MPLS network. The LSPID is composed of the ingress LSR router ID (or any of its own IPv4 addresses) and a locally unique CR-LSPID to that LSR. The LSPID also includes an action indicator flag, which indicates the action that should be taken if the LSP already exists on the LSR in question. For example, we might want to modify the traffic parameters associated with a given LSR (such as add more bandwidth), in which case the action indicator flag would indicate that the LSP is to be modified. If set to 0, the action indicator flag indicates that this is an initial LSP establishment. The ability to indicate

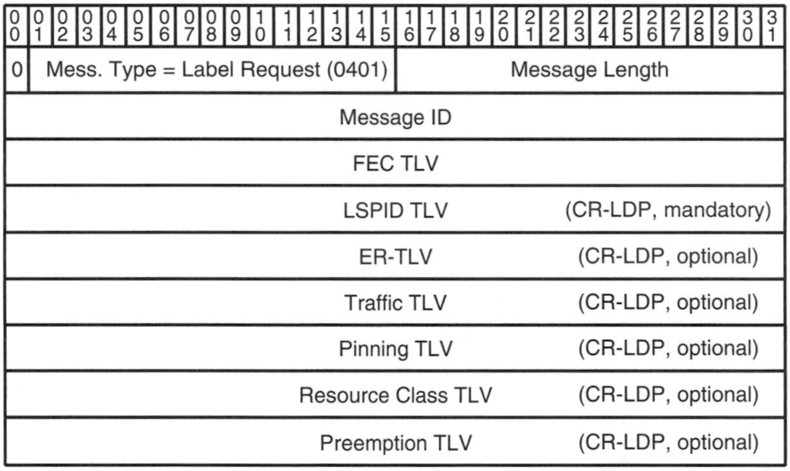

Figure 8-30
CR-LDP Label Request message

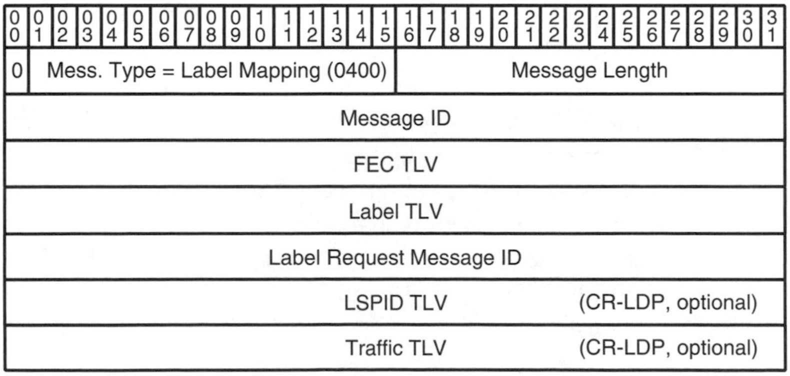

Figure 8-31
CR-LDP Label Mapping message

the required action means that the parameters associated with an LSP can be changed without having to release and reestablish the LSP.

The ER TLV specifies the path to be taken through the network. The TLV is composed of a number of ER-hop TLVs. Each ER-hop TLV indicates the type of hop (T), the length of the hop address (L), and the value of the address associated with the hop (V). Four types of ER hops are defined: an IPv4 address, an IPv6 address, an autonomous system number, and an LSPID. For each ER-hop TLV, a single bit specifies whether this is a loose hop or a strict hop. If this is a strict hop, then the address in the ER-hop TLV must match the LSR that has received this Label Request message. Otherwise, we have an error situation. If this is a loose hop, then the LSR receiving the message must pass the request on towards the LSP indicated in the address part of the ER-hop TLV.

The ability to indicate an existing LSPID as a hop to be traversed is quite useful and enables us to easily establish MPLS tunnels within tunnels. Recall our earlier example of a transatlantic VoIP operator that purchased international IP transport from an international IP service provider. Let's assume that the international IP provider establishes an MPLS tunnel for the VoIP operator to use. That tunnel could have a specific LSPID. If the VoIP operator wants to establish multiple LSPs through that tunnel (one LSP for bearer traffic, one for signaling traffic, one for admin traffic, and so on), the establishment of each of the LSPs could specify the LSPID of the overall tunnel as one of the ER-hop TLVs. This action would effectively establish each new LSP as a tunnel within the overall tunnel allocated to the VoIP operator.

The Pinning TLV is used with loose ERs. It includes a means of indicating whether or not a path may be changed at a given LSR if a better next hop becomes available later. Of course, such changes cannot apply to a strict ER.

Enabling Traffic Engineering and QoS with CR-LDP The foregoing description of CR-LDP indicates how we can apply traffic requirements and explicit routing requirements to the establishment of an LSP. By ensuring that an LSP is established according to certain criteria, we can ensure that the traffic carried on that LSP always gets the resources it needs. For example, we can set up an LSP that meets certain resource requirements (such as bandwidth) and reserve those resources so that they are dedicated to that FEC's traffic. Moreover, we can specifically indicate the route that a given LSP needs to take through the network.

Imagine a traffic-engineering tool used to engineer an MPLS network. The tool has a *graphical user interface* (GUI) that shows the topology of the

IP network, as depicted in Figure 8-32. The tool is provisioned with data regarding the network resources (such as router configurations and link bandwidths) and has a command interface to the various LSRs in the network.

To establish an LSP, the network architect would simply "drag" the LSP from one side of the network to the other on the tool's GUI interface. The tool would then prompt the user for LSP constraints such as bandwidth requirements, explicit routing requirements, and so on. Once the user has entered the required data, the tool would send the appropriate command to the first LSR, which would trigger the LSR to send a CR-LDP Label Request message through the network with the various parameters. The various routers would reserve the necessary resources and send the necessary CR-LDP Label Mapping messages to set up the LSP. Not only would the network have established the LSP, the tool would also store information about the LSP.

If the network architect wants to set up another LSP, then he or she would repeat the process. This time, however, the tool would be aware of

Figure 8-32
Example GUI for an MPLS engineering tool

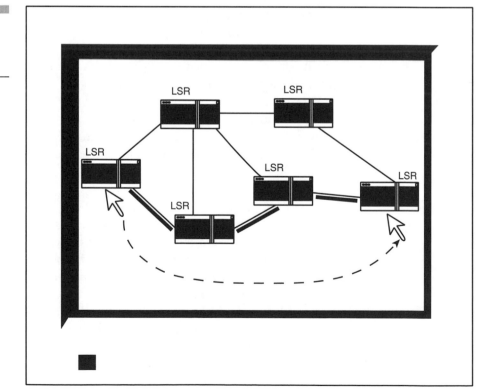

committed resources on the network and command the edge router to send a CR-LDP Label Request message that meets the requirements for the required LSP while also maintaining the integrity of the first LSP.

RSVP Traffic Engineering (RSVP-TE)

Our discussion of RSVP earlier in this chapter focused on the usage of RSVP for the reservation of resources on a session-by-session basis. When used in such a manner, RSVP has difficulty scaling to support large systems. With the appropriate extensions, however, RSVP can be easily adapted to resource reservation for traffic trunks rather than individual sessions. In other words, we can reserve resources for traffic of a given class from one edge of the network to the other rather than for each individual session.

RSVP-TE is defined in RFC 3209, "Extensions to RSVP for LSP Tunnels." This specification describes the additions made to RSVP messages (PATH and RESV) to enable RSVP to be used as the label distribution and traffic-engineering protocol for MPLS networks.

In order to support LSP establishment using RSVP, a number of new objects are added to RSVP messages: the Label_Request object, the Label object, the Explicit_Route object, the Record_Route object, and the Session_Attribute object.

The Label_Request object can have three formats. One has no specific label range, one includes the maximum and minimum VPI and VCI values and would be used with ATM interfaces, and one includes the maximum and minimum DLCI values and would be used with Frame Relay interfaces.

The Label object is a four-octet parameter with the same structure as shown in the label portion of Figure 8-19 earlier in this chapter.

The Explicit_Route object is a list of addresses that a particular LSP must follow and is similar to the ER TLV of CR-LDP, which is composed of a number of ER-hop elements. Like the ER-hop elements of CR-LDP, the Explicit_Route addresses of RSVP-TE have a loose/strict indicator bit.

The Record_Route object is a list of IPv4 or IPv6 addresses for the nodes that a given RSVP message has passed through. The object can be used to detect loops in RSVP signaling and it can also be used as an input to the Explicit_Route object. In some ways, this object is similar to the Path Vector TLV of LDP.

The Session_Attribute object is similar to the Preemption TLV of CR-LDP, and it includes a setupPriority and holdingPriority for an LSP.

RSVP establishes LSPs by using these new objects with the existing RSVP PATH and RESV messages. The PATH message is enhanced to include the Label_Request object and optionally the Explicit_Route and Session_Attribute objects. The PATH message already includes traffic data (sender TSpec, ADSpec, and so on), but the traffic data in this case would apply to a traffic trunk rather than an individual session between individual users. With the addition of the new objects, the PATH message now includes much of the same information that is included in a CR-LDP Label Request message and can be used for the same purpose.

The RESV message is enhanced to include the Label object and optionally the Record_Route object. These new objects are attached to a filter spec in an RESV message. Recall that the filter spec in RSVP indicates a specific traffic flow. If a Filter spec is applied to a traffic trunk between two points, then the filter spec is equivalent to an identification of an FEC. If we use RSVP to assign a label to that filter spec, then we have effectively assigned a label to the FEC.

Based on the foregoing, we can see that LSP establishment using RSVP-TE would work as follows. The ingress LSR would send an RSVP PATH message with traffic parameters for the traffic trunk in question with a Label_Request object, and potentially one or more of the Explicit_Route, Record_Route, and Session_Attribute objects. That message would propagate through the network to the point of egress. At that point, an RSVP RESV message would be returned, including a Label object and optionally a Record_Route object. The net effect of this process is the same as that of CR-LDP. An LSP is established through the network according to specific traffic requirements and potentially along a specific, predefined path.

Combining QoS Solutions

The QoS solutions described in this chapter take different approaches and each has its advantages and disadvantages. Although RSVP at the session level is very powerful, it has the drawback that it requires a path state to be maintained in each router, something that can be difficult and costly in a large network. DiffServ is simpler, but it is more of a prioritization technique than a resource-guarantee mechanism. MPLS offers great promise as an overall solution, but requires significant changes to all routers that want to use it. Therefore, each of the solutions has its drawbacks and no single solution is likely to be a silver bullet.

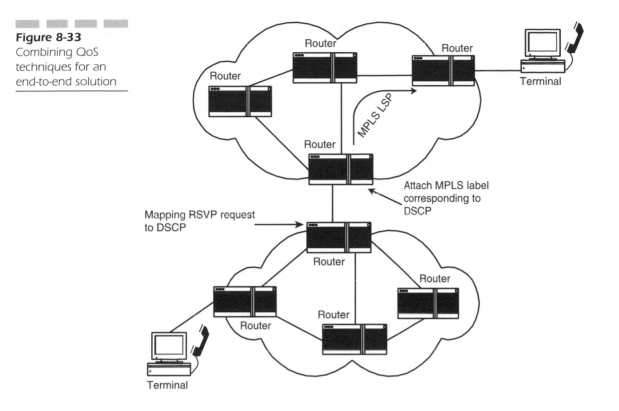

Figure 8-33
Combining QoS
techniques for an
end-to-end solution

Combining the different solutions in smart ways, however, can allow them to be used in different parts of the network where they can be put to best use. For example, if RSVP is used in one domain and DiffServ is used in another, then it is possible to map an RSVP service request (such as guaranteed service) to an appropriate DiffServ PHB (such as EF). Similarly, it might be possible to map a particular DiffServ behavior aggregate to an MPLS FEC at an MPLS edge router. Figure 8-33 gives one example of how different solutions might be combined.

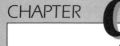

Designing a Voice over IP Network

In general, the design of any network involves striking a balance between three requirements: meeting or exceeding the capacity needed to handle the projected demand, minimizing the capital and operational cost of the network, and ensuring high network reliability and availability. In short, we can refer to these three issues as cost, capacity, and quality. Meeting one or more of the requirements often means making sacrifices elsewhere, such that it is impossible to divorce one network design consideration from any of the others. For example, a lower cost might well mean a lower network capacity or a lower network quality. Thus, we will never get a network that is remarkably cheap to implement and operate while still offering high capacity and high quality. Instead, we must aim to establish some happy medium where we satisfy at least the most important criteria to an acceptable degree. Of course, the question is, "What is an acceptable degree?"

If we want to design a carrier-grade network, one design criteria to include is an overall availability of at least 99.999 percent (five-nines), which corresponds to a down time of no more than about 5 minutes per year. Other design criteria will be based upon projected usage of the network and business considerations. For example, network capacity will be driven by expected traffic demand. Network cost will usually be the result of a negotiation between the carrier's engineering organization and the financial/commercial department. At the end of the day, the cost of the network must be sufficiently low to ensure that it costs less to build and operate than the revenue it will generate.

In general, designing a network has no "one-size-fits-all" solution. A number of general rules and an overall approach, however, can be applied in most network design efforts. The overall approach includes the following aspects:

- Understanding the expected traffic demand, such as where traffic will come from and go to, and what typical per-subscriber usage is expected
- Establishing network design criteria, such as build-ahead, voice-coding schemes, network technology (such as softswitch versus H.323), whether to use silence suppression, whether to include network-level redundancy, and so on
- Vendor and product selection, including the development of a *Request for Information* (RFI) or *Request for Proposal* (RFP) and the comparison of vendor responses
- Network topology, connectivity, and bandwidth requirements
- Physical connectivity

Often the network design process is iterative. For example, one might apply certain design criteria to develop a draft design that meets capacity requirements only to find that such a design would be too expensive. In such a case, one might need to revise the design assumptions in order to design a more cost-effective network.

In order to illustrate the various steps in the process, we focus much of this chapter on a fictitious *Voice over IP* (VoIP) network. We will design this network from scratch, with an explanation of each step along the way. For some of the steps, we will perform all the detailed calculations. The fictitious network will be required to provide long-distance service in North America. Our network will provide service to subscribers in 12 cities, City 1 through City 12, as shown in Figure 9-1.

Design Criteria

In order to develop a network design, we must establish certain design criteria. Among other things, we need to specify whether we will have a build-ahead or some capacity buffer. For example, it is pointless to design a network that will only support the immediate demand. Taking such an

Figure 9-1
The cities for the VoIP example network design

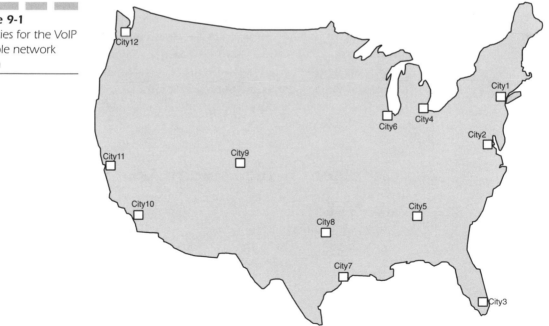

approach, however, would mean that we would constantly be redesigning the network as traffic increases.

Build-Ahead or Capacity Buffer

Rather than design for the immediate expected load, we could include a certain build-ahead in the design process. For example, we could include a one-year build-ahead, which means that the initial network implementation will be designed for the traffic load expected one year after launch. (We establish example traffic projections later in this chapter.)

A build-ahead serves two purposes. First, it avoids the necessity for constant redesigning as traffic demand increases. Second, it provides a buffer in case traffic demand increases faster than expected. Given that our network design will be based on a number of input assumptions, and given that assumptions can be incorrect, such a buffer is wise.

Although a build-ahead is a wise approach for a completely new network, a simple design buffer might be better if we do not expect sudden, drastic traffic growth. Imagine, for example, a situation where we have a traditional circuit-switched network and we want to replace that network with a new VoIP network. In such a case, we should have a good understanding of the traffic demand and it is possible that forecasted growth might not be very great. Nonetheless, it is wise to ensure that the network has some extra capacity to cater for sudden demand changes (as might occur if the marketing department decides to suddenly offer some new pricing deal). One way or another, a network should be sized such that the expected traffic demand is less than some percentage of network capacity (say 70 or 80 percent). For example purposes, in our example design, we will assume a 12-month build ahead.

Fundamental Technology Assumptions

As described in earlier chapters, VoIP has several technology solutions. For example, we could choose to deploy an H.323 architecture or a softswitch-based architecture. Given that the softswitch architecture is becoming the most accepted VoIP approach, let's apply that architecture to our network design.

The technology choice does not end there, however. We must also decide on the type of signaling solution to deploy. Should we use the *Media Gateway Control Protocol* (MGCP) or MEGACO? MEGACO is probably the best choice as it is the media gateway control protocol of the future. Should we use external *signaling gateways* (SGs) with Sigtran between *media gateway controllers* (MGCs) and SFs or should we deploy MGCs that support *Signaling System 7* (SS7) directly? In many cases, we will find that we need to make a trade-off between what we would ideally like and what vendors can provide at a good price.

For example, we might want MEGACO, but the overall best-value vendor solution might support only MGCP in the near-term. Similarly, we might want to deploy SGs, but the overall best solution on the market might include SS7 links directly on the MGC. In such cases, the agreement with the vendor should include a commitment by the vendor to implement the desired functionality within a specific time period.

Later in this chapter we delve further into the issues surrounding product and vendor selection. In the meantime, let's assume that our solution will use MEGACO between MGCs and *media gateways* (MGs) and the *Session Initiation Protocol* (SIP) between MGCs. Let's also assume that we use separate signaling gateways, with *SS7 MTP3-User Adaptation Layer* (M3UA) over *Stream Control Transmission Protocol* (SCTP) between the MGCs and the SGs.

Network-Level Redundancy

If we are to design a carrier-grade network, then all the products in the network must include significant levels of redundancy. From a network design perspective, however, we must decide whether to include redundancy at the network level.

We saw in Chapter 6, "Media Gateway Control and the Softswitch Architecture," that an MG can be handed off to a secondary MGC if a primary MGC fails. If the secondary MGC normally carries traffic of its own, we should make sure that the secondary MGC has the capacity to handle some (or all) of the load of the new MG in addition to existing traffic. In other words, we must design the network so that excess capacity is available in the secondary MGC. We must also decide how much excess capacity we will include. For example, we might decide that the secondary MGC should have the capacity to handle its normal expected load (including build-ahead), plus only 50 percent of the load that would be carried by the failed MGC.

Implementing network-level redundancy can be done in several ways. If, for example, we have 3 MGCs (MGC-A, MGC-B, and MGC-C), then each could back up another. Thus, MGC-B could be a backup for some or all of MGC-A's traffic, MGC-C could be a backup for some or all of MGC-B's traffic, and MGC-A could be a backup for some or all of MGC-C's traffic. Alternatively, we might choose to have a single redundant MGC for the whole network. That MGC would begin handling traffic in the event of failure of any MGC on the network.

Again, the choice of redundancy technique depends on the capabilities of the vendor. For example, although a single MGC acting as a backup for all others might be the most cost-effective solution, it might require that the backup MGC be loaded with all the port/termination configuration data related to all the MGs on the network. The on-board availability of such data might be required in order for the backup MGC to quickly assume control in the event of a failure.

The number of supported ports might be a capacity-limiting factor on the MGC, however, which might mean that a single backup MGC for the whole network would not be possible. You might think that we should not have such a problem. After all, MEGACO enables an MGC to query an MG as to the capabilities and current status of its terminations. Thus, the back-up MGC could automatically load data for the MG(s) that it must support without having data in advance for all terminations on all MGs. MEGACO does support such capabilities, but whether a given MGC supports that aspect of MEGACO is another matter and could be vendor dependent. In our example, we will assume that we can implement a single backup MGC to support the failure of any active MGC.

Voice Coder/Decoder (Codec) Selection Issues

We must make a number of decisions regarding voice coding. First, we must choose the actual *coder/decoder* (codec) to use. Second, we must select the packetization interval. Third, we must decide whether or not to use silence suppression. These decisions can have a profound impact on the *quality of service* (QoS) offered to our subscribers. As with many of our decisions, however, a trade-off takes place between the quality we offer, the cost of the network, and the capabilities of our vendors.

Clearly, G.711 is the simplest (and highest-quality) codec we can deploy and should be the first choice if we want to ensure superior voice quality. G.711 uses higher bandwidth than other codecs, however. Of course, we can choose any number of coding schemes (G.723.1, G.729, G.726, and so on), many of which provide good quality and that enable lower-bandwidth consumption than G.711. The network architecture should at least perform a laboratory test of each of the codec choices under ideal and degraded circumstances before making any firm decision. Those tests should include tests with silence suppression turned on and turned off.

The packetization interval is also important. A longer packetization interval means a net lower overhead for the *Real-Time Transport Protocol* (RTP), *User Datagram Protocol* (UDP), and *Internet Protocol* (IP) headers, which means lower bandwidth consumption overall. On the other hand, a longer packetization interval means a longer delay, which impacts voice quality. Typically, we will find that a packetization interval of about 20 to 40 milliseconds provides a good balance between delay and bandwidth consumption. The exact packetization delay will depend on the chosen codec. For example, G.729 operates on 10-millisecond speech samples, so we could not choose a 25-millisecond packetization interval with G.729. Similarly, G.723.1 operates on 30-millisecond speech samples, which means that we could not choose a 20-millisecond packetization delay with that codec. For our example network design, we will assume that we use G.729B (G.729 with silence suppression) and with a packetization interval of 20 milliseconds.

Voice Activity Factor   Since we have decided that we will use silence suppression, our VoIP MGs will transmit little or no IP voice traffic during periods of silence. Consequently, we can reduce the amount of bandwidth that we will require on the network. Therefore, we must decide how much of a typical conversation is actual voice rather than silence. Clearly, if a typical voice conversation is 80 percent voice and 20 percent silence, greater bandwidth will be required compared to 60 percent voice and 40 percent silence.

In fact, we consider voice activity on a one-way basis only. This is because transmission facilities are generally symmetrical. For example, if we deploy a DS3 (approximately 45 Mbps), then we have about 45 Mbps of bandwidth in each direction. If we typically have 40 percent voice activity per person in a call, then there is voice activity for about 80 percent of the time and silence for 20 percent of the time. In each direction, however, we are consuming approximately 40 percent of the bandwidth that would be consumed in a call where each party is speaking 100 percent of the time. In other words, we are consuming 40 percent in the direction from A to B and 40 percent in the direction from B to A, which amounts to 40 percent overall.

Statistics show that in a typical conversation each party speaks for approximately 30 to 40 percent of the time, with the remainder of the time consumed by silence. For the purposes of our example network design, we will assume a voice activity factor of 40 percent.

Blocking Probability

The Erlang is the standard measure of traffic on a circuit-switched network. Since our VoIP network will have interfaces to external circuit-switched networks (such as the *Public Switched Telephone Network* [PSTN]), we will need to use Erlangs to dimension the interfaces to those networks.

One Erlang corresponds to a channel being occupied for one hour. Alternatively, we can think of an Erlang as representing one hour of conversation. This one hour could be a single conversation that lasts 1 hour, 2 conversations that last 30 minutes each, 5 conversations that last 12 minutes each, and so on. At first glance, one would think that a single channel can carry one Erlang. That is not the case. Imagine 5 callers that each make a 12-minute call. In order for those five calls to all be carried on the same channel in a given one-hour period, then one call would need to begin exactly when another call ends. Different users will not be synchronized in such a manner. Therefore, we will likely find that one of our five callers will attempt to call when a different caller is using the channel. That call attempt will fail.

Depending on the number of available channels and the amount of offered traffic, there is a statistical probability that a channel will be available when a user wants to make a call. If we want to ensure that there will always be a channel available to carry a call, we would need an infinite number of channels, which is obviously not feasible. Instead, we want a high probability that a channel will be available. Alternatively, we want a low probability that a call will be blocked due to a lack of available channels. This probability is known as *blocking probability*.

For interfaces towards the PSTN, we must decide which blocking probability we want. The lower the blocking probability, the greater the service quality and the greater the number of required channels (and hence, greater cost).

Appendix A, "Table of Erlang B," provides a table of Erlang values for combinations of channel counts and blocking probabilities. For our network design, we will assume a blocking probability of 0.1 percent on interfaces between the VoIP network and external circuit-switched networks.

QoS Protocol Considerations and Layer 2 Protocol Choices

As described in Chapter 8, "Quality of Service," we can choose a number of protocols to help ensure high QoS on the network. The simplest approach is to use *Differentiated Service* (DiffServ), which is a form of traffic prioritization. The *Resource Reservation Protocol* (RSVP) comes closest to circuit emulation, but can have difficulty scaling if resources are reserved on a per-session basis. MPLS can provide the best of both worlds by enabling us to reserve resources for traffic trunks. For the purposes of our example network, we will assume the use of MPLS to help ensure high QoS.

This assumption does not necessarily mean that we will totally disregard DiffServ or RSVP. For example, we might choose to employ RSVP-*traffic engineering* (RSUP-TE) for *Multiprotocol Label Switching* (MPLS) label distribution and traffic engineering purposes. Moreover, we can choose to mark voice packets from an MG with a specific *DiffServ Code Point* (DSCP) value, which a *label edge router* (LER) will map to a particular *forwarding equivalence class* (FEC). Alternatively, we might want the MG to label the packet directly.

The ultimate choice will depend on the capabilities of the MG we choose and the routers we choose. It will likely be more effective for the MG to mark packets with a particular DSCP and let a separate LER take care of label assignment. Since we will probably find that one or more other nodes (such as an MGC, SG, or *element management system* [EMS]) will be unable to apply labels directly, we will need a separate LER to perform that function. Given that such an LER exists, we can allow it to function at the edge of the MPLS network for all nodes in a location, which means that we do not need to restrict an MG selection to those products that natively support MPLS.

Clearly, we will need to design a wide area IP network to support traffic to and from the various cities. For that *wide area network* (WAN), we must select a Layer 2 protocol (the layer below IP). We have several choices, such as Frame Relay, *Asynchronous Transfer Mode* (ATM), or the *Point-to-Point Protocol* (PPP). Considering our choice of MPLS for QoS purposes, we do not need ATM for QoS purposes. Therefore, let's assume that we use PPP at Layer 2 in our WAN.

Product and Vendor Selection

Having established our overall network design criteria, and understanding our traffic demand, we must choose vendors and products that provide the functionality, capacity, and quality we desire at a cost-effective price. Traffic forecasts are described later in this chapter. For now, let's assume that we have already determined our estimated traffic demand and that we will use the expected demand as an input to our product and vendor selection process.

Generally, the selection process will involve the preparation of a *Request for Information* (RFI) or a *Request for Proposal* (RFP). Such documents will address all the functionality and capacity required of the network elements, management, and billing systems. In addition, the RFI/RFP should solicit general information regarding the vendor (such as the company size, revenues, revenues from VoIP solutions, and the existing installed base of the vendor's VoIP solution). The last thing we want to do is take a risk on a vendor that is not financially sound or not sufficiently experienced. Finally, the RFI/RFP should request information regarding interworking that has been performed between the vendor's equipment and that of other suppliers. In general, we want a solution that is standards compliant and that can interwork with products from other vendors.

Generic VoIP Product Requirements

In addition to the requirements that are specific to the network in question (such as support for a specific SS7 or *multifrequency* [MF] signaling variant or for specific codecs), a number of requirements are critical to the performance of any carrier-grade network.

Node-Level Redundancy Every node in the network should include redundancy within itself. Typically, we will require N + 1 redundancy for all internal components, management components, media-handling components, power supplies, fans, and disk drives. For example, if an MG requires one active management card to maintain the health of the platform, then we should have at least two such cards, with one as a standby in case the primary fails. If a gateway can be equipped with, say, up to eight media-handling cards, then seven should normally be active, with one spare in case an active card fails.

Typically, an MG will be based on a backplane or midplane design, as shown in Figure 9-2. In a backplane-based system, it is common for media

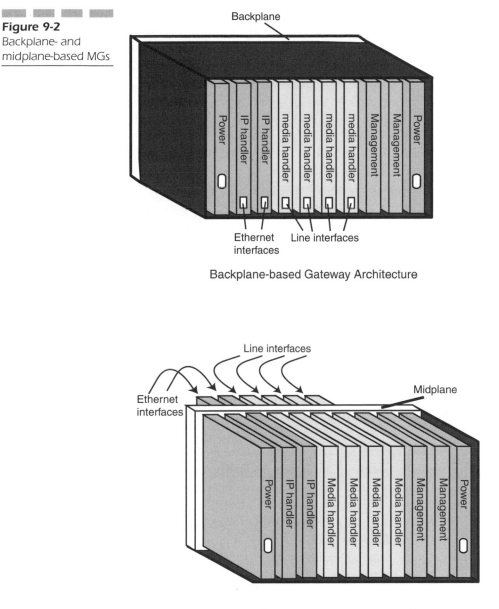

Figure 9-2
Backplane- and
midplane-based MGs

Backplane-based Gateway Architecture

Midplane-based Gateway Architecture

cards to handle both the media manipulation (such as voice coding and echo cancelling) and the physical interface to the outside world. Alternatively, we may find that certain cards are dedicated to line interface functions, while other cards are dedicated to media-handling functions. Moreover, one might need different cards depending on the coding scheme. Such a situation is generally undesirable. If different cards are needed for different coding schemes, and if we want to implement a variety of coding schemes, then we will need to populate multiple different cards, thereby using up cards that could be used for interfacing with the outside network. The result can be an overall reduction in the call-handling capacity of the network.

A midplane design is often preferable to a backplane design. With a midplane design, the handling of information that is sent on an interface is separated from the handling of the interface itself. In a midplane architecture, the physical interfaces to the outside world typically enter at the back of the chassis, while the cards that process the data on those interfaces are at the front of the chassis. In a midplane architecture, we can designate a spare media-handling card at the front of the chassis. Thus, if a media-handling card fails, the spare can take over and be linked to the interface card that the failed media-handling card was supporting. This arrangement allows for N + 1 redundancy of media-handling cards, something that is often impossible with a backplane design.

Of course, redundancy does not apply only to an MG; it also applies to other nodes such as MGCs, SGs, and so on. For each of these nodes, we require redundant interfaces to other nodes (such as redundant Ethernet interfaces). In the case of an SG, we will have SS7 links to separate *signal transfer points* (STPs).

Node Availability Each node on the network should provide at least 99.999 percent availability. One way to gauge the availability of a node is for the vendor to provide *Mean Time Between Failure* (MTBF) values for each component of a given node as well as for the node as a whole. Often, we will find that MTBF values are in tens of years, which means that we could expect decades to pass before a given component fails. Although this sounds like an impressive duration, do not be overly impressed. The MTBF statistics for a complete node will be much lower than the lowest MTBF value for a single component.

Assume, for example, that a gateway has five cards, with MTBF values of 5 years, 10 years, 10 years, 20 years, and 20 years. Then in a given 100-year period, we can expect a total of 50 failures (20 + 10 + 10 + 5 + 5). Thus, our overall MTBF is two years.

Although hardware failures can certainly occur, we will most likely find that software failures are more common and in some cases can be catastrophic. A software bug that occurs only under some strange call scenario might not be a major problem, particularly if it means that only the call in question fails. There can be, however, problems that affect all nodes at the same time. For example, a software problem that occurs only during the changeover to daylight savings time would strike all nodes simultaneously and could disable a complete network. For these reasons, one should require that a vendor provide a list of known software bugs plus copies of tests (and test results) that have been run on the software load being considered.

Alarms and Statistics   As a general rule, a given node should produce an alarm whenever a fault situation arises or whenever operations and maintenance staff need to be informed of a specific condition. Typically, alarms will fall into one of three categories: minor, major, and critical. There might also be a fourth: observational, which means the occurrence of a condition that is not a fault but that needs to be communicated to operations personnel.

Not only should nodes provide alarms when faults occur, but alarms should also be generated whenever the load on a network element reaches some percentage of available capacity. For example, an MGC might issue an alarm when total call handling reaches 75 percent capacity, with a further alarm when the load reaches 90 percent capacity. The alarm threshold levels should be settable by the network operator.

Every node should be able to produce a great deal of statistics regarding carried traffic, node performance, and so on. For example, an MGC should be able to provide statistics and reports regarding traffic patterns (including when the busy hour occurs, traffic distribution, and *mean holding time* [MHT]). Equally, an MG should provide reports and statistics regarding IP network performance (such as packets lost, average latency, and jitter) and circuit-switched network performance (such as trunk hold times). A large range of statistics and reports is vital for the network operator to fully understand the performance of the network.

Element Management

How we will manage the network elements is a major consideration in vendor selection. Often, for example, element management considerations might force us to choose a single vendor for all network elements rather than mix and match MGCs, SGs, and so on. Although we might be tempted to pick the best in class for each type of network element, such an approach might lead to difficulty, as each type of network might need to be managed separately.

Imagine, for example, that we need to provision a new trunk group on the network. We need to provision data on an MG, an MGC, and potentially on an SG (for example, if we have an M3UA application server associated with the new trunk group). It would be nice to have a single EMS that would be used to provision all the correct data for the various nodes, ensuring the consistency of data between the nodes. If we have to use separate provisioning systems to provision each type of node, we run the risk of making errors and introducing data inconsistencies across network elements.

Typically, the interface between the network elements and the EMS will use the *Simple Network Management Protocol* (SNMP) version 2 or higher. Above the EMS, we might require or already have a *Network Management System* (NMS). This situation is often the case where a VoIP network is being introduced as part of a larger network that includes circuit-switched equipment, signaling equipment, transmission equipment, and so on. The EMS will need to have a northbound interface towards the EMS. That interface should be standardized based on, for example, SNMP or the *Common Object Request Broker Architecture* (CORBA).

Traffic Forecasts

Obviously, we need to design a network that will support the projected traffic demand. Consequently, projecting subscriber usage is a critical first step in the network design process. This projection often involves a certain amount of up-front guesswork, particularly if this is the first network of a given type in a given market. If one is building a network in competition with another provider, then one can forecast subscriber growth based on the competitor's subscriber numbers, which are often publicly available. For example, one might assume that the new network will attract a certain percentage of subscribers of a given type (for example, 10 percent of the residential users and 10 percent of the business users in the first year, greater percentages in the second year, and so on). Typically, a marketing or forecasting group will be responsible for determining the subscriber forecast and providing that information to the network designers.

Let's assume the subscriber counts shown in Table 9-1.

Voice Usage Forecast

Let's assume that we intend to provide service only to residential users (which is done by many of the 1010-XXX long-distance service providers). Let's also assume (for simplicity) that the average user in each city makes

Table 9-1

Subscriber
projections

	Launch + 3 months	Launch + 6 months	Launch + 9 months	Launch + 1 year	Launch + 18 months	Launch + 2 years	Launch + 3 years
City1	15,000	48,000	98,800	159,900	301,200	427,600	627,200
City2	8,300	26,400	54,400	88,000	165,700	235,200	345,000
City3	9,000	28,800	59,300	96,000	180,800	256,600	376,400
City4	7,800	25,000	51,400	83,200	156,700	222,400	326,200
City5	9,800	31,200	64,300	104,000	195,800	278,000	407,700
City6	12,600	40,400	83,000	134,400	253,100	359,200	526,900
City7	11,700	37,500	77,100	124,800	235,000	333,600	489,300
City8	11,300	36,000	74,100	120,000	225,900	320,700	470,400
City9	7,500	24,000	49,400	80,000	150,600	213,800	313,600
City10	13,800	44,200	90,900	147,200	277,200	393,400	577,100
City11	12,000	38,400	79,100	128,000	241,000	342,100	501,800
City12	10,500	33,600	69,200	112,000	210,900	299,400	439,100

about the same number of calls. Thus, we assume that any differences in overall traffic demand will be based on differences in subscriber numbers between cities rather than differences in average per-user demand. Since we typically bill per minute and on a monthly basis, we will commonly find that traffic forecasts are often provided in terms of *minutes of use* (MoU) per subscriber per month.

Let's assume our minutes of use per subscriber are as shown in Table 9-2. These figures apply to originating calls only.

The first job of the network designer is to convert such a forecast into measurements that are more meaningful in network design. In particular, the network designer must establish the traffic demand during the busy hour and the MHT per call.

Example: Average User with 120 MoUs per Month of Domestic Long-distance Traffic Assume, for example, that 60 percent of the traffic occurs during weekdays (that is, 40 percent on weekends) and that there are 21 work days per month. Assume that on a given day, 20 percent of the

Table 9-2

Per-subscriber
minutes of use
projections

	Launch + 3 months	Launch + 6 months	Launch + 9 months	Launch + 1 year	Launch + 18 months	Launch + 2 years	Launch + 3 years
City1	110	115	120	122	125	130	135
City2	110	115	120	122	125	130	135
City3	110	115	120	122	125	130	135
City4	110	115	120	122	125	130	135
City5	110	115	120	122	125	130	135
City6	110	115	120	122	125	130	135
City7	110	115	120	122	125	130	135
City8	110	115	120	122	125	130	135
City9	110	115	120	122	125	130	135
City10	110	115	120	122	125	130	135
City11	110	115	120	122	125	130	135
City12	110	115	120	122	125	130	135

voice traffic occurs during the busy hour. For a domestic long-distance service, much of the traffic is likely to occur in the evening, as opposed to business-generated long distance, which occurs during work hours. Then the average busy-hour usage (in minutes of use) per subscriber is given by

(MoUs per month) × (fraction during work days) × (percentage in busy hour) / (work days per month)

Thus, in our example, we get 120 × 0.6 × 0.20/21 = 0.686 MoU/sub/busy hour.

We convert this number to Erlangs, which is the standard measurement of voice network traffic. An Erlang is equivalent to one hour of usage. Therefore, one Erlang is equivalent to 60 MoUs. Thus, busy-hour Erlangs/sub = 0.686/60 = 0.0114 Erlangs/sub/busy hour.

Next, we need to understand the average call length, known as the MHT. The combination of Erlangs and MHT will enable us to determine the number of calls in the busy hour as well as the total traffic. The quantity we seek is the number of *busy-hour call attempts* (BHCA). Although busy-hour Erlangs is often a driving factor in the design of a network, so is the BHCA. Typically, we will find that Erlangs are the driving factor for those network elements that reside in the bearer path, while BHCA tends to be the critical factor for call-control entities such as MGCs. Let's assume that the average call length is 5 minutes (300 seconds).

BHCA/subscriber is given by

$$BHCA = Erlangs \times 3600/MHT$$

Thus, we have BHCA/sub = 0.0114×3600/300 = 0.137.

In other words, we assume that a subscriber with 120 MoUs per month will make 0.137 calls each busy hour.

If we apply the above busy hour and MHT assumptions (60 percent in the work week, 20 percent in the busy hour, 21 work days in a month, and a 300-second MIIT), we can use simple spreadsheet formulas to convert the assumptions of Tables 9-1 and 9-2 into total busy-hour Erlangs and BHCA requirements for each city. The busy-hour Erlangs per city are shown in Table 9-3 and the BHCA values per city are shown in Table 9-4. As was the case for Table 9-2, the figures in Tables 9-3 and 9-4 apply only to calls originated by our subscribers, not calls made to our subscribers.

Note that the input assumptions of Tables 9-1 and 9-2 should be revisited on a regular basis to make sure they are still valid.

Traffic Distribution Forecast

Knowing how much traffic will originate in each city is one thing. Knowing where that traffic will go is something else. Every call that enters the network must also leave. If a given call enters the network at City 1 and leaves at City 2, then resources will be needed in each city, not just at the point of ingress. Therefore, we need to understand the traffic distribution. In other words, for each city, we need to know where the calls will go.

Table 9-5 provides an example of traffic distribution for each city in our network. We can see that the table includes data for traffic that is destined for a city not served by the network. The fact is that our example network is rather simple, as it involves only 12 cities. If our network were to be deployed, the network operator would need to partner with some other operator to terminate calls outside network footprint. In fact, due to the limited scope of our example, most of the calls handed to the network in a given city would likely be destined for a location not served by the network and

Table 9-3

Busy-hour erlangs (originating) per city

	Launch + 3 months	Launch + 6 months	Launch + 9 months	Launch + 1 year	Launch + 18 months	Launch + 2 years	Launch + 3 years
City1	157.1	525.7	1129.1	1857.9	3585.7	5294.1	8064.0
City2	87.0	289.1	621.7	1022.5	1972.6	2912.0	4435.7
City3	94.3	315.4	677.7	1115.4	2152.4	3177.0	4839.4
City4	81.7	273.8	587.4	966.7	1865.5	2753.5	4194.0
City5	102.7	341.7	734.9	1208.4	2331.0	3441.9	5241.9
City6	132.0	442.5	948.6	1561.6	3013.1	4447.2	6774.4
City7	122.6	410.7	881.1	1450.1	2797.6	4130.3	6291.0
City8	118.4	394.3	846.9	1394.3	2689.3	3970.6	6048.0
City9	78.6	262.9	564.6	929.5	1792.9	2647.0	4032.0
City10	144.6	484.1	1038.9	1710.3	3300.0	4870.7	7419.9
City11	125.7	420.6	904.0	1487.2	2869.0	4235.5	6451.7
City12	110.0	368.0	790.9	1301.3	2510.7	3706.9	5645.6

Table 9-4

Originating BHCA per city

	Launch + 3 months	Launch + 6 months	Launch + 9 months	Launch + 1 year	Launch + 18 months	Launch + 2 years	Launch + 3 years
City1	1,886	6,309	13,550	22,295	43,029	63,529	96,768
City2	1,043	3,470	7,461	12,270	23,671	34,944	53,229
City3	1,131	3,785	8,133	13,385	25,829	38,123	58,073
City4	981	3,286	7,049	11,600	22,386	33,042	50,328
City5	1,232	4,101	8,818	14,501	27,971	41,303	62,902
City6	1,584	5,310	11,383	18,739	36,157	53,367	81,293
City7	1,471	4,929	10,574	17,401	33,571	49,563	75,492
City8	1,421	4,731	10,162	16,731	32,271	47,647	72,576
City9	943	3,154	6,775	11,154	21,514	31,765	48,384
City10	1,735	5,809	12,466	20,524	39,600	58,448	89,038
City11	1,509	5,047	10,848	17,847	34,429	50,826	77,421
City12	1,320	4,416	9,490	15,616	30,129	44,482	67,747

would need to be handed off to a partner network immediately. Therefore, the distribution in Table 9-5 is not realistic, but it will serve the purpose of illustrating the type of information needed.

Node Locations and Bandwidth Requirements

When we have established our expected traffic demand and traffic distribution, developed our design criteria, and selected our vendors and products, the next step is to establish the initial network topology. The topology of the network will specify how many network elements of a given type will be in each location and the bandwidth requirements between those network elements and the outside world. Initially, we will deal with logical connectivity. Later, once we are satisfied with the logical connectivity (such as the bandwidth required between nodes), then we can develop greater detail regarding the actual connections. That greater detail will involve the specifications of physical connections to routers, such as point-to-point versus ring transport, and so on.

MG Locations and PSTN Trunk Dimensioning

Clearly, we will need to place at least 1 MG in each of the 12 cities where we will provide service. Given that our MGs will connect to the PSTN, our first task is to determine the size of the trunk groups to the PSTN. To do so, we use the Erlang values of Table 9-3 and the traffic distribution statistics of Table 9-5. The Erlang values of Table 9-3 indicate the amount of traffic that we will receive from the PSTN in each city. From the traffic distribution percentages of Table 9-5, we can calculate how much traffic will be sent to the PSTN in each city. Once we have the total Erlang counts, we can use standard Erlang tables to determine the number of DS0s needed. First, however, we must understand how many individual trunk groups we will have to the PSTN (and to our partner *long distance* [LD] carrier) in each city.

Erlang calculations are not linear. As the number of channels in a trunk group increases, the overall traffic per channel increases. For example, a trunk group with 200 DS0s can carry more than twice the traffic of a trunk group with 100 DS0s for the same blocking probability. Because of this fact, we cannot simply take the total traffic in and out of a gateway, look up an

Table 9-5

Projected traffic distribution

To→	City1	City2	City3	City4	City5	City6	City7	City8	City9	City10	City11	City12	Other
From → City1	0	10%	8%	7%	9%	12%	8%	7%	5%	11%	8%	7%	8%
City2	12%	0	8%	5%	8%	11%	9%	7%	6%	12%	9%	5%	8%
City3	12%	8%	0	5%	10%	10%	7%	6%	5%	11%	9%	7%	10%
City4	11%	5%	5%	0	9%	15%	9%	8%	6%	10%	7%	6%	9%
City5	11%	8%	9%	6%	0	9%	7%	7%	5%	10%	9%	8%	11%
City6	11%	9%	6%	10%	8%	0	7%	7%	6%	9%	8%	7%	12%
City7	11%	8%	7%	6%	8%	9%	0	13%	8%	9%	7%	5%	9%
City8	10%	8%	7%	6%	8%	10%	13%	0	8%	9%	8%	5%	8%
City9	10%	6%	5%	5%	7%	8%	10%	10%	0	9%	8%	8%	14%
City10	11%	8%	6%	5%	6%	10%	8%	8%	7%	0	13%	9%	9%
City11	10%	7%	6%	5%	6%	10%	8%	7%	7%	14%	0	10%	10%
City12	11%	7%	5%	5%	6%	10%	8%	8%	8%	11%	10%	0	11%
Other	0	0	0	0	0	0	0	0	0	0	0	0	0

Erlang table, and determine the number of channels required. Instead, we must allocate the total traffic across the various trunks groups and perform an Erlang table lookup (or Erlang calculation) for each trunk group individually.

In a real network, we would need to consider the PSTN networks in each city and determine what type of connectivity we need to various PSTN switches. For example, in one city we might choose to simply interface to one or two *Local Exchange Carrier* (LEC) tandem switches. In other cities, we might choose to include some direct end-office trunks, which can result in cheaper interconnection under certain circumstances. For simplicity, however, let's just assume that we connect to three separate LEC tandems in each city and also assume that the traffic is evenly spread across those tandems.

Therefore, in each city we will have three both-way trunk groups to and from separate LEC tandem switches. In addition, we will have a one-way trunk group from our MG(s) to a partner LD carrier. Recall that some of the incoming traffic from the LEC will be destined for a city that our network does not serve. We will need to send that traffic directly to a partner LD carrier (the Other column in Table 9-5). Consequently, we will need a trunk group to that partner network. Since we do not expect the partner network to send traffic to our network, we can assume that the trunk groups to the partner network will be one-way.

To determine the size of the trunks to the LEC and our partner network in each city, we combine the information of Tables 9-3 and 9-5. Table 9-5 tells us how much traffic we will receive from the LEC in a given city. Table 9-5 also tells us how much traffic we will send to the LEC in each city and our partner LD carrier in each city. Table 9-6 provides the result of combining Tables 9-3 and 9-5 and tells how much traffic will be carried on each of the trunk groups to and from each gateway location.

The quantities in Table 9-6 are calculated from Tables 9-3 and 9-5. For each city, Table 9-3 provides a projected traffic volume towards the network. Since we are assuming a 12-month build-ahead, we assume the traffic load specified in the Launch + 1 year column of Table 9-3. Consider City 1. Table 9-3 tells us that we expect to receive 1857.9 Erlangs of traffic from the PSTN. Clearly, the MG(s) in City 1 must handle that traffic. Moreover, we have already assumed that the LEC will have three trunk groups and that the traffic will be spread evenly across those trunks groups. Thus, we receive 619.3 Erlangs on each trunk group from the LEC.

From Table 9-5, we can see that 8 percent of the 1857.9 Erlangs received from the LEC must be immediately passed on a trunk group to our LD partner (the LD carrier that carries traffic to destinations that our network does

Table 9-6

Erlang demand per city and trunk group

| | LEC Trunk Group 1 | | | LEC Trunk Group 2 | | | LEC Trunk Group 3 | | | LD Partner Trunk Group | | |
	From	To	Total	From	To	Total	From	To	Total	From	To	Total
City1	619.3	513.2	1132.5	619.3	513.2	1132.5	619.3	513.2	1132.5	0	148.6	148.6
City2	340.8	392.0	732.8	340.8	392.0	732.8	340.8	392.0	732.8	0	81.8	81.8
City3	371.8	327.9	699.7	371.8	327.9	699.7	371.8	327.9	699.7	0	111.5	111.5
City4	322.2	302.6	624.8	322.2	302.6	624.8	322.2	302.6	624.8	0	87.0	87.0
City5	402.8	378.3	781.1	402.8	378.3	781.1	402.8	378.3	781.1	0	132.9	132.9
City6	520.5	498.3	1018.8	520.5	498.3	1018.8	520.5	498.3	1018.8	0	187.4	187.4
City7	483.4	411.3	894.6	483.4	411.3	894.6	483.4	411.3	894.6	0	130.5	130.5
City8	464.8	388.8	853.5	464.8	388.8	853.5	464.8	388.8	853.5	0	111.5	111.5
City9	309.8	325.9	635.7	309.8	325.9	635.7	309.8	325.9	635.7	0	130.1	130.1
City10	570.1	499.6	1069.7	570.1	499.6	1069.7	570.1	499.6	1069.7	0	153.9	153.9
City11	495.7	427.4	923.2	495.7	427.4	923.2	495.7	427.4	923.2	0	148.7	148.7
City12	433.8	347.5	781.3	433.8	347.5	781.3	433.8	347.5	781.3	0	143.1	143.1

not serve). Thus, we must send 148.6 Erlangs from City 1 to our LD partner. Also from Table 9-5, we can see that 12 percent of the traffic that arrives from the LEC in City 2 is destined for City 1. This corresponds to 122.7 Erlangs, which must be handled by the MG(s) in City 1 and passed to the LEC in City 1. Similarly, Table 9-5 tells us that a certain percentage of traffic from the various other cities will be destined for City 1 and must be handled by the MG(s) in City 1. We simply sum these amounts to determine the total traffic that must be passed from our MG(s) in City 1 to the LEC in City 1. We have assumed that such traffic will be evenly spread across the three LEC trunk groups.

The same calculations are performed for each city served by our network. As for many of the calculations so far, the calculations can easily be performed with a spreadsheet program.

Based on the values calculated in Table 9-6, we use the Erlang table of Appendix A to determine the number of DS0 channels needed for each trunk group, assuming a blocking probability of 0.1 percent. Alternatively, we can use the Visual Basic® code of Appendix B as a module in a spreadsheet so that the Erlang calculations can be performed automatically. From the number of DS0s, we can determine the number of DS1 interfaces required (if our MGs use DS1 interfaces) or the number of DS3 interfaces required (if our MGs uses DS3 interfaces). The results are shown in Table 9-7.

A single DS3 corresponds to 28 DS1s. Table 9-7 assumes that a given DS3 card in an MG can handle DS1s belonging to different trunk groups. This is a normal situation. Not only does Table 9-7 specify how many DS1s are required on each interface, but it also specifies the total number of DS3 cards, including the allocation of an extra DS3 card to provide N + 1 redundancy. Based on the information in Table 9-7, the interfaces to our external networks will be as shown in Figure 9-3.

MGC, SG, and EMS Dimensioning and Placement

As shown in Figure 9-3, we will place an MG in each of the 12 cities in our example network. We must also determine the placement of other network elements such as MGCs, SGs, and our EMS. Let's begin with the MGCs.

Table 9-7

DS0, DS1, and DS3 interfaces and cards required per MG location

	LEC trunk group 1		LEC trunk group 2		LEC trunk group 3		LD partner trunk group		DS1 Total	DS3 Total	DS3 cards (N + 1)
	DS0	DS1	DS0	DS1	DS0	DS1	DS0	DS1			
City1	1208	51	1208	51	1208	51	183	8	161	6	7
City2	798	34	798	34	798	34	109	5	107	4	5
City3	763	32	763	32	763	32	142	6	102	4	5
City4	686	29	686	29	686	29	114	5	92	4	5
City5	847	36	847	36	847	36	166	7	115	5	6
City6	1093	46	1093	46	1093	46	224	10	148	6	7
City7	965	41	965	41	965	41	163	7	130	5	6
City8	922	39	922	39	922	39	142	6	123	5	6
City9	697	30	697	30	697	30	162	7	97	4	5
City10	1145	48	1145	48	1145	48	189	8	152	6	7
City11	993	42	993	42	993	42	183	8	134	5	6
City12	847	36	847	36	847	36	176	8	116	5	6

Figure 9-3
PSTN DS1
connections for the
example VoIP
network

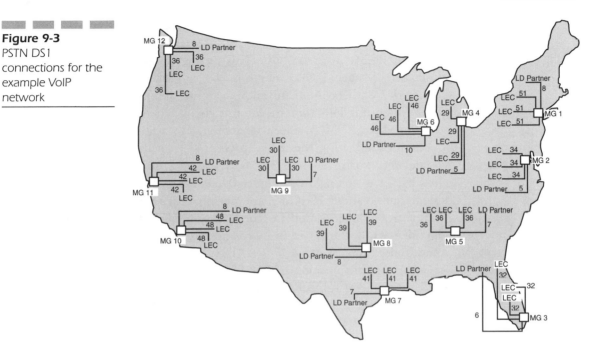

MGC Quantities and Placement   The capacity of an MGC is usually limited by the total BHCA supported. In some cases, we might also find that the total number of MG ports (such as MEGACO terminations) supported is a limiting factor. For our network, let's assume that the BHCA is the limiting factor of the MGC and that the BHCA capacity of an MGC is 150,000 BHCA.

Table 9-4 provides the BHCA projections for our example network. Since we are planning the network with a 12-month build-ahead, we need to consider the BHCA values in the Launch + 1 year column of Table 9-4. At first glance, we might believe that we simply need to divide the BHCA total by the MGC BHCA capacity to determine the number of MGCs required. Unfortunately, the solution might not be that simple. If a call passes between two MGs controlled by the same MGC, then the call consumes capacity on just one MGC. If, however, a call passes between two MGs that are controlled by different MGCs, then that single call consumes capacity on two MGCs. Consequently, the number of MGCs we require depends on which MGs a given MGC is to control and the distribution of traffic to and from those MGs. Consequently, determining the number and location of MGCs can be an iterative process.

We could start by dividing the total BHCA projection by the BHCA capacity of an MGC to develop an initial estimate of the number of MGCs. We then allocate MGs to MGCs. Next we determine the total BHCA to be supported by each MGC, including the impact of calls that cross MGC boundaries, and see if our initial MGC allocation still fits within the MGC BHCA limit. If it does, then we can proceed. If it does not, then we may need to change the MG-MGC allocation, potentially adding MGC capacity.

When establishing the initial MG-MGC allocation, we should arrange the network such that a given MGC will control MGs that are relatively close by. If we look again at Figure 9-1, we can see that it would be preferable to have a single MGC control MGs in Cities 8, 9, 10, 11, and 12, rather than Cities 1, 3, 8, 11, and 12. The former arrangement is simply easier to visualize, while the latter is less intuitive, which could lead to confusion among operations staff.

Given our assumption of 150,000 BHCA capacity per MGC and the BHCA demand of Table 9-4, we make an initial assumption that two MGCs should be able to handle the total demand. Let's therefore assume that MGC1 supports MGs in Cities 1 through 6 and that MGC2 supports MGs in Cities 7 through 12. In terms of placement, let's assume that MGC1 is located in City 2, while MGC 2 is located in City 9.

In order to make sure that this allocation will work correctly, we need to revisit the BHCA demand of Table 9-3 and the traffic distribution of Table 9-5 to make sure that the total BHCA per MGC is less than the MGC BHCA capacity limit. An analysis of Tables 9-4 and 9-5 provides the following information:

- BHCA for calls within footprint of MGC1 = 50,369.7 (a)
- BHCA for calls within footprint of MGC2 = 53,810.5 (b)
- BHCA for calls from MGC1 to MGC2 = 42,420.0 (c)
- BHCA for calls from MGC2 to MGC1 = 45,462.6 (d)
- BHCA load for MGC1 is (a) + (c) + (d) = 138,252.4 BHCA
- BHCA load for MGC2 is (b) + (c) + (d) = 141,693.2 BHCA

Since both of these values are below the BHCA capacity limit of a single MGC, our initial MGC allocation is feasible. We have also assumed that we will have a backup MGC. Let's call it MGC3 and place it in City 8.

SG Quantities, Dimensioning, and Placement   Like MGCs, SGs will typically be limited by BHCA capacity. In addition, we might find that the number of SS7 links supported also limits the capacity of an SG. Let's assume that our SGs will be BHCA limited, with a BHCA capacity of 500,000 BHCA (not an unreasonable number for a carrier-grade machine).

From Table 9-4, we can see that a single SG would have the capacity to support the whole network. We would never allow a single point of failure on our network, however. Therefore, we will deploy two SGs, each of which will be connected to both MGCs. Given that we have placed our two primary MGCs at cities 2 and 9, it is reasonable to place our SGs at the same locations.

From the foregoing discussions, our MG-MGC allocation and SG-MGC allocation are as shown in Figure 9-4. Figure 9-4 also assumes that we have chosen to place our EMS in City 8.

Calculating VoIP Bandwidth Requirements

So far, we have determined the connectivity between our MGs and external networks. We have also determined the types and numbers of interface cards required in each of our MGs. In this section we will calculate the bandwidth required between MGs for VoIP traffic. Later, we will calculate the

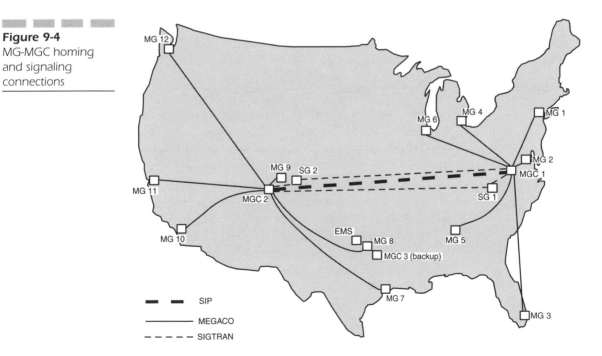

Figure 9-4
MG-MGC homing and signaling connections

additional network bandwidth needed for IP signaling (such as MEGACO) and *Operation, Administration & Maintenance* (OA&M) purposes.

The bandwidth required for a single call depends on the following factors:

- Voice-coding scheme
- Packetization interval
- The use of silence suppression
- Probability of excessive packet collision if silence suppression is used

Impact of Voice-Coding Schemes   Different voice-coding schemes generate different numbers of bits for a certain duration of speech. For example, G.711 operates at 64 Kbps, while G.729 operates at 8 Kbps. For a 20-millisecond sample, G.711 generates 1,280 bits (160 octets), while G.729 generates 160 bits (20 octets).

IP, UDP, and RTP Overhead   For every voice packet that we send, we have a good deal of overhead that is sent in addition to the coded voice itself. Recall that voice is carried using RTP over UDP over IP. Each of these protocols has a header that needs to be transported in addition to the voice payload. The IPv4 header is at least 20 octets long (assuming that no optional fields are included). The UDP header is 8 octets long, and the RTP header is at least 12 octets long. Together, IP, UDP, and RTP add 40 octets. Therefore, for a 20-millisecond sample of G.729-coded voice, we have a total of 60 octets. If we use a 10-millisecond sample, then we have a total of 50 octets. Clearly, the smaller the sample size, the greater the net overhead.

Silence Suppression   Silence suppression is a valuable tool for reducing the total bandwidth demand. If we assume a voice activity factor of 40 percent, then we can save approximately 60 percent bandwidth compared to a network that does not use silence suppression. As described in Chapter 3, "Speech-Coding Techniques," however, periods of silence do not mean a complete cessation in the transmission of voice-related packets. During periods of silence, the voice coder will periodically generate periodic *silence descriptor* (SID) frames. A SID frame is quite small compared to a speech frame, however. Moreover, significant time gaps typically take place between SID frames during periods of silence. Consequently, the bandwidth occupied by SID frames is small and can be disregarded.

Peaks in the Number of Simultaneous Speakers   Unfortunately, a voice activity factor of 40 percent does not mean that we can simply calculate the bandwidth needed without silence suppression and then multiply by 40 percent to determine the actual required bandwidth. Although a voice activity factor of 40 percent does imply that an average user is speaking for only 40 percent of the time, there might be brief periods where many more people are speaking. If we design the network based only on an average of 40 percent voice activity, then we risk packet delay or loss if brief periods occur when more than 40 percent of users are speaking at the same time. Therefore, we need to include extra bandwidth to accommodate such situations. The question is how much extra bandwidth is needed. The answer is given by a binomial distribution function.

Consider n speakers. If voice activity is 40 percent, then the probability of an individual user speaking at a given instant is 40 percent. If we want to determine the probability that exactly x subscribers are speaking at a given time, then the probability is given by

$$Pa(x) = \left\{ \begin{array}{c} n \\ x \end{array} \right\} p^x (1 - p)^{n-x}, \text{ where p} = 0.4 \text{ in our example.}$$

The probability that there are no more than x speakers at a time is given by

$$Pb(x) = \sum_{y=0}^{x} Pa(y)$$

We need to determine the value x, such that the probability is very high that no more than x speakers are active at a given time. Let's assume that we want a probability of 0.1 percent or less that packets will be lost or delayed as a result of too many speakers talking at one time. This is equivalent to saying that we need to find the value of x such that $Pb(x) = 0.999$ or greater. From an algorithmic perspective, we start at $x = 0$ and determine $Pa(0) + Pa(1) + Pa(2) + \ldots$, until the running total exceeds 0.999. The value of x at that point is the value we seek. Fortunately, most spreadsheet programs already have built-in formulas that perform these calculations.

Let's assume a voice activity factor of 40 percent. If, for example, we have up to 1,000 simultaneous calls and we want to make sure that no more than 0.1 percent of the packets are dropped due to too many simultaneous speakers, then the value of x is 448. In other words, we need to dimension the network for 448 simultaneous speakers in a given direction.

Number of Simultaneous Calls for Our Example Network From Table 9-7, we know the number of DS1s that we will deploy to external networks in each city. We can use this information to determine the maximum number of simultaneous calls in and out of the IP network at each MG. Recall, however, that the DS1s to our LD partner network represent traffic that arrives at an MG from the LEC and is routed directly to the LD partner. Therefore, when considering the bandwidth to the IP network at an MG, we should take the total number of DS1s to and from the LEC and subtract the total number of DS1s to our partner network. The resulting number of DS1 enables us to determine the maximum number of simultaneous calls to and from the IP network at each gateway.

As described previously, we cannot simply multiply the maximum number of simultaneous calls by the voice activity factor (40 percent) and use the result as a basis for calculating our VoIP bandwidth demand. We should use a binomial probability function to accommodate for a situation where more than 40 percent of the users are speaking at a given instant. The results of these calculations are shown in Table 9-8. The column "Max. calls for VoIP bandwidth planning (x)" is calculated based on a voice activity factor of 40 percent and a probability of 99.9 percent that no more than x per-

Table 9-8

DS1 demand to/from IP network

	Total DS1s to/from LEC	DS1s to LD partner	Net DS1s to/from IP network	Max. simultaneous calls to/from IP network	Max. calls for VoIP bandwidth planning (x)
City1	153	8	145	3,480	1,482
City2	102	5	97	2,328	1,005
City3	96	6	90	2,160	935
City4	87	5	82	1,968	855
City5	108	7	101	2,424	1,045
City6	138	10	128	3,072	1,313
City7	123	7	116	2,784	1,194
City8	117	6	111	2,664	1,144
City9	90	7	83	1,992	865
City10	144	8	136	3,264	1,393
City11	126	8	118	2,832	1,214
City12	108	8	100	2,400	1,035

sons will be speaking in a given direction at a given instant. When we speak about direction, we mean either the direction from the MG towards the IP network or from the IP network towards the MG. We can see that the values of x are greater than the voice activity factor times the maximum number of simultaneous calls.

As mentioned previously, popular spreadsheet programs include binomial probability functions for the types of calculations required to generate the information in Table 9-8. One drawback of those functions, however, is the fact that binomial calculations include a good deal of factorials, which can lead to extremely high or extremely low interim results prior to the final result. If the input data values are large, the spreadsheet calculations can generate a floating point overflow as the algorithm is being executed. If you detect such a situation, you should be aware that a normal distribution closely approximates a binomial distribution for large values. Therefore, if you find that a spreadsheet binomial function fails due to a large input value, you can use a normal distribution function instead and still achieve a very accurate result.

VoIP Bandwidth for Our Example Network The per-call bandwidth for a VoIP network is calculated as follows:

Voice packet size + 40 octets (for IP, UDP, and RTP) +
WAN layer 2 overhead + MPLS overhead (if applicable)

In our case, the voice packet size is based on a 20-millisecond packet duration with G.729 coding and corresponds to 20 octets. For our WAN network, we will use PPP, which has a two-octet header size. In addition, we have stated that we intend to use MPLS within our WAN for QoS purposes. Since we will use PPP at layer 2, we will need to use a generic MPLS label structure, with four octets per label. Assuming just a single MPLS label is in the stack, the use of MPLS and PPP will add a total of six octets to our packet size. Therefore, our packet size is 66 octets. Since each packet corresponds to 20 milliseconds of speech, our total bandwidth per call is 3,300 KBps or 26.4 Kbps. We can clearly see that this is far greater than the native 8 Kbps of G.729, with the extra bandwidth caused by the various overheads we need to accommodate.

Recall from Chapter 2, "Transporting Voice by Using IP," that we use *RTP Control Protocol* (RTCP) in conjunction with RTP for quality feedback purposes. Recall also that RTCP bandwidth should be limited to about 5 percent of the actual VoIP bandwidth. Therefore, we need to add 5 percent to the VoIP bandwidth just calculated. Consequently, the IP network bandwidth per call is (26.4 Kbps) $\times$ 1.05 = 27.72 Kbps. We multiply this rate by the value x in Table 9-8 to determine the bandwidth required between gateways in the IP network. We then use the traffic distribution of Table 9-5 to determine the amount of bandwidth required between pairs of MGs. As for most of the calculations described in this chapter, simple spreadsheet formulas can be used to make the task relatively easy. The resulting bandwidth requirement between MGs is shown in Table 9-9 in *megabits per second* (Mbps).

Signaling and OA&M Bandwidth   Not only will traffic take place between MGs, but traffic will exist between MGs and their controlling MGCs. As a rule of thumb, we allocate the same bandwidth for MG-MGC signaling per call as we allocate for RTCP traffic. Unlike RTCP traffic, however, we need to allocate our MG-MGC bandwidth for all calls, not just those that happen to have active speech at a given instant.

Next, we allocate bandwidth for MGC-SG signaling. As another rule of thumb, we will allocate the same bandwidth for MGC-SG signaling as we allocate for MG-MGC signaling.

In addition to calculating bandwidth for signaling between SGs and MGCs, we must also calculate the number of SS7 links required between our SGs and the STPs of the networks to which we connect. First, we must remember that our SGs will appear to an external network as an STP pair. For connection to an external network, each of our two SGs will connect to two STPs of the external network. Such an approach will create the standard SS7 quad arrangement described in Chapter 7, "VoIP and SS7" (see Figure 7-4). Second, we will design our SS7 links to run at 40 percent load under normal conditions. By applying such a design rule, we create an environment where we can lose an SS7 link, and a different link can handle the traffic while running at no more than an 80 percent load. Moreover, with a quad arrangement, we can lose a complete STP and still send traffic through the alternative STP of the destination network without overload.

An SS7 link that carries *ISDN User Part* (ISUP) traffic can handle about 30 calls per second. This equates to over 100,000 BHCA. Knowing this number enables us to determine how many SS7 links we will need from our SG pair to each pair of STPs that we connect to.

Note that we do not necessarily need to connect to separate STPs or even have separate SS7 links for each trunk group on our network. We need to have separate SS7 links and connect to separate STPs only for traffic to different networks. For example, the LEC in City 1 and the LEC in City 2 might happen to be the same network. In such a case, we might connect to a single STP pair with SS7 links that carry traffic related to both City 1 and

Table 9-9

VoIP bandwidth required between MGs (Mbps)

	City1	City2	City3	City4	City5	City6	City7	City8	City9	City10	City11
City1	—	—	—	—	—	—	—	—	—	—	—
City2	3.96	—	—	—	—	—	—	—	—	—	—
City3	3.62	2.24	—	—	—	—	—	—	—	—	—
City4	3.05	1.31	1.37	—	—	—	—	—	—	—	—
City5	3.86	2.34	2.88	2.10	—	—	—	—	—	—	—
City6	5.00	3.26	2.64	3.90	3.01	—	—	—	—	—	—
City7	3.95	2.71	2.33	2.28	2.61	3.08	—	—	—	—	—
City8	3.45	2.39	2.14	2.11	2.55	3.19	4.79	—	—	—	—
City9	2.43	1.56	1.35	1.40	1.67	2.19	2.76	2.70	—	—	—
City10	4.96	3.34	2.90	2.36	2.88	3.94	3.42	3.35	2.64	—	—
City11	3.79	2.55	2.45	1.85	2.57	3.49	2.84	2.78	2.34	5.48	—
City12	3.51	1.86	1.87	1.62	2.28	3.08	2.30	2.26	2.36	3.81	3.61

City 2. In a real deployment, the exact SS7 connectivity will be agreed between the network operators as part of an interconnect agreement.

We must also allocate bandwidth for MGC-MGC signaling. We again apply the same rule of thumb. For each call that will pass between MGs controlled by different MGCs, we will assume that the MGC-MGC signaling bandwidth is equal to the per-call bandwidth between a single MGC and an MG.

Finally, we need to allocate some bandwidth for OA&M traffic between each network element and the EMS. The amount of bandwidth allocated to such traffic will depend on many factors, including the number and frequency of reports and statistics, the number of and frequency of call data records (for billing purposes), the number of commands issued to network elements, and the number of alarms generated. No general rule exists for determining the bandwidth needed for such functions. For our network, we will assume that each network element will require approximately 128 Kbps of bandwidth for such functions.

Physical Connectivity

The previous sections of this chapter have shown how we calculate the bandwidth requirements between the various network elements and the outside world. We have performed the detailed calculations required to determine the MG-MG bandwidth and we have explained how to calculate other bandwidth requirements. Knowing how much bandwidth will be required between different nodes does not, however, specify how the network will actually be laid out. After all, we are unlikely to build a fully meshed network, because to do so would be wasteful. Therefore, the next step is to determine how we will connect the different cities together to provide the bandwidth we need between various points. Moreover, we should design the connectivity so that each city has an alternative path to every other city. Such a design will ensure that our network does not fail, even if we lose some transport facilities.

If we consider Figure 9-4, we see that City 2 (MGC1 location) forms a natural hubbing location for the eastern part of our network. Similarly, City 9 (MGC2 location) forms a natural hubbing point for the western part of our network. As a first step, we could assume a network layout where all traffic to and from Cities 1 through 6 passes through City 2 and all traffic to and from Cities 7 through 12 passes through City 9. With a connection between City 2 and City 9, we would have a network layout as shown in Figure 9-5.

Unfortunately, such a network design is not sufficiently reliable. For example, a break in the connection between City 2 and City 9 would disable all connections between the east and west sides of the network. We need to have two paths from each location to each other location. Such an approach will mean that the failure of any given transport facility or the complete loss of a location (such as City 2) will not jeopardize a great portion of the network. Figure 9-6 is one way we could provide the network redundancy and diversity we need.

Finally, we must consider the connectivity within a given location. After all, there is no point in having redundancy and diversity across the network if we have single points of failure within a given location. In general, our VoIP network elements will have *Fast Ethernet* (FE) or *Gigabit Ethernet* (Gig-E) interfaces. In order to provide sufficient diversity, the FE or Gig-E interfaces from our network element should be connected to a pair of Ethernet switches. These Ethernet switches should then be connected to a pair of routers, as shown in Figure 9-7.

Figure 9-5
Minimal network connectivity

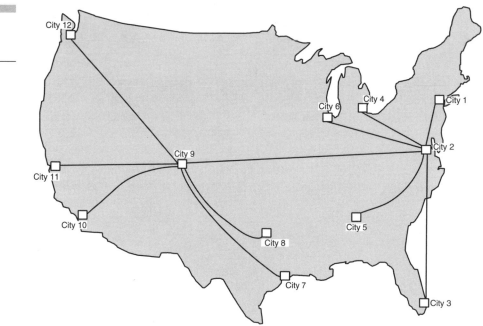

Figure 9-6
Robust network
connectivity

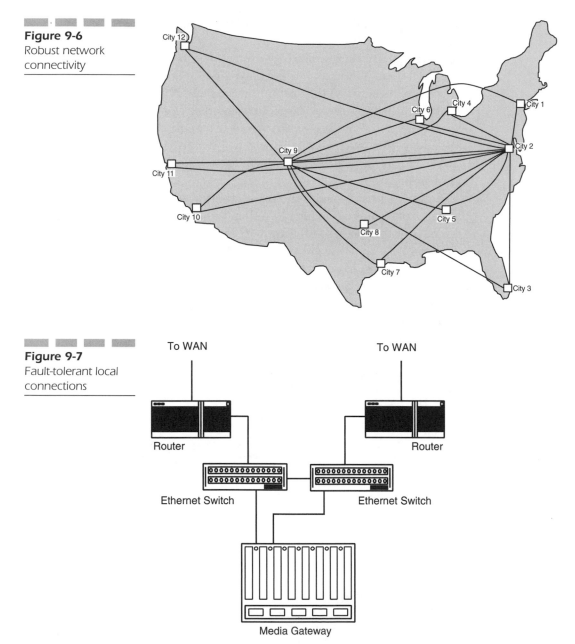

Figure 9-7
Fault-tolerant local
connections

Figure 9-7 shows a fault-tolerant arrangement for conncting an MG to the IP network. A similar arrangement should apply to other network elements such as an MGC, SG or EMS.

APPENDIX A

Table of Erlang B

The following table specifies the number of channels required for a given blocking probability and traffic demand.

In order to determine the required number of channels for a given blocking probability, select the column corresponding to the blocking probability in question. Within that column, find the Erlang value that provides the closest match to the required Erlang value. The leftmost column of the same row provides the required number of channels.

channels	Blocking probability															
	0.01%	0.05%	0.1%	0.2%	0.3%	0.4%	0.5%	0.6%	0.7%	0.8%	0.9%	1.0%	2.0%	3.0%	4.0%	5.0%
1	0.0	0.0	0.0	0.0	0.0	0.0	0.0	0.0	0.0	0.0	0.0	0.0	0.0	0.0	0.0	0.1
2	0.0	0.0	0.0	0.1	0.1	0.1	0.1	0.1	0.1	0.1	0.1	0.2	0.2	0.3	0.3	0.4
3	0.1	0.2	0.2	0.2	0.3	0.3	0.3	0.4	0.4	0.4	0.4	0.5	0.6	0.7	0.8	0.9
4	0.2	0.4	0.4	0.5	0.6	0.7	0.7	0.7	0.8	0.8	0.8	0.9	1.1	1.3	1.4	1.5
5	0.5	0.6	0.8	0.9	1.0	1.1	1.1	1.2	1.2	1.3	1.3	1.4	1.7	1.9	2.1	2.2
6	0.7	1.0	1.1	1.3	1.4	1.5	1.6	1.7	1.8	1.8	1.9	1.9	2.3	2.5	2.8	3.0
7	1.1	1.4	1.6	1.8	1.9	2.1	2.2	2.2	2.3	2.4	2.4	2.5	2.9	3.2	3.5	3.7
8	1.4	1.8	2.1	2.3	2.5	2.6	2.7	2.8	2.9	3.0	3.0	3.1	3.6	4.0	4.3	4.5
9	1.8	2.3	2.6	2.9	3.1	3.2	3.3	3.4	3.5	3.6	3.7	3.8	4.3	4.7	5.1	5.4
10	2.3	2.8	3.1	3.4	3.6	3.8	4.0	4.1	4.2	4.3	4.4	4.5	5.1	5.5	5.9	6.2
11	2.7	3.3	3.7	4.0	4.3	4.5	4.6	4.7	4.9	5.0	5.1	5.2	5.8	6.3	6.7	7.1
12	3.2	3.9	4.2	4.6	4.9	5.1	5.3	5.4	5.6	5.7	5.8	5.9	6.6	7.1	7.6	7.9
13	3.7	4.4	4.8	5.3	5.6	5.8	6.0	6.1	6.3	6.4	6.5	6.6	7.4	8.0	8.4	8.8
14	4.2	5.0	5.4	5.9	6.2	6.5	6.7	6.8	7.0	7.1	7.2	7.4	8.2	8.8	9.3	9.7
15	4.8	5.6	6.1	6.6	6.9	7.2	7.4	7.6	7.7	7.9	8.0	8.1	9.0	9.6	10.2	10.6
16	5.3	6.2	6.7	7.3	7.6	7.9	8.1	8.3	8.5	8.6	8.7	8.9	9.8	10.5	11.1	11.5
17	5.9	6.9	7.4	7.9	8.3	8.6	8.8	9.0	9.2	9.4	9.5	9.7	10.7	11.4	12.0	12.5
18	6.5	7.5	8.0	8.6	9.0	9.3	9.6	9.8	10.0	10.1	10.3	10.4	11.5	12.2	12.9	13.4
19	7.1	8.2	8.7	9.4	9.8	10.1	10.3	10.6	10.7	10.9	11.1	11.2	12.3	13.1	13.8	14.3
20	7.7	8.8	9.4	10.1	10.5	10.8	11.1	11.3	11.5	11.7	11.9	12.0	13.2	14.0	14.7	15.2
21	8.3	9.5	10.1	10.8	11.2	11.6	11.9	12.1	12.3	12.5	12.7	12.8	14.0	14.9	15.6	16.2
22	8.9	10.2	10.8	11.5	12.0	12.3	12.6	12.9	13.1	13.3	13.5	13.7	14.9	15.8	16.5	17.1
23	9.6	10.9	11.5	12.3	12.7	13.1	13.4	13.7	13.9	14.1	14.3	14.5	15.8	16.7	17.4	18.1
24	10.2	11.6	12.2	13.0	13.5	13.9	14.2	14.5	14.7	14.9	15.1	15.3	16.6	17.6	18.4	19.0
25	10.9	12.3	13.0	13.8	14.3	14.7	15.0	15.3	15.5	15.7	15.9	16.1	17.5	18.5	19.3	20.0
26	11.5	13.0	13.7	14.5	15.1	15.5	15.8	16.1	16.3	16.6	16.8	17.0	18.4	19.4	20.2	20.9
27	12.2	13.7	14.4	15.3	15.8	16.3	16.6	16.9	17.2	17.4	17.6	17.8	19.3	20.3	21.2	21.9
28	12.9	14.4	15.2	16.1	16.6	17.1	17.4	17.7	18.0	18.2	18.4	18.6	20.1	21.2	22.1	22.9
29	13.6	15.1	15.9	16.8	17.4	17.9	18.2	18.5	18.8	19.1	19.3	19.5	21.0	22.1	23.0	23.8
30	14.2	15.9	16.7	17.6	18.2	18.7	19.0	19.4	19.6	19.9	20.1	20.3	21.9	23.1	24.0	24.8

channels	0.01%	0.05%	0.1%	0.2%	0.3%	0.4%	0.5%	Blocking probability 0.6%	0.7%	0.8%	0.9%	1.0%	2.0%	3.0%	4.0%	5.0%
31	14.9	16.6	17.4	18.4	19.0	19.5	19.9	20.2	20.5	20.7	21.0	21.2	22.8	24.0	24.9	25.8
32	15.6	17.3	18.2	19.2	19.8	20.3	20.7	21.0	21.3	21.6	21.8	22.0	23.7	24.9	25.9	26.7
33	16.3	18.1	19.0	20.0	20.6	21.1	21.5	21.9	22.2	22.4	22.7	22.9	24.6	25.8	26.8	27.7
34	17.0	18.8	19.7	20.8	21.4	21.9	22.3	22.7	23.0	23.3	23.5	23.8	25.5	26.8	27.8	28.7
35	17.8	19.6	20.5	21.6	22.2	22.7	23.2	23.5	23.8	24.1	24.4	24.6	26.4	27.7	28.8	29.7
36	18.5	20.3	21.3	22.4	23.0	23.6	24.0	24.4	24.7	25.0	25.3	25.5	27.3	28.6	29.7	30.7
37	19.2	21.1	22.1	23.2	23.9	24.4	24.8	25.2	25.6	25.9	26.1	26.4	28.3	29.6	30.7	31.6
38	19.9	21.9	22.9	24.0	24.7	25.2	25.7	26.1	26.4	26.7	27.0	27.3	29.2	30.5	31.6	32.6
39	20.6	22.6	23.7	24.8	25.5	26.1	26.5	26.9	27.3	27.6	27.9	28.1	30.1	31.5	32.6	33.6
40	21.4	23.4	24.4	25.6	26.3	26.9	27.4	27.8	28.1	28.5	28.7	29.0	31.0	32.4	33.6	34.6
41	22.1	24.2	25.2	26.4	27.2	27.8	28.2	28.6	29.0	29.3	29.6	29.9	31.9	33.3	34.5	35.6
42	22.8	25.0	26.0	27.2	28.0	28.6	29.1	29.5	29.9	30.2	30.5	30.8	32.8	34.3	35.5	36.6
43	23.6	25.7	26.8	28.1	28.8	29.4	29.9	30.4	30.7	31.1	31.4	31.7	33.8	35.3	36.5	37.6
44	24.3	26.5	27.6	28.9	29.7	30.3	30.8	31.2	31.6	31.9	32.3	32.5	34.7	36.2	37.5	38.6
45	25.1	27.3	28.4	29.7	30.5	31.1	31.7	32.1	32.5	32.8	33.1	33.4	35.6	37.2	38.4	39.6
46	25.8	28.1	29.3	30.5	31.4	32.0	32.5	33.0	33.4	33.7	34.0	34.3	36.5	38.1	39.4	40.5
47	26.6	28.9	30.1	31.4	32.2	32.9	33.4	33.8	34.2	34.6	34.9	35.2	37.5	39.1	40.4	41.5
48	27.3	29.7	30.9	32.2	33.1	33.7	34.2	34.7	35.1	35.5	35.8	36.1	38.4	40.0	41.4	42.5
49	28.1	30.5	31.7	33.0	33.9	34.6	35.1	35.6	36.0	36.4	36.7	37.0	39.3	41.0	42.3	43.5
50	28.9	31.3	32.5	33.9	34.8	35.4	36.0	36.5	36.9	37.2	37.6	37.9	40.3	41.9	43.3	44.5
51	29.6	32.1	33.3	34.7	35.6	36.3	36.9	37.3	37.8	38.1	38.5	38.8	41.2	42.9	44.3	45.5
52	30.4	32.9	34.2	35.6	36.5	37.2	37.7	38.2	38.6	39.0	39.4	39.7	42.1	43.9	45.3	46.5
53	31.2	33.7	35.0	36.4	37.3	38.0	38.6	39.1	39.5	39.9	40.3	40.6	43.1	44.8	46.3	47.5
54	31.9	34.5	35.8	37.2	38.2	38.9	39.5	40.0	40.4	40.8	41.2	41.5	44.0	45.8	47.2	48.5
55	32.7	35.3	36.6	38.1	39.0	39.8	40.4	40.9	41.3	41.7	42.1	42.4	44.9	46.7	48.2	49.5
56	33.5	36.1	37.5	38.9	39.9	40.6	41.2	41.7	42.2	42.6	43.0	43.3	45.9	47.7	49.2	50.5
57	34.3	36.9	38.3	39.8	40.8	41.5	42.1	42.6	43.1	43.5	43.9	44.2	46.8	48.7	50.2	51.5
58	35.1	37.7	39.1	40.6	41.6	42.4	43.0	43.5	44.0	44.4	44.8	45.1	47.8	49.6	51.2	52.6
59	35.8	38.6	40.0	41.5	42.5	43.3	43.9	44.4	44.9	45.3	45.7	46.0	48.7	50.6	52.2	53.6
60	36.6	39.4	40.8	42.4	43.4	44.1	44.8	45.3	45.8	46.2	46.6	46.9	49.6	51.6	53.2	54.6

channels		Blocking probability														
	0.01%	0.05%	0.1%	0.2%	0.3%	0.4%	0.5%	0.6%	0.7%	0.8%	0.9%	1.0%	2.0%	3.0%	4.0%	5.0%
61	37.4	40.2	41.6	43.2	44.2	45.0	45.6	46.2	46.7	47.1	47.5	47.9	50.6	52.5	54.2	55.6
62	38.2	41.0	42.5	44.1	45.1	45.9	46.5	47.1	47.6	48.0	48.4	48.8	51.5	53.5	55.1	56.6
63	39.0	41.9	43.3	44.9	46.0	46.8	47.4	48.0	48.5	48.9	49.3	49.7	52.5	54.5	56.1	57.6
64	39.8	42.7	44.2	45.8	46.8	47.6	48.3	48.9	49.4	49.8	50.2	50.6	53.4	55.4	57.1	58.6
65	40.6	43.5	45.0	46.6	47.7	48.5	49.2	49.8	50.3	50.7	51.1	51.5	54.4	56.4	58.1	59.6
66	41.4	44.4	45.8	47.5	48.6	49.4	50.1	50.7	51.2	51.6	52.0	52.4	55.3	57.4	59.1	60.6
67	42.2	45.2	46.7	48.4	49.5	50.3	51.0	51.6	52.1	52.5	53.0	53.4	56.3	58.4	60.1	61.6
68	43.0	46.0	47.5	49.2	50.3	51.2	51.9	52.5	53.0	53.4	53.9	54.3	57.2	59.3	61.1	62.6
69	43.8	46.8	48.4	50.1	51.2	52.1	52.8	53.4	53.9	54.4	54.8	55.2	58.2	60.3	62.1	63.7
70	44.6	47.7	49.2	51.0	52.1	53.0	53.7	54.3	54.8	55.3	55.7	56.1	59.1	61.3	63.1	64.7
71	45.4	48.5	50.1	51.8	53.0	53.8	54.6	55.2	55.7	56.2	56.6	57.0	60.1	62.3	64.1	65.7
72	46.2	49.4	50.9	52.7	53.9	54.7	55.5	56.1	56.6	57.1	57.5	58.0	61.0	63.2	65.1	66.7
73	47.0	50.2	51.8	53.6	54.7	55.6	56.4	57.0	57.5	58.0	58.5	58.9	62.0	64.2	66.1	67.7
74	47.8	51.0	52.7	54.5	55.6	56.5	57.3	57.9	58.4	58.9	59.4	59.8	62.9	65.2	67.1	68.7
75	48.6	51.9	53.5	55.3	56.5	57.4	58.2	58.8	59.3	59.8	60.3	60.7	63.9	66.2	68.1	69.7
76	49.4	52.7	54.4	56.2	57.4	58.3	59.1	59.7	60.3	60.8	61.2	61.7	64.9	67.2	69.1	70.8
77	50.2	53.6	55.2	57.1	58.3	59.2	60.0	60.6	61.2	61.7	62.1	62.6	65.8	68.1	70.1	71.8
78	51.1	54.4	56.1	58.0	59.2	60.1	60.9	61.5	62.1	62.6	63.1	63.5	66.8	69.1	71.1	72.8
79	51.9	55.3	56.9	58.8	60.1	61.0	61.8	62.4	63.0	63.5	64.0	64.4	67.7	70.1	72.1	73.8
80	52.7	56.1	57.8	59.7	61.0	61.9	62.7	63.3	63.9	64.4	64.9	65.4	68.7	71.1	73.1	74.8
81	53.5	56.9	58.7	60.6	61.8	62.8	63.6	64.2	64.8	65.4	65.8	66.3	69.6	72.1	74.1	75.8
82	54.3	57.8	59.5	61.5	62.7	63.7	64.5	65.2	65.7	66.3	66.8	67.2	70.6	73.0	75.1	76.9
83	55.1	58.6	60.4	62.4	63.6	64.6	65.4	66.1	66.7	67.2	67.7	68.2	71.6	74.0	76.1	77.9
84	56.0	59.5	61.3	63.2	64.5	65.5	66.3	67.0	67.6	68.1	68.6	69.1	72.5	75.0	77.1	78.9
85	56.8	60.4	62.1	64.1	65.4	66.4	67.2	67.9	68.5	69.1	69.6	70.0	73.5	76.0	78.1	79.9
86	57.6	61.2	63.0	65.0	66.3	67.3	68.1	68.8	69.4	70.0	70.5	70.9	74.5	77.0	79.1	80.9
87	58.4	62.1	63.9	65.9	67.2	68.2	69.0	69.7	70.3	70.9	71.4	71.9	75.4	78.0	80.1	82.0
88	59.3	62.9	64.7	66.8	68.1	69.1	69.9	70.6	71.3	71.8	72.3	72.8	76.4	78.9	81.1	83.0
89	60.1	63.8	65.6	67.7	69.0	70.0	70.8	71.6	72.2	72.8	73.3	73.7	77.3	79.9	82.1	84.0
90	60.9	64.6	66.5	68.6	69.9	70.9	71.8	72.5	73.1	73.7	74.2	74.7	78.3	80.9	83.1	85.0

channels	\multicolumn{16}{c}{Blocking probability}															
	0.01%	0.05%	0.1%	0.2%	0.3%	0.4%	0.5%	0.6%	0.7%	0.8%	0.9%	1.0%	2.0%	3.0%	4.0%	5.0%
91	61.8	65.5	67.4	69.4	70.8	71.8	72.7	73.4	74.0	74.6	75.1	75.6	79.3	81.9	84.1	86.0
92	62.6	66.3	68.2	70.3	71.7	72.7	73.6	74.3	75.0	75.5	76.1	76.6	80.2	82.9	85.1	87.1
93	63.4	67.2	69.1	71.2	72.6	73.6	74.5	75.2	75.9	76.5	77.0	77.5	81.2	83.9	86.1	88.1
94	64.2	68.1	70.0	72.1	73.5	74.5	75.4	76.2	76.8	77.4	77.9	78.4	82.2	84.9	87.1	89.1
95	65.1	68.9	70.9	73.0	74.4	75.5	76.3	77.1	77.7	78.3	78.9	79.4	83.1	85.8	88.1	90.1
96	65.9	69.8	71.7	73.9	75.3	76.4	77.2	78.0	78.7	79.3	79.8	80.3	84.1	86.8	89.1	91.1
97	66.8	70.7	72.6	74.8	76.2	77.3	78.2	78.9	79.6	80.2	80.7	81.2	85.1	87.8	90.1	92.2
98	67.6	71.5	73.5	75.7	77.1	78.2	79.1	79.8	80.5	81.1	81.7	82.2	86.0	88.8	91.1	93.2
99	68.4	72.4	74.4	76.6	78.0	79.1	80.0	80.8	81.4	82.0	82.6	83.1	87.0	89.8	92.1	94.2
100	69.3	73.2	75.2	77.5	78.9	80.0	80.9	81.7	82.4	83.0	83.5	84.1	88.0	90.8	93.1	95.2
102	70.9	75.0	77.0	79.3	80.7	81.8	82.7	83.5	84.2	84.8	85.4	85.9	89.9	92.8	95.2	97.3
104	72.6	76.7	78.8	81.1	82.5	83.7	84.6	85.4	86.1	86.7	87.3	87.8	91.8	94.8	97.2	99.3
106	74.3	78.5	80.5	82.8	84.3	85.5	86.4	87.2	87.9	88.6	89.2	89.7	93.8	96.7	99.2	101.4
108	76.0	80.2	82.3	84.6	86.2	87.3	88.3	89.1	89.8	90.5	91.1	91.6	95.7	98.7	101.2	103.4
110	77.7	81.9	84.1	86.4	88.0	89.2	90.1	90.9	91.7	92.3	92.9	93.5	97.7	100.7	103.2	105.5
112	79.4	83.7	85.8	88.3	89.8	91.0	92.0	92.8	93.5	94.2	94.8	95.4	99.6	102.7	105.3	107.5
114	81.1	85.4	87.6	90.1	91.6	92.8	93.8	94.7	95.4	96.1	96.7	97.3	101.6	104.7	107.3	109.6
116	82.8	87.2	89.4	91.9	93.5	94.7	95.7	96.5	97.3	98.0	98.6	99.2	103.5	106.7	109.3	111.7
118	84.5	89.0	91.2	93.7	95.3	96.5	97.5	98.4	99.2	99.9	100.5	101.1	105.5	108.7	111.3	113.7
120	86.2	90.7	93.0	95.5	97.1	98.4	99.4	100.3	101.0	101.7	102.4	103.0	107.4	110.6	113.4	115.8
122	87.9	92.5	94.7	97.3	98.9	100.2	101.2	102.1	102.9	103.6	104.3	104.9	109.4	112.6	115.4	117.8
124	89.6	94.2	96.5	99.1	100.8	102.1	103.1	104.0	104.3	105.5	106.2	106.8	111.3	114.6	117.4	119.9
126	91.3	96.0	98.3	100.9	102.6	103.9	105.0	105.9	106.7	107.4	108.1	108.7	113.3	116.6	119.4	121.9
128	93.1	97.8	100.1	102.7	104.5	105.8	106.8	107.7	108.5	109.3	109.9	110.6	115.2	118.6	121.5	124.0
130	94.8	99.5	101.9	104.6	106.3	107.6	108.7	109.6	110.4	111.2	111.8	112.5	117.2	120.6	123.5	126.1
132	96.5	101.3	103.7	106.4	108.1	109.5	110.5	111.5	112.3	113.1	113.7	114.4	119.1	122.6	125.5	128.1
134	98.2	103.1	105.5	108.2	110.0	111.3	112.4	113.4	114.2	115.0	115.6	116.3	121.1	124.6	127.6	130.2
136	100.0	104.9	107.3	110.0	111.8	113.2	114.3	115.2	116.1	116.8	117.5	118.2	123.1	126.6	129.6	132.3
138	101.7	106.6	109.1	111.9	113.7	115.0	116.2	117.1	118.0	118.7	119.4	120.1	125.0	128.6	131.6	134.3
140	103.4	108.4	110.9	113.7	115.5	116.9	118.0	119.0	119.9	120.6	121.4	122.0	127.0	130.6	133.7	135.4

channels	0.01%	0.05%	0.1%	0.2%	0.3%	0.4%	0.5%	0.6%	0.7%	0.8%	0.9%	1.0%	2.0%	3.0%	4.0%	5.0%
									Blocking probability							
142	105.1	110.2	112.7	115.5	117.4	118.7	119.9	120.9	121.8	122.5	123.3	123.9	128.9	132.6	135.7	138.4
144	106.9	112.0	114.5	117.4	119.2	120.6	121.8	122.8	123.6	124.4	125.2	125.8	130.9	134.6	137.7	140.5
146	108.6	113.8	116.3	119.2	121.1	122.5	123.6	124.6	125.5	126.3	127.1	127.7	132.9	136.6	139.8	142.6
148	110.4	115.5	118.1	121.0	122.9	124.3	125.5	126.5	127.4	128.2	129.0	129.7	134.8	138.6	141.8	144.6
150	112.1	117.3	119.9	122.9	124.8	126.2	127.4	128.4	129.3	130.1	130.9	131.6	136.8	140.6	143.8	146.7
152	113.8	119.1	121.8	124.7	126.6	128.1	129.3	130.3	131.2	132.0	132.8	133.5	138.8	142.6	145.9	148.8
154	115.6	120.9	123.6	126.5	128.5	129.9	131.2	132.2	133.1	133.9	134.7	135.4	140.7	144.6	147.9	150.8
156	117.3	122.7	125.4	128.4	130.3	131.8	133.0	134.1	135.0	135.9	136.6	137.3	142.7	146.6	149.9	152.9
158	119.1	124.5	127.2	130.2	132.2	133.7	134.9	136.0	136.9	137.8	138.5	139.2	144.7	148.6	152.0	155.0
160	120.8	126.3	129.0	132.1	134.0	135.6	136.8	137.9	138.8	139.7	140.4	141.2	146.6	150.6	154.0	157.0
162	122.6	128.1	130.8	133.9	135.9	137.4	138.7	139.8	140.7	141.6	142.4	143.1	148.6	152.7	156.1	159.1
164	124.3	129.9	132.7	135.8	137.8	139.3	140.6	141.7	142.6	143.5	144.3	145.0	150.6	154.7	158.1	161.2
166	126.1	131.7	134.5	137.6	139.6	141.2	142.5	143.5	144.5	145.4	146.2	146.9	152.6	156.7	160.1	163.3
168	127.9	133.5	136.3	139.4	141.5	143.1	144.3	145.4	146.4	147.3	148.1	148.9	154.5	158.7	162.2	165.3
170	129.6	135.3	138.1	141.3	143.4	144.9	146.2	147.3	148.3	149.2	150.0	150.8	156.5	160.7	164.2	167.4
172	131.4	137.1	139.9	143.1	145.2	146.8	148.1	149.2	150.2	151.1	151.9	152.7	158.5	162.7	166.3	169.5
174	133.1	138.9	141.8	145.0	147.1	148.7	150.0	151.1	152.1	153.0	153.9	154.6	160.4	164.7	168.3	171.5
176	134.9	140.7	143.6	146.9	149.0	150.6	151.9	153.0	154.0	155.0	155.8	156.6	162.4	166.7	170.3	173.6
178	136.7	142.5	145.4	148.7	150.8	152.4	153.8	154.9	156.0	156.9	157.7	158.5	164.4	168.7	172.4	175.7
180	138.4	144.3	147.3	150.6	152.7	154.3	155.7	156.8	157.9	158.8	159.6	160.4	166.4	170.7	174.4	177.8
182	140.2	146.1	149.1	152.4	154.6	156.2	157.6	158.7	159.8	160.7	161.6	162.3	168.3	172.8	176.5	179.8
184	142.0	147.9	150.9	154.3	156.4	158.1	159.5	160.6	161.7	162.6	163.5	164.3	170.3	174.8	178.5	181.9
186	143.7	149.8	152.8	156.1	158.3	160.0	161.4	162.5	163.6	164.5	165.4	166.2	172.3	176.8	180.6	184.0
188	145.5	151.6	154.6	158.0	160.2	161.9	163.3	164.4	165.5	166.5	167.3	168.1	174.3	178.8	182.6	186.1
190	147.3	153.4	156.4	159.8	162.1	163.8	165.2	166.4	167.4	168.4	169.3	170.1	176.3	180.8	184.7	188.1
192	149.1	155.2	158.3	161.7	163.9	165.6	167.0	168.3	169.3	170.3	171.2	172.0	178.2	182.8	186.7	190.2
194	150.8	157.0	160.1	163.6	165.8	167.5	168.9	170.2	171.2	172.2	173.1	173.9	180.2	184.8	188.8	192.3
196	152.6	158.8	161.9	165.4	167.7	169.4	170.8	172.1	173.2	174.1	175.0	175.9	182.2	186.9	190.8	194.4
198	154.4	160.7	163.8	167.3	169.6	171.3	172.7	174.0	175.1	176.1	177.0	177.8	184.2	188.9	192.8	196.4
200	156.2	162.5	165.6	169.2	171.4	173.2	174.6	175.9	177.0	178.0	178.9	179.7	186.2	190.9	194.9	198.5

channels	0.01%	0.05%	0.1%	0.2%	0.3%	0.4%	0.5%	0.6%	0.7%	0.8%	0.9%	1.0%	2.0%	3.0%	4.0%	5.0%
									Blocking probability							
205	160.6	167.0	170.2	173.8	176.1	177.9	179.4	180.7	181.8	182.8	183.7	184.6	191.1	195.9	200.0	203.7
210	165.1	171.6	174.8	178.5	180.9	182.7	184.2	185.4	186.6	187.6	188.6	189.4	196.1	201.0	205.1	208.9
215	169.6	176.2	179.5	183.2	185.6	187.4	188.9	190.2	191.4	192.4	193.4	194.3	201.0	206.0	210.3	214.1
220	174.0	180.7	184.1	187.8	190.3	192.1	193.7	195.0	196.2	197.2	198.2	199.1	206.0	211.1	215.4	219.3
225	178.5	185.3	188.7	192.5	195.0	196.9	198.5	199.8	201.0	202.1	203.1	204.0	211.0	216.1	220.5	224.5
230	183.0	189.9	193.3	197.2	199.7	201.6	203.2	204.6	205.8	206.9	207.9	208.8	215.9	221.2	225.6	229.7
235	187.5	194.5	198.0	201.9	204.4	206.4	208.0	209.4	210.6	211.7	212.8	213.7	220.9	226.2	230.8	234.9
240	192.0	199.1	202.6	206.6	209.2	211.2	212.8	214.2	215.4	216.6	217.6	218.6	225.9	231.3	235.9	240.1
245	196.5	203.7	207.3	211.3	213.9	215.9	217.6	219.0	220.3	221.4	222.5	223.4	230.8	236.4	241.0	245.3
250	201.0	208.3	211.9	216.0	218.7	220.7	222.4	223.8	225.1	226.2	227.3	228.3	235.8	241.4	246.2	250.5
255	205.6	212.9	216.6	220.7	223.4	225.5	227.2	228.6	229.9	231.1	232.2	233.2	240.8	246.5	251.3	255.7
260	210.1	217.5	221.2	225.4	228.1	230.2	232.0	233.4	234.7	235.9	237.0	238.0	245.8	251.5	256.5	260.9
265	214.6	222.1	225.9	230.1	232.9	235.0	236.7	238.3	239.6	240.8	241.9	242.9	250.8	256.6	261.6	266.1
270	219.1	226.8	230.6	234.9	237.6	239.8	241.5	243.1	244.4	245.6	246.8	247.3	255.8	261.7	266.7	271.3
275	223.7	231.4	235.2	239.6	242.4	244.6	246.4	247.9	249.3	250.5	251.6	252.7	260.7	266.7	271.9	276.5
280	228.2	236.0	239.9	244.3	247.2	249.3	251.2	252.7	254.1	255.4	256.5	257.6	265.7	271.8	277.0	281.8
285	232.8	240.6	244.6	249.0	251.9	254.1	256.0	257.5	258.9	260.2	261.4	262.5	270.7	276.9	282.2	287.0
290	237.3	245.3	249.3	253.8	256.7	258.9	260.8	262.4	263.8	265.1	266.2	267.3	275.7	282.0	287.3	292.2
295	241.9	249.9	254.0	258.5	261.5	263.7	265.6	267.2	268.6	269.9	271.1	272.2	280.7	287.0	292.5	297.4
300	246.4	254.6	258.6	263.2	266.2	268.5	270.4	272.0	273.5	274.8	276.0	277.1	285.7	292.1	297.6	302.6
305	251.0	259.2	263.3	268.0	271.0	273.3	275.2	276.9	278.3	279.7	280.9	282.0	290.7	297.2	302.8	307.8
310	255.6	263.9	268.0	272.7	275.8	278.1	280.0	281.7	283.2	284.5	285.8	286.9	295.7	302.3	307.9	313.1
315	260.1	268.5	272.7	277.5	280.5	282.9	284.9	286.6	288.1	289.4	290.7	291.8	300.7	307.4	313.1	318.3
320	264.7	273.2	277.4	282.2	285.3	287.7	289.7	291.4	292.9	294.3	295.5	296.7	305.7	312.4	318.2	323.5
325	269.3	277.8	282.1	287.0	290.1	292.5	294.5	296.2	297.8	299.2	300.4	301.6	310.7	317.5	323.4	328.7
330	273.9	282.5	286.8	291.7	294.9	297.3	299.4	301.1	302.6	304.0	305.3	306.5	315.7	322.6	328.5	333.9
335	278.5	287.2	291.6	296.5	299.7	302.1	304.2	305.9	307.5	308.9	310.2	311.4	320.7	327.7	333.7	339.2
340	283.1	291.8	296.3	301.2	304.5	307.0	309.0	310.8	312.4	313.8	315.1	316.3	325.7	332.8	338.8	344.4
345	287.7	296.5	301.0	306.0	309.3	311.8	313.9	315.6	317.2	318.7	320.0	321.2	330.7	337.9	344.0	349.6
350	292.3	301.2	305.7	310.8	314.1	316.6	318.7	320.5	322.1	323.6	324.9	326.2	335.7	342.9	349.1	354.8

channels	0.01%	0.05%	0.1%	0.2%	0.3%	0.4%	0.5%	0.6%	0.7%	0.8%	0.9%	1.0%	2.0%	3.0%	4.0%	5.0%
								Blocking probability								
355	296.9	305.9	310.4	315.5	318.9	321.4	323.5	325.4	327.0	328.5	329.8	331.1	340.7	348.0	354.3	360.0
360	301.5	310.6	315.1	320.3	323.7	326.2	328.4	330.2	331.9	333.3	334.7	336.0	345.8	353.1	359.5	365.3
365	306.1	315.2	319.9	325.1	328.4	331.1	333.2	335.1	336.7	338.2	339.6	340.9	350.8	358.2	364.6	370.5
370	310.7	319.9	324.6	329.8	333.3	335.9	338.1	339.9	341.6	343.1	344.5	345.8	355.8	363.3	369.8	375.7
375	315.3	324.6	329.3	334.6	338.1	340.7	342.9	344.8	346.5	348.0	349.4	350.7	360.8	368.4	374.9	380.9
380	319.9	329.3	334.0	339.4	342.9	345.5	347.8	349.7	351.4	352.9	354.3	355.6	365.8	373.5	380.1	386.2
385	324.5	334.0	338.8	344.2	347.7	350.4	352.6	354.5	356.2	357.8	359.2	360.6	370.8	378.6	385.3	391.4
390	329.1	338.7	343.5	348.9	352.5	355.2	357.5	359.4	361.1	362.7	364.1	365.5	375.8	383.7	390.4	396.6
395	333.8	343.4	348.3	353.7	357.3	360.0	362.3	364.3	366.0	367.6	369.1	370.4	380.9	388.8	395.6	401.9
400	338.4	348.1	353.0	358.5	362.1	364.9	367.2	369.1	370.9	372.5	374.0	375.3	385.9	393.9	400.8	407.1
410	347.7	357.5	362.5	368.1	371.7	374.5	376.9	378.9	380.7	382.3	383.8	385.2	395.9	404.1	411.1	417.5
420	356.9	366.9	372.0	377.7	381.4	384.2	386.6	388.6	390.5	392.1	393.6	395.0	406.0	414.2	421.4	428.0
430	366.2	376.4	381.5	387.2	391.0	393.9	396.3	398.4	400.2	401.9	403.5	404.9	416.0	424.4	431.8	438.5
440	375.5	385.8	391.0	396.8	400.7	403.6	406.0	408.2	410.0	411.7	413.3	414.8	426.1	434.6	442.1	448.9
450	384.8	395.2	400.5	406.4	410.3	413.3	415.8	417.9	419.8	421.6	423.2	424.6	436.1	444.8	452.4	459.4
460	394.1	404.7	410.0	416.0	420.0	423.0	425.5	427.7	429.6	431.4	433.0	434.5	446.2	455.1	462.8	469.9
470	403.4	414.1	419.5	425.6	429.6	432.7	435.3	437.5	439.4	441.2	442.9	444.4	456.2	465.3	473.1	480.3
480	412.7	423.6	429.1	435.3	439.3	442.4	445.0	447.2	449.2	451.0	452.7	454.3	466.3	475.5	483.5	490.8
490	422.1	433.1	438.6	444.9	449.0	452.1	454.8	457.0	459.1	460.9	462.6	464.1	476.4	485.7	493.8	501.3
500	431.4	442.5	448.2	454.5	458.7	461.9	464.5	466.8	468.9	470.7	472.4	474.0	486.4	495.9	504.1	511.8
510	440.8	452.0	457.7	464.1	468.3	471.6	474.3	476.6	478.7	480.6	482.3	483.9	496.5	506.1	514.5	522.2
520	450.1	461.5	467.3	473.8	478.0	481.3	484.0	486.4	488.5	490.4	492.2	493.8	506.6	516.3	524.8	532.7
530	459.5	471.0	476.8	483.4	487.7	491.0	493.8	496.2	498.3	500.3	502.1	503.7	516.7	526.6	535.2	543.2
540	468.8	480.5	486.4	493.1	497.4	500.8	503.6	506.0	508.2	510.1	511.9	513.6	526.7	536.8	545.5	553.7
550	478.2	490.0	496.0	502.7	507.1	510.5	513.4	515.8	518.0	520.0	521.8	523.5	536.8	547.0	555.9	564.1
560	487.6	499.5	505.5	512.4	516.8	520.3	523.1	525.6	527.8	529.8	531.7	533.4	546.9	557.2	566.3	574.6
570	497.0	509.0	515.1	522.0	526.5	530.0	532.9	535.4	537.7	539.7	541.6	543.3	557.0	567.4	576.6	585.1
580	506.4	518.6	524.7	531.7	536.2	539.8	542.7	545.3	547.5	549.6	551.5	553.2	567.1	577.7	587.0	595.6
590	515.8	528.1	534.3	541.3	546.0	549.5	552.5	555.1	557.4	559.4	561.4	563.2	577.2	587.9	597.3	606.1
600	525.2	537.6	543.9	551.0	555.7	559.3	562.3	564.9	567.2	569.3	571.3	573.1	587.2	598.1	607.7	616.5

channels	0.01%	0.05%	0.1%	0.2%	0.3%	0.4%	0.5%	Blocking probability 0.6%	0.7%	0.8%	0.9%	1.0%	2.0%	3.0%	4.0%	5.0%
610	534.6	547.1	553.5	560.7	565.4	569.1	572.1	574.7	577.1	579.2	581.2	583.0	597.3	608.4	618.0	627.0
620	544.0	556.7	563.1	570.4	575.1	578.8	581.9	584.5	586.9	589.1	591.1	592.9	607.4	618.6	628.4	637.5
630	553.4	566.2	572.7	580.0	584.9	588.6	591.7	594.4	596.8	599.0	601.0	602.8	617.5	628.8	638.8	648.0
640	562.8	575.8	582.3	589.7	594.6	598.4	601.5	604.2	606.6	608.8	610.9	612.8	627.6	639.1	649.1	658.5
650	572.3	585.3	591.9	599.4	604.3	608.1	611.3	614.0	616.5	618.7	620.8	622.7	637.7	649.3	659.5	669.0
660	581.7	594.9	601.6	609.1	614.1	617.9	621.1	623.9	626.4	628.6	630.7	632.6	647.8	659.5	669.9	679.5
670	591.2	604.5	611.2	618.8	623.8	627.7	630.9	633.7	636.2	638.5	640.6	642.6	657.9	669.8	680.2	690.0
680	600.6	614.0	620.8	628.5	633.6	637.5	640.7	643.6	646.1	648.4	650.5	652.5	668.0	680.0	690.6	700.4
690	610.1	623.6	630.4	638.2	643.3	647.3	650.6	653.4	656.0	658.3	660.4	662.4	678.1	690.3	701.0	710.9
700	619.5	633.2	640.1	647.9	653.1	657.1	660.4	663.3	665.8	668.2	670.3	672.4	688.2	700.5	711.3	721.4
710	629.0	642.7	649.7	657.6	662.8	666.8	670.2	673.1	675.7	678.1	680.3	682.3	698.3	710.7	721.7	731.9
720	638.4	652.3	659.3	667.3	672.6	676.6	680.0	683.0	685.6	688.0	690.2	692.2	708.4	721.0	732.1	742.4
730	647.9	661.9	669.0	677.0	682.3	686.4	689.8	692.8	695.5	697.9	700.1	702.2	718.5	731.2	742.4	752.9
740	657.4	671.5	678.6	686.7	692.1	696.2	699.7	702.7	705.3	707.8	710.0	712.1	728.6	741.5	752.8	763.4
750	666.8	681.1	688.3	696.5	701.9	706.0	709.5	712.5	715.2	717.7	720.0	722.1	738.8	751.7	763.2	773.9
760	676.3	690.7	697.9	706.2	711.6	715.8	719.3	722.4	725.1	727.6	729.9	732.0	748.9	762.0	773.6	784.4
770	685.8	700.3	707.6	715.9	721.4	725.6	729.2	732.3	735.0	737.5	739.8	742.0	759.0	772.2	783.9	794.9
780	695.3	709.9	717.2	725.6	731.2	735.4	739.0	742.1	744.9	747.4	749.7	751.9	769.1	782.5	794.3	805.4
790	704.8	719.5	726.9	735.4	740.9	745.2	748.8	752.0	754.8	757.3	759.7	761.9	779.2	792.7	804.7	815.9
800	714.3	729.1	736.6	745.1	750.7	755.1	758.7	761.8	764.7	767.2	769.6	771.8	789.3	803.0	815.1	826.4
810	723.8	738.7	746.2	754.8	760.5	764.9	768.5	771.7	774.6	777.2	779.6	781.8	799.4	813.2	825.4	836.9
820	733.3	748.3	755.9	764.6	770.3	774.7	778.4	781.6	784.5	787.1	789.5	791.7	809.6	823.5	835.8	847.4
830	742.8	757.9	765.6	774.3	780.0	784.5	788.2	791.5	794.4	797.0	799.4	801.7	819.7	833.7	846.2	857.9
840	752.3	767.5	775.2	784.0	789.8	794.3	798.1	801.3	804.3	806.9	809.4	811.7	829.8	844.0	856.6	868.4
850	761.8	777.1	784.9	793.8	799.6	804.1	807.9	811.2	814.2	816.8	819.3	821.6	839.9	854.2	867.0	878.9
860	771.3	786.8	794.6	803.5	809.4	814.0	817.8	821.1	824.1	826.8	829.3	831.6	850.1	864.5	877.3	889.4
870	780.8	796.4	804.3	813.3	819.2	823.8	827.6	831.0	834.0	836.7	839.2	841.5	860.2	874.8	887.7	899.9
880	790.4	806.0	814.0	823.0	829.0	833.6	837.5	840.9	843.9	846.6	849.1	851.5	870.3	885.0	898.1	910.4
890	799.9	815.6	823.6	832.8	838.8	843.4	847.3	850.7	853.8	856.5	859.1	861.5	880.4	895.3	908.5	920.9
900	809.4	825.3	833.3	842.5	848.6	853.3	857.2	860.6	863.7	866.5	869.0	871.5	890.6	905.5	918.9	931.4

channels								Blocking probability								
	0.01%	0.05%	0.1%	0.2%	0.3%	0.4%	0.5%	0.6%	0.7%	0.8%	0.9%	1.0%	2.0%	3.0%	4.0%	5.0%
910	818.9	834.9	843.0	852.3	858.4	863.1	867.1	870.5	873.6	876.4	879.0	881.4	900.7	915.8	929.2	941.9
920	828.5	844.5	852.7	862.0	868.2	872.9	876.9	880.4	883.5	886.3	888.9	891.4	910.8	926.0	939.6	952.4
930	838.0	854.2	862.4	871.8	878.0	882.8	886.8	890.3	893.4	896.3	898.9	901.4	920.9	936.3	950.0	962.9
940	847.5	863.8	872.1	881.5	887.8	892.6	896.6	900.2	903.3	906.2	908.9	911.3	931.1	946.6	960.4	973.4
950	857.1	873.5	881.8	891.3	897.6	902.4	906.5	910.1	913.2	916.1	918.8	921.3	941.2	956.8	970.8	983.9
960	866.6	883.1	891.5	901.1	907.4	912.3	916.4	920.0	923.2	926.1	928.8	931.3	951.3	967.1	981.1	994.4
970	876.2	892.8	901.2	910.8	917.2	922.1	926.3	929.8	933.1	936.0	938.7	941.3	961.5	977.3	991.5	1004.9
980	885.7	902.4	910.9	920.6	927.0	932.0	936.1	939.7	943.0	945.9	948.7	951.2	971.6	987.6	1001.9	1015.4
990	895.3	912.1	920.6	930.4	936.8	941.8	946.0	949.6	952.9	955.9	958.6	961.2	981.7	997.9	1012.3	1025.9
1000	904.8	921.7	930.3	940.1	946.6	951.7	955.9	959.5	962.8	965.8	968.6	971.2	991.9	1008.1	1022.7	1036.4
1010	914.4	931.4	940.0	949.9	956.4	961.5	965.7	969.4	972.7	975.8	978.6	981.2	1002.0	1018.4	1033.1	1046.9
1020	923.9	941.0	949.7	959.7	966.2	971.3	975.6	979.3	982.7	985.7	988.5	991.2	1012.1	1028.7	1043.5	1057.4
1030	933.5	950.7	959.5	969.4	976.1	981.2	985.5	989.2	992.6	995.7	998.5	1001.1	1022.3	1038.9	1053.8	1067.9
1040	943.1	960.4	969.2	979.2	985.9	991.0	995.4	999.1	1002.5	1005.6	1008.5	1011.1	1032.4	1049.2	1064.2	1078.4
1050	952.6	970.0	978.9	989.0	995.7	1000.9	1005.2	1009.0	1012.4	1015.5	1018.4	1021.1	1042.5	1059.5	1074.6	1088.9
1060	962.2	979.7	988.6	998.8	1005.5	1010.7	1015.1	1018.9	1022.4	1025.5	1028.4	1031.1	1052.7	1069.7	1085.0	1099.4
1070	971.8	989.4	998.3	1008.6	1015.3	1020.6	1025.0	1028.9	1032.3	1035.4	1038.4	1041.1	1062.8	1080.0	1095.4	1109.9
1080	981.4	999.0	1008.1	1018.3	1025.2	1030.5	1034.9	1038.8	1042.2	1045.4	1048.3	1051.1	1072.9	1090.3	1105.8	1120.4
1090	990.9	1008.7	1017.8	1028.1	1035.0	1040.3	1044.8	1048.7	1052.2	1055.3	1058.3	1061.1	1083.1	1100.5	1116.2	1130.9
1100	1000.5	1018.4	1027.5	1037.9	1044.8	1050.2	1054.7	1058.6	1062.1	1065.3	1068.3	1071.1	1093.2	1110.8	1126.6	1141.4
1110	1010.1	1028.1	1037.2	1047.7	1054.6	1060.0	1064.5	1068.5	1072.0	1075.2	1078.2	1081.0	1103.4	1121.1	1137.0	1152.0
1120	1019.7	1037.7	1047.0	1057.5	1064.5	1069.9	1074.4	1078.4	1082.0	1085.2	1088.2	1091.0	1113.5	1131.3	1147.3	1162.5
1130	1029.3	1047.4	1056.7	1067.3	1074.3	1079.8	1084.3	1088.3	1091.9	1095.2	1098.2	1101.0	1123.6	1141.6	1157.7	1173.0
1140	1038.8	1057.1	1066.4	1077.1	1084.1	1089.6	1094.2	1098.2	1101.8	1105.1	1108.2	1111.0	1133.8	1151.9	1168.1	1183.5
1150	1048.4	1066.8	1076.2	1086.9	1094.0	1099.5	1104.1	1108.1	1111.8	1115.1	1118.1	1121.0	1143.9	1162.2	1178.5	1194.0
1160	1058.0	1076.5	1085.9	1096.6	1103.8	1109.3	1114.0	1118.1	1121.7	1125.0	1128.1	1131.0	1154.1	1172.4	1188.9	1204.5
1170	1067.6	1086.2	1095.6	1106.4	1113.6	1119.2	1123.9	1128.0	1131.6	1135.0	1138.1	1141.0	1164.2	1182.7	1199.3	1215.0
1180	1077.2	1095.9	1105.4	1116.2	1123.5	1129.1	1133.8	1137.9	1141.6	1144.9	1148.1	1151.0	1174.4	1193.0	1209.7	1225.5
1190	1086.8	1105.5	1115.1	1126.0	1133.3	1138.9	1143.7	1147.8	1151.5	1154.9	1158.0	1161.0	1184.5	1203.2	1220.1	1236.0
1200	1096.4	1115.2	1124.8	1135.8	1143.1	1148.8	1153.6	1157.7	1161.5	1164.9	1168.0	1171.0	1194.7	1213.5	1230.5	1246.5

channels	Blocking probability															
	0.01%	0.05%	0.1%	0.2%	0.3%	0.4%	0.5%	0.6%	0.7%	0.8%	0.9%	1.0%	2.0%	3.0%	4.0%	5.0%
1210	1106.0	1124.9	1134.6	1145.6	1153.0	1158.7	1163.5	1167.6	1171.4	1174.8	1178.0	1181.0	1204.8	1223.8	1240.9	1257.0
1220	1115.6	1134.6	1144.3	1155.4	1162.8	1168.5	1173.4	1177.6	1181.3	1184.8	1188.0	1191.0	1214.9	1234.1	1251.3	1267.5
1230	1125.2	1144.3	1154.1	1165.2	1172.6	1178.4	1183.3	1187.5	1191.3	1194.7	1198.0	1201.0	1225.1	1244.3	1261.6	1278.1
1240	1134.8	1154.0	1163.8	1175.0	1182.5	1188.3	1193.2	1197.4	1201.2	1204.7	1207.9	1211.0	1235.2	1254.6	1272.0	1288.6
1250	1144.4	1163.7	1173.6	1184.8	1192.3	1198.2	1203.1	1207.3	1211.2	1214.7	1217.9	1221.0	1245.4	1264.9	1282.4	1299.1
1260	1154.0	1173.4	1183.3	1194.6	1202.2	1208.0	1213.0	1217.3	1221.1	1224.6	1227.9	1231.0	1255.5	1275.1	1292.8	1309.6
1270	1163.6	1183.1	1193.1	1204.5	1212.0	1217.9	1222.9	1227.2	1231.1	1234.6	1237.9	1241.0	1265.7	1285.4	1303.2	1320.1
1280	1173.3	1192.8	1202.8	1214.3	1221.9	1227.8	1232.8	1237.1	1241.0	1244.6	1247.9	1251.0	1275.8	1295.7	1313.6	1330.6
1290	1182.9	1202.5	1212.6	1224.1	1231.7	1237.7	1242.7	1247.0	1251.0	1254.5	1257.9	1261.0	1286.0	1306.0	1324.0	1341.1
1300	1192.5	1212.2	1222.3	1233.9	1241.6	1247.5	1252.6	1257.0	1260.9	1264.5	1267.9	1271.0	1296.1	1316.2	1334.4	1351.6
1310	1202.1	1221.9	1232.1	1243.7	1251.4	1257.4	1262.5	1266.9	1270.8	1274.5	1277.8	1281.0	1306.3	1326.5	1344.8	1362.1
1320	1211.7	1231.7	1241.8	1253.5	1261.3	1267.3	1272.4	1276.8	1280.8	1284.4	1287.8	1291.0	1316.4	1336.8	1355.2	1372.7
1330	1221.3	1241.4	1251.6	1263.3	1271.1	1277.2	1282.3	1286.7	1290.7	1294.4	1297.8	1301.0	1326.6	1347.1	1365.6	1383.2
1340	1231.0	1251.1	1261.4	1273.1	1281.0	1287.1	1292.2	1296.7	1300.7	1304.4	1307.8	1311.0	1336.7	1357.4	1376.0	1393.7
1350	1240.6	1260.8	1271.1	1282.9	1290.8	1297.0	1302.1	1306.6	1310.7	1314.4	1317.8	1321.0	1346.9	1367.6	1386.4	1404.2
1360	1250.2	1270.5	1280.9	1292.8	1300.7	1306.8	1312.0	1316.5	1320.6	1324.3	1327.8	1331.0	1357.0	1377.9	1396.8	1414.7
1370	1259.8	1280.2	1290.6	1302.6	1310.5	1316.7	1321.9	1326.5	1330.6	1334.3	1337.8	1341.0	1367.2	1388.2	1407.2	1425.2
1380	1269.5	1289.9	1300.4	1312.4	1320.4	1326.6	1331.8	1336.4	1340.5	1344.3	1347.8	1351.0	1377.4	1398.5	1417.6	1435.7
1390	1279.1	1299.7	1310.2	1322.2	1330.2	1336.5	1341.7	1346.3	1350.5	1354.2	1357.8	1361.0	1387.5	1408.7	1428.0	1446.2
1400	1288.7	1309.4	1319.9	1332.0	1340.1	1346.4	1351.7	1356.3	1360.4	1364.2	1367.7	1371.1	1397.7	1419.0	1438.4	1456.7
1410	1298.4	1319.1	1329.7	1341.9	1349.9	1356.3	1361.6	1366.2	1370.4	1374.2	1377.7	1381.1	1407.8	1429.3	1448.8	1467.3
1420	1308.0	1328.8	1339.5	1351.7	1359.8	1366.2	1371.5	1376.1	1380.3	1384.2	1387.7	1391.1	1418.0	1439.6	1459.1	1477.8
1430	1317.6	1338.5	1349.2	1361.5	1369.7	1376.0	1381.4	1386.1	1390.3	1394.2	1397.7	1401.1	1428.1	1449.9	1469.5	1488.3
1440	1327.3	1348.3	1359.0	1371.3	1379.5	1385.9	1391.3	1396.0	1400.3	1404.1	1407.7	1411.1	1438.3	1460.1	1479.9	1498.8
1450	1336.9	1358.0	1368.8	1381.1	1389.4	1395.8	1401.2	1406.0	1410.2	1414.1	1417.7	1421.1	1448.4	1470.4	1490.3	1509.3
1460	1346.6	1367.7	1378.6	1391.0	1399.3	1405.7	1411.1	1415.9	1420.2	1424.1	1427.7	1431.1	1458.6	1480.7	1500.7	1519.8
1470	1356.2	1377.4	1388.3	1400.8	1409.1	1415.6	1421.1	1425.8	1430.1	1434.1	1437.7	1441.1	1468.7	1491.0	1511.1	1530.3
1480	1365.8	1387.2	1398.1	1410.6	1419.0	1425.5	1431.0	1435.8	1440.1	1444.0	1447.7	1451.2	1478.9	1501.3	1521.5	1540.8
1490	1375.5	1396.9	1407.9	1420.5	1428.8	1435.4	1440.9	1445.7	1450.1	1454.0	1457.7	1461.2	1489.1	1511.5	1531.9	1551.4
1500	1385.1	1406.6	1417.7	1430.3	1438.7	1445.3	1450.8	1455.7	1460.0	1464.0	1467.7	1471.2	1499.2	1521.8	1542.3	1561.9

channels	0.01%	0.05%	0.1%	0.2%	0.3%	0.4%	0.5%	0.6%	0.7%	0.8%	0.9%	1.0%	2.0%	3.0%	4.0%	5.0%
1510	1394.8	1416.4	1427.4	1440.1	1448.6	1455.2	1460.7	1465.6	1470.0	1474.0	1477.7	1481.2	1509.4	1532.1	1552.7	1572.4
1520	1404.4	1426.1	1437.2	1449.9	1458.4	1465.1	1470.7	1475.5	1479.9	1484.0	1487.7	1491.2	1519.5	1542.4	1563.1	1582.9
1530	1414.1	1435.8	1447.0	1459.8	1468.3	1475.0	1480.6	1485.5	1489.9	1493.9	1497.7	1501.2	1529.7	1552.7	1573.5	1593.4
1540	1423.7	1445.6	1456.8	1469.6	1478.2	1484.9	1490.5	1495.4	1499.9	1503.9	1507.7	1511.3	1539.9	1562.9	1583.9	1603.9
1550	1433.4	1455.3	1466.5	1479.4	1488.0	1494.8	1500.4	1505.4	1509.8	1513.9	1517.7	1521.3	1550.0	1573.2	1594.3	1614.4
1560	1443.0	1465.0	1476.3	1489.3	1497.9	1504.7	1510.3	1515.3	1519.8	1523.9	1527.7	1531.3	1560.2	1583.5	1604.7	1625.0
1570	1452.7	1474.8	1486.1	1499.1	1507.8	1514.6	1520.3	1525.3	1529.8	1533.9	1537.7	1541.3	1570.3	1593.8	1615.1	1635.5
1580	1462.4	1484.5	1495.9	1508.9	1517.7	1524.5	1530.2	1535.2	1539.7	1543.9	1547.7	1551.3	1580.5	1604.1	1625.5	1646.0
1590	1472.0	1494.3	1505.7	1518.8	1527.5	1534.4	1540.1	1545.2	1549.7	1553.8	1557.7	1561.3	1590.7	1614.4	1635.9	1656.5
1600	1481.7	1504.0	1515.5	1528.6	1537.4	1544.3	1550.0	1555.1	1559.7	1563.8	1567.7	1571.4	1600.8	1624.6	1646.3	1667.0
1610	1491.3	1513.8	1525.3	1538.5	1547.3	1554.2	1560.0	1565.1	1569.6	1573.8	1577.7	1581.4	1611.0	1634.9	1656.7	1677.5
1620	1501.0	1523.5	1535.0	1548.3	1557.2	1564.1	1569.9	1575.0	1579.6	1583.8	1587.7	1591.4	1621.1	1645.2	1667.1	1688.0
1630	1510.7	1533.2	1544.8	1558.1	1567.0	1574.0	1579.8	1585.0	1589.6	1593.8	1597.7	1601.4	1631.3	1655.5	1677.5	1698.6
1640	1520.3	1543.0	1554.6	1568.0	1576.9	1583.9	1589.8	1594.9	1599.5	1603.8	1607.7	1611.4	1641.5	1665.8	1687.9	1709.1
1650	1530.0	1552.7	1564.4	1577.8	1586.8	1593.8	1599.7	1604.9	1609.5	1613.8	1617.7	1621.5	1651.6	1676.1	1698.3	1719.6
1660	1539.7	1562.5	1574.2	1587.7	1596.7	1603.7	1609.6	1614.8	1619.5	1623.8	1627.7	1631.5	1661.8	1686.3	1708.7	1730.1
1670	1549.3	1572.2	1584.0	1597.5	1606.5	1613.6	1619.5	1624.8	1629.4	1633.7	1637.8	1641.5	1671.9	1696.6	1719.1	1740.6
1680	1559.0	1582.0	1593.8	1607.3	1616.4	1623.5	1629.5	1634.7	1639.4	1643.7	1647.8	1651.5	1682.1	1706.9	1729.5	1751.1
1690	1568.7	1591.7	1603.6	1617.2	1626.3	1633.4	1639.4	1644.7	1649.4	1653.7	1657.8	1661.6	1692.3	1717.2	1739.9	1761.7
1700	1578.3	1601.5	1613.4	1627.0	1636.2	1643.3	1649.3	1654.6	1659.4	1663.7	1667.8	1671.6	1702.4	1727.5	1750.3	1772.2
1710	1588.0	1611.2	1623.2	1636.9	1646.0	1653.2	1659.3	1664.6	1669.3	1673.7	1677.8	1681.6	1712.6	1737.8	1760.7	1782.7
1720	1597.7	1621.0	1633.0	1646.7	1655.9	1663.1	1669.2	1674.5	1679.3	1683.7	1687.8	1691.6	1722.8	1748.0	1771.1	1793.2
1730	1607.4	1630.7	1642.8	1656.6	1665.8	1673.0	1679.1	1684.5	1689.3	1693.7	1697.8	1701.7	1732.9	1758.3	1781.5	1803.7
1740	1617.0	1640.5	1652.6	1666.4	1675.7	1683.0	1689.1	1694.4	1699.3	1703.7	1707.8	1711.7	1743.1	1768.6	1791.9	1814.2
1750	1626.7	1650.3	1662.4	1676.3	1685.6	1692.9	1699.0	1704.4	1709.2	1713.7	1717.8	1721.7	1753.3	1778.9	1802.3	1824.8
1760	1636.4	1660.0	1672.1	1686.1	1695.5	1702.8	1708.9	1714.3	1719.2	1723.7	1727.8	1731.7	1763.4	1789.2	1812.7	1835.3
1770	1646.1	1669.8	1681.9	1696.0	1705.3	1712.7	1718.9	1724.3	1729.2	1733.7	1737.8	1741.8	1773.6	1799.5	1823.1	1845.8
1780	1655.7	1679.5	1691.7	1705.8	1715.2	1722.6	1728.8	1734.3	1739.2	1743.7	1747.8	1751.8	1783.7	1809.8	1833.5	1856.3
1790	1665.4	1689.3	1701.6	1715.7	1725.1	1732.5	1738.7	1744.2	1749.1	1753.7	1757.9	1761.8	1793.9	1820.0	1843.9	1866.8
1800	1675.1	1699.0	1711.4	1725.5	1735.0	1742.4	1748.7	1754.2	1759.1	1763.6	1767.9	1771.8	1804.1	1830.3	1854.3	1877.3

Blocking probability

channels	0.01%	0.05%	0.1%	0.2%	0.3%	0.4%	0.5%	0.6%	0.7%	0.8%	0.9%	1.0%	2.0%	3.0%	4.0%	5.0%
									Blocking probability							
1810	1684.8	1708.8	1721.2	1735.4	1744.9	1752.3	1758.6	1764.1	1769.1	1773.6	1777.9	1781.9	1814.2	1840.6	1864.7	1887.9
1820	1694.5	1718.6	1731.0	1745.2	1754.8	1762.3	1768.6	1774.1	1779.1	1783.6	1787.9	1791.9	1824.4	1850.9	1875.1	1898.4
1830	1704.2	1728.3	1740.8	1755.1	1764.7	1772.2	1778.5	1784.0	1789.0	1793.6	1797.9	1801.9	1834.6	1861.2	1885.5	1908.9
1840	1713.8	1738.1	1750.6	1764.9	1774.5	1782.1	1788.4	1794.0	1799.0	1803.6	1807.9	1812.0	1844.7	1871.5	1895.9	1919.4
1850	1723.5	1747.9	1760.4	1774.8	1784.4	1792.0	1798.4	1804.0	1809.0	1813.6	1817.9	1822.0	1854.9	1881.8	1906.3	1929.9
1860	1733.2	1757.6	1770.2	1784.6	1794.3	1801.9	1808.3	1813.9	1819.0	1823.6	1827.9	1832.0	1865.1	1892.1	1916.8	1940.4
1870	1742.9	1767.4	1780.0	1794.5	1804.2	1811.8	1818.2	1823.9	1829.0	1833.6	1838.0	1842.0	1875.2	1902.3	1927.2	1951.0
1880	1752.6	1777.2	1789.8	1804.4	1814.1	1821.7	1828.2	1833.8	1838.9	1843.6	1848.0	1852.1	1885.4	1912.6	1937.6	1961.5
1890	1762.3	1786.9	1799.6	1814.2	1824.0	1831.7	1838.1	1843.8	1848.9	1853.6	1858.0	1862.1	1895.6	1922.9	1948.0	1972.0
1900	1772.0	1796.7	1809.4	1824.1	1833.9	1841.6	1848.1	1853.8	1858.9	1863.6	1868.0	1872.1	1905.7	1933.2	1958.4	1982.5
1910	1781.7	1806.5	1819.2	1833.9	1843.8	1851.5	1858.0	1863.7	1868.9	1873.6	1878.0	1882.2	1915.9	1943.5	1968.8	1993.0
1920	1791.4	1816.2	1829.0	1843.8	1853.7	1861.4	1868.0	1873.7	1878.8	1883.6	1888.1	1892.2	1926.1	1953.8	1979.2	2003.5
1930	1801.0	1826.0	1838.8	1853.6	1863.6	1871.3	1877.9	1883.7	1888.8	1893.6	1898.1	1902.2	1936.3	1964.1	1989.6	2014.1
1940	1810.7	1835.8	1848.7	1863.5	1873.5	1881.3	1887.8	1893.6	1898.8	1903.6	1908.1	1912.3	1946.4	1974.4	2000.0	2024.6
1950	1820.4	1845.5	1858.5	1873.4	1883.3	1891.2	1897.8	1903.6	1908.8	1913.6	1918.1	1922.3	1956.6	1984.6	2010.4	2035.1
1960	1830.1	1855.3	1868.3	1883.3	1893.2	1901.1	1907.7	1913.5	1918.8	1923.6	1928.1	1932.3	1966.8	1994.9	2020.8	2045.6
1970	1839.8	1865.1	1878.1	1893.1	1903.1	1911.0	1917.7	1923.5	1928.8	1933.6	1938.1	1942.4	1976.9	2005.2	2031.2	2056.1
1980	1849.5	1874.9	1887.9	1902.9	1913.0	1920.9	1927.6	1933.5	1938.8	1943.6	1948.1	1952.4	1987.1	2015.5	2041.6	2066.6
1990	1859.2	1884.6	1897.7	1912.8	1922.9	1930.9	1937.6	1943.4	1948.7	1953.6	1958.2	1962.4	1997.3	2025.8	2052.0	2077.2
2000	1868.9	1894.4	1907.5	1922.7	1932.8	1940.8	1947.5	1953.4	1958.7	1963.6	1968.2	1972.5	2007.4	2036.1	2062.4	2087.7
2010	1878.6	1904.2	1917.4	1932.5	1942.7	1950.7	1957.4	1963.4	1968.7	1973.6	1978.2	1982.5	2017.6	2046.4	2072.8	2098.2
2020	1888.3	1914.0	1927.2	1942.4	1952.6	1960.6	1967.4	1973.3	1978.7	1983.6	1988.2	1992.5	2027.8	2056.7	2083.2	2108.7
2030	1898.0	1923.7	1937.0	1952.3	1962.5	1970.6	1977.3	1983.3	1988.7	1993.6	1998.2	2002.6	2037.9	2067.0	2093.6	2119.2
2040	1907.7	1933.5	1946.8	1962.1	1972.4	1980.5	1987.3	1993.3	1998.7	2003.6	2008.3	2012.6	2048.1	2077.2	2104.0	2129.8
2050	1917.4	1943.3	1956.6	1972.0	1982.3	1990.4	1997.2	2003.2	2008.7	2013.6	2018.3	2022.6	2058.3	2087.5	2114.4	2140.3
2060	1927.1	1953.1	1966.4	1981.9	1992.2	2000.3	2007.2	2013.2	2018.6	2023.6	2028.3	2032.7	2068.5	2097.8	2124.8	2150.8
2070	1936.9	1962.9	1976.3	1991.7	2002.1	2010.3	2017.1	2023.2	2028.6	2033.6	2038.3	2042.7	2078.6	2108.1	2135.2	2161.3
2080	1946.6	1972.6	1986.1	2001.6	2012.0	2020.2	2027.1	2033.1	2038.6	2043.6	2048.3	2052.8	2088.8	2118.4	2145.6	2171.8
2090	1956.3	1982.4	1995.9	2011.5	2021.9	2030.1	2037.0	2043.1	2048.6	2053.7	2058.4	2062.8	2099.0	2128.7	2156.0	2182.3
2100	1966.0	1992.2	2005.7	2021.3	2031.8	2040.0	2047.0	2053.1	2058.6	2063.7	2068.4	2072.8	2109.1	2139.0	2166.4	2192.9

channels								Blocking probability								
	0.01%	0.05%	0.1%	0.2%	0.3%	0.4%	0.5%	0.6%	0.7%	0.8%	0.9%	1.0%	2.0%	3.0%	4.0%	5.0%
2110	1975.7	2002.0	2015.5	2031.2	2041.7	2050.0	2056.9	2063.1	2068.6	2073.7	2078.4	2082.9	2119.3	2149.3	2176.9	2203.4
2120	1985.4	2011.8	2025.4	2041.1	2051.6	2059.9	2066.9	2073.0	2078.6	2083.7	2088.4	2092.9	2129.5	2159.6	2187.3	2213.9
2130	1995.1	2021.6	2035.2	2050.9	2061.5	2069.8	2076.8	2083.0	2088.6	2093.7	2098.4	2102.9	2139.7	2169.9	2197.7	2224.4
2140	2004.8	2031.3	2045.0	2060.8	2071.4	2079.7	2086.8	2093.0	2098.5	2103.7	2108.5	2113.0	2149.8	2180.2	2208.1	2234.9
2150	2014.5	2041.1	2054.8	2070.7	2081.3	2089.7	2096.7	2102.9	2108.5	2113.7	2118.5	2123.0	2160.0	2190.4	2218.5	2245.5
2160	2024.2	2050.9	2064.7	2080.6	2091.2	2099.6	2106.7	2112.9	2118.5	2123.7	2128.5	2133.1	2170.2	2200.7	2228.9	2256.0
2170	2034.0	2060.7	2074.5	2090.4	2101.1	2109.5	2116.6	2122.9	2128.5	2133.7	2138.5	2143.1	2180.4	2211.0	2239.3	2266.5
2180	2043.7	2070.5	2084.3	2100.3	2111.0	2119.5	2126.6	2132.9	2138.5	2143.7	2148.6	2153.1	2190.5	2221.3	2249.7	2277.0
2190	2053.4	2080.3	2094.1	2110.2	2120.9	2129.4	2136.5	2142.8	2148.5	2153.7	2158.6	2163.2	2200.7	2231.6	2260.1	2287.5
2200	2063.1	2090.1	2104.0	2120.0	2130.8	2139.3	2146.5	2152.8	2158.5	2163.7	2168.6	2173.2	2210.9	2241.9	2270.5	2298.1
2210	2072.8	2099.8	2113.8	2129.9	2140.7	2149.3	2156.4	2162.8	2168.5	2173.7	2178.6	2183.3	2221.0	2252.2	2280.9	2308.6
2220	2082.5	2109.6	2123.6	2139.8	2150.6	2159.2	2166.4	2172.7	2178.5	2183.7	2188.7	2193.3	2231.2	2262.5	2291.3	2319.1
2230	2092.3	2119.4	2133.5	2149.7	2160.6	2169.1	2176.4	2182.7	2188.5	2193.8	2198.7	2203.3	2241.4	2272.8	2301.7	2329.6
2240	2102.0	2129.2	2143.3	2159.5	2170.5	2179.0	2186.3	2192.7	2198.5	2203.8	2208.7	2213.4	2251.6	2283.1	2312.1	2340.1
2250	2111.7	2139.0	2153.1	2169.4	2180.4	2189.0	2196.3	2202.7	2208.5	2213.8	2218.7	2223.4	2261.7	2293.4	2322.5	2350.6
2260	2121.4	2148.8	2162.9	2179.3	2190.3	2198.9	2206.2	2212.6	2218.4	2223.8	2228.8	2233.5	2271.9	2303.7	2332.9	2361.2
2270	2131.1	2158.6	2172.8	2189.2	2200.2	2208.8	2216.2	2222.6	2228.4	2233.8	2238.8	2243.5	2282.1	2313.9	2343.3	2371.7
2280	2140.9	2168.4	2182.6	2199.0	2210.1	2218.8	2226.1	2232.6	2238.4	2243.8	2248.8	2253.5	2292.3	2324.2	2353.7	2382.2
2290	2150.6	2178.2	2192.4	2208.9	2220.0	2228.7	2236.1	2242.6	2248.4	2253.8	2258.8	2263.6	2302.4	2334.5	2364.2	2392.7
2300	2160.3	2188.0	2202.3	2218.8	2229.9	2238.6	2246.0	2252.5	2258.4	2263.8	2268.9	2273.6	2312.6	2344.8	2374.6	2403.2
2310	2170.0	2197.8	2212.1	2228.7	2239.8	2248.6	2256.0	2262.5	2268.4	2273.8	2278.9	2283.7	2322.8	2355.1	2385.0	2413.8
2320	2179.8	2207.6	2221.9	2238.5	2249.7	2258.5	2266.0	2272.5	2278.4	2283.8	2288.9	2293.7	2333.0	2365.4	2395.4	2424.3
2330	2189.5	2217.4	2231.8	2248.4	2259.6	2268.5	2275.9	2282.5	2288.4	2293.9	2299.0	2303.8	2343.1	2375.7	2405.8	2434.8
2340	2199.2	2227.2	2241.6	2258.3	2269.5	2278.4	2285.9	2292.4	2298.4	2303.9	2309.0	2313.8	2353.3	2386.0	2416.2	2445.3
2350	2208.9	2237.0	2251.4	2268.2	2279.5	2288.3	2295.8	2302.4	2308.4	2313.9	2319.0	2323.8	2363.5	2396.3	2426.6	2455.8
2360	2218.7	2246.8	2261.3	2278.1	2289.4	2298.3	2305.8	2312.4	2318.4	2323.9	2329.0	2333.9	2373.7	2406.6	2437.0	2466.4
2370	2228.4	2256.6	2271.1	2287.9	2299.3	2308.2	2315.7	2322.4	2328.4	2333.9	2339.1	2343.9	2383.8	2416.9	2447.4	2476.9
2380	2238.1	2266.3	2280.9	2297.8	2309.2	2318.1	2325.7	2332.4	2338.4	2343.9	2349.1	2354.0	2394.0	2427.2	2457.8	2487.4
2390	2247.8	2276.1	2290.8	2307.7	2319.1	2328.1	2335.7	2342.3	2348.4	2353.9	2359.1	2364.0	2404.2	2437.5	2468.2	2497.9
2400	2257.6	2285.9	2300.6	2317.6	2329.0	2338.0	2345.6	2352.3	2358.4	2364.0	2369.2	2374.1	2414.4	2447.7	2478.6	2508.4

Blocking probability

channels	0.01%	0.05%	0.1%	0.2%	0.3%	0.4%	0.5%	0.6%	0.7%	0.8%	0.9%	1.0%	2.0%	3.0%	4.0%	5.0%
2410	2267.3	2295.7	2310.4	2327.5	2338.9	2347.9	2355.6	2362.3	2368.4	2374.0	2379.2	2384.1	2424.5	2458.0	2489.0	2519.0
2420	2277.0	2305.5	2320.3	2337.4	2348.8	2357.9	2365.5	2372.3	2378.4	2384.0	2389.2	2394.2	2434.7	2468.3	2499.4	2529.5
2430	2286.8	2315.3	2330.1	2347.2	2358.8	2367.8	2375.5	2382.3	2388.4	2394.0	2399.2	2404.2	2444.9	2478.6	2509.8	2540.0
2440	2296.5	2325.2	2340.0	2357.1	2368.7	2377.8	2385.5	2392.2	2398.4	2404.0	2409.3	2414.2	2455.1	2488.9	2520.3	2550.5
2450	2306.2	2335.0	2349.8	2367.0	2378.6	2387.7	2395.4	2402.2	2408.4	2414.0	2419.3	2424.3	2465.2	2499.2	2530.7	2561.0
2460	2316.0	2344.8	2359.6	2376.9	2388.5	2397.6	2405.4	2412.2	2418.4	2424.0	2429.3	2434.3	2475.4	2509.5	2541.1	2571.6
2470	2325.7	2354.6	2369.5	2386.8	2398.4	2407.6	2415.3	2422.2	2428.4	2434.1	2439.4	2444.4	2485.6	2519.8	2551.5	2582.1
2480	2335.4	2364.4	2379.3	2396.7	2408.3	2417.5	2425.3	2432.2	2438.4	2444.1	2449.4	2454.4	2495.8	2530.1	2561.9	2592.6
2490	2345.2	2374.2	2389.2	2406.5	2418.2	2427.5	2435.3	2442.1	2448.4	2454.1	2459.4	2464.5	2506.0	2540.4	2572.3	2603.1
2500	2354.9	2384.0	2399.0	2416.4	2428.2	2437.4	2445.2	2452.1	2456.4	2464.1	2469.5	2474.5	2516.1	2550.7	2582.7	2613.6

APPENDIX B

Visual Basic® Code for Erlang Calculations

The following code can be used in a spreadsheet macro to enable the automatic calculation of Erlang values. The function GetBlocking generates a blocking value based on two inputs: Erlangs and circuits. The function Get-Circuits generates a number of circuits based on two inputs: Erlangs and blocking. The function GetErlangs calculates the number of Erlangs based on two inputs: circuits and blocking.

```
Attribute VB_Name = "Erlangs"
Option Explicit
Function Ceil(Value As Double) As Integer
    Dim i As Integer
    i = Value
    If i < Value Then
        i = i + 1
    End If
    Ceil = i
End Function

Function GetBlocking(Erlangs As Double, Circuits As Integer) As
Double
    Dim Blocking As Double
    Dim i As Integer
    If (Erlangs = 0) Or (Circuits = 0) Then
        GetBlocking = 0
        Exit Function
    End If
    Blocking = 0
    For i = 1 To Circuits
        Blocking = (1 + Blocking) * (i / Erlangs)
    Next
    GetBlocking = 1 / (1 + Blocking)
End Function

Function GetCircuits(Blocking As Double, Erlangs As Double) As
Integer
    Dim fR As Double, fMid As Double
    Dim l As Integer, r As Integer, mid As Integer
  If (Blocking < 0.0001) Or (Blocking > 0.9999) Or (Erlangs = 0)
Then
        GetCircuits = 0
        Exit Function
    End If
    l = 0
    r = Ceil(Erlangs)
    fR = GetBlocking(Erlangs, r)
    While (fR > Blocking)
        l = r
        r = r + 32
        fR = GetBlocking(Erlangs, r)
```

```
        Wend
        While ((r - l) > 1)
            mid = Ceil((l + r) / 2)
            fMid = GetBlocking(Erlangs, mid)
            If (fMid > Blocking) Then
                l = mid
            Else
                r = mid
            End If
        Wend
        GetCircuits = r
    End Function


    Function GetErlangs(Blocking As Double, Circuits As Integer) As
    Double
        Dim l As Double, r As Double, i As Double, Delta As Double

        If (Blocking < 0.0001) Or (Blocking > 0.9999) Or (Circuits = 0)
    Then
            GetErlangs = 0
            Exit Function
        End If

        i = 0
        Do
            i = i + 1
        Loop While (Blocking > GetBlocking(Circuits * i, Circuits))

        r = i * Circuits
        l = (i - 1) * Circuits
        Delta = r - l

        While (Delta > 0.001)
            Delta = (r - l) / 2
            If (Blocking > GetBlocking(l + Delta, Circuits)) Then
                l = l + Delta
            Else
                r = l + Delta
            End If
        Wend

        GetErlangs = l
    End Function
```

GLOSSARY OF ACRONYMS

ABNF	Augmented Backus-Naur Form
ACELP	Algebraic Code-Excited Linear Prediction
ACF	AdmissionConfirm message
ACM	Address Complete message
ADPCM	Adaptive Differential Pulse Code Modulation
ADSL	Asymmetric Digital Subscriber Line
AF	Assured Forwarding
AMI	Alternate Mark Inversion
AMR	Adaptive Multi-Rate
ANM	Answer message
ANSI	American National Standards Institute
API	Application Programming Interface
ARJ	AdmissionReject message
ARQ	AdmissionRequest message
AS	Application server
ASN.1	Abstract Syntax Notation 1
ASP	Application Server Process
ATM	Asynchronous Transfer Mode
AUCX	AuditConnection (MGCP command)
AUEP	AuditEndpoint (MGCP command)
BCF	BandwidthConfirm message
BCP	Best Current Practice
BGP	Border Gateway Protocol
BHCA	Busy-hour call attempts
BRI	Basic Rate Interface
BRJ	BandwidthReject message
BRQ	BandwidthRequest message
CAP	Competitive Access Provider
CAS	Channel-Associated Signaling
CBR	Constant bit rate
CBS	Committed burst size
CCBS	Call Completion to Busy Subscriber
CCS	Common Channel Signaling
CDMA	Code Division Multiple Access
CDR	Committed data rate
CELP	Code-Excited Linear Prediction
CIC	Circuit Identification Code
CLEC	Competitive Local Exchange Carrier
CNG	Comfort noise generation

COPS	Common Open Policy Service
CORBA	Common Object Request Broker Architecture
CPE	Customer Premises Equipment
CPG	Call Progress
CRCX	CreateConnection (MGCP command)
CRLF	Carriage return and line feed
CR-LSP	Constraint-Based LSP
CS-1	Capability Set 1
CSCF	Call State Control Function
CSeq	Command sequence
CSRC	Contributing Source
DCF	DisengageConfirm message
DiffServ	Differentiated Services
DLCX	DeleteConnection (MGCP Command)
DNS	Domain Name System
DPC	Destination Point Code
DPCM	Differential Pulse Code Modulation
DRJ	DisengageReject message
DRQ	DisengageRequest message
DS	DiffServ
DS0	Digital Signal 0 (equivalent to 64 Kbps)
DS1	Digital Signal 1 (equivalent to 1.544 Mbps)
DS3	Digital Signal 3 (equivalent to 44.736 Mbps)
DSCP	DiffServ codepoint
DSL	Digital Subscriber Line
DSP	Digital signal processor
DTMF	Dual-tone multifrequency
DTX	Discontinuous transmission
DWDM	Dense Wave Division Multiplexing
EBS	Excess burst size
EC	Echo cancellation
EF	Expedited Forwarding
EFR	Enhanced Full Rate
EMS	Element Management System
EPCF	EndpointConfiguration (MGCP command)
ER	Explicit route
ESP	Enhanced service provider
ETSI	European Telecommunications Standards Institute
FCS	Frame Check Sequence
FDM	Frequency Division Multiplexing

FE	Fast Ethernet
FEC	Forwarding Equivalence Class
FISU	Fill-In Signal Unit
FoIP	Fax over IP
FR	Full Rate
GCF	GatekeeperConfirm message
Gig-E	Gigabit Ethernet
GMSK	Gaussian Minimum Shift Key
GRJ	GatekeeperReject message
GRQ	GatekeeperRequest message
GSM	Global System for Mobile communications
GSTN	General Switched Telephone Network
GT	Global Title
HDLC	High-level Data Link Control
HDSL	High-bit-rate Digital Subscriber Line
HTTP	Hypertext Transfer Protocol
IAB	Internet Architecture Board
IACK	InfoRequestAck message
IAM	Initial Address message
IANA	Internet Assigned Numbers Authority
ICMP	Internet Control Message Protocol
IESG	Internet Engineering Steering Group
IETF	Internet Engineering Task Force
IGMP	Internet Group Message Protocol
IGRP	Internet Gateway Routing Protocol
IN	Intelligent Network
INAK	InfoRequestNak message
INAP	Intelligent Network Application Part
IP	Internet Protocol
IRQ	InformationRequest message
IRR	InformationRequestResponse message
ISDN	Integrated Services Digital Network
ISDN-UP or ISUP	ISDN User Part
ISP	Internet service provider
ITSP	IP telephony service provider
ITU	International Telecommunication Union
ITU-T	International Telecommunication Union-Telecommunications Standardization Sector (formerly CCITT)
IUA	ISDN User Adaptation Layer

IVR	Interactive Voice Response
JPEG	Joint Photographic Experts Group
LAN	Local area network
LCF	LocationConfirm message
LD	Long distance
LD-CELP	Low-Delay Code-Excited Linear Prediction
LDP	Label Distribution Protocol
LEC	Local Exchange Carrier
LER	Label edge router
LMDS	Local Multipoint Distribution Service
LPC	Linear Prediction Coding
LRJ	LocationReject message
LRQ	LocationRequest message
LSP	Label-switched path
LSR	Label-switching router
LSSU	Link Status Signal Unit
M2UA	MTP2 User Adaptation Layer
M2PA	MTP2 Peer-to-Peer Adaptation Layer
M3UA	MTP3 User Adaptation Layer
MAP	Mobile Application Part
MC	Multipoint controller
MCF	MessageConfirm message
MCU	Multipoint controller unit
MDCX	ModifyConnection (MGCP command)
MF	Multifrequency
MFC	Multifrequency Compelled
MG	Media gateway
MGC	Media gateway controller
MGCP	Media Gateway Control Protocol
MHT	Mean holding time
MIME	Multipurpose Internet Mail Extension
MIPS	Million instructions per second
MMUSIC	Multiparty Multimedia Session Control
MOS	Mean Opinion Score
MoU	Minutes of use
MP	Multipoint processor
MPE	Multi-Pulse Excited
MPLS	Multiprotocol Label Switching
MP-MLQ	Multi-Pulse Maximum Likelihood Quantization
MRF	Multimedia Resource Function
MSU	Message signal unit

MTA	Media terminal adapter
MTBF	Mean Time Between Failures
MTP	Message Transfer Part
MTU	Maximum transmission unit
NGW	Network gateway
NHLFE	Next Hop Level Forwarding Entry
NIF	Nodal Interworking Function
NIST	National Institute of Standards and Technology
NMS	Network Management System
NSF	Non-Standard Facilities
NSM	Non-Standard Message
NTFY	Notify (MGCP command)
NTP	Network Time Protocol
OA&M	Operations, Administration, and Maintenance
OPC	Origination point code
OSI	Open Systems Interconnection
OSPF	Open Shortest Path First
PBS	Peak burst size
PBX	Private branch exchange
PC	Personal computer
PC	Point code
PCM	Pulse Code Modulation
PDD	Post-Dial Delay
PDP	Policy Decision Point
PDR	Peak data rate
PEP	Policy Enforcement Point
PHB	Per-hop behavior
PINT	PSTN and Internet Interworking
POTS	Plain Old Telephone Service
PPP	Point-to-Point Protocol
PQ	Priority queue
PRACK	ProvisionalResponseAcknowledgment message
PSTN	Public Switched Telephone Network
PT	Payload Type
PVC	Permanent virtual circuit
QCIF	Quarter Common Intermediate Format
QoS	Quality of service
RAC	ResourceAvailableConfirm message
RAck	ResponseAcknowledgment Header
RAI	ResourceAvailableIndicate message
RAS	Registration, Admission, and Status

RCF	RegistrationConfirm message
REL	Release message
RFC	Request for Comments
RFI	Request for Information
RFP	Request for Proposal
RIP	Routing Information Protocol
RIP	Request in Progress
RLC	Release Complete message
RPE-LTP	Regular Pulse-Excited Long term Prediction
RQNT	NotificationRequest (MGCP command)
RRJ	RegistrationReject message
RRQ	RegistrationRequest message
RSeq	Response sequence
RSIP	RestartInProgress (MGCP command)
RSVP	Resource Reservation Protocol
RSVP-TE	RSVP Traffic Engineering
RTCP	Real-Time Control Protocol
RTP	Real-Time Transport Protocol
RTSP	Real-Time Streaming Protocol
SAP	Session Announcement Protocol
SCCP	Signaling Connection Control Part
SCP	Service control point
SCTP	Stream Control Transmission Protocol
SDES	Source Description
SDP	Session Description Protocol
SG	Signaling gateway
SID	Silence Insertion Description
SIF	Signaling Information Field
SIMPLE	SIP for Instant Message and Presence Leveraging Extensions
SIO	Service Information Octet
SIP	Session Initiation Protocol
SLA	Service Level Agreement
SLS	Signaling Link Selection
SNMP	Simple Network Management Protocol
SOHO	Small Office/Home Office
SONET	Synchronous Optical Network
SP	Signaling point
SPC	Signaling point code
SS7	Signaling System Number 7
SSN	Subsystem Number

SSP	Service switching point
SSRC	Synchronization Source
STP	Signal transfer point
SUA	SCCP User Adaptation Layer
TA	Terminal adapter
TCAP	Transaction Capabilities Application Part
TCP	Transmission Control Protocol
TDM	Time Division Multiplexing
TDMA	Time Division Multiple Access
TFA	Transfer Allowed
TFP	Transfer Prohibited
TLV	Type-Length-Value
ToS	Type of Service
TSAP	Transport Service Access Point
TTL	Time to live
UCF	UnregistrationConfirm message
UDP	User Datagram Protocol
URI	Uniform Resource Identifier
URJ	UnregistrationReject message
URL	Uniform Resource Locator
URQ	UnregistrationRequest message
UTF	Unicode Standard Transformation Format
V5UA	V5.2 User Adaptation Layer
VAD	Voice activity detection
VBR	Variable bit rate
VoATM	Voice over ATM
VoIP	Voice over IP
WAN	Wide area network
WRR	Weighted Round-Robin
XRS	UnknownMessageResponse message

REFERENCES

Chapter 2

Braden, R. "Requirements for Internet Hosts—Communication Layers." RFC 1122, October 1989.

Braden, R. "Requirements for Internet Hosts—Application and Support." RFC 1123, October 1989.

Bradner, S. "The Internet Standards Process—Revision 3." RFC 2026, October 1996.

Deering, S. "Host Extensions for IP Multicasting." RFC 1112, August 1989.

Deering, S. and R. Hinden. "IP Version 6 Addressing Architecture." RFC 2373, July 1998.

Fenner, W. "Internet Group Management Protocol, Version 2." RFC 2236, November 1997.

Mills, D. "Network Time Protocol (Version 3) Specification, Implementation, and Analysis." RFC 1305, March 1992.

Mogul, J. "Broadcasting Internet Datagrams." RFC 919, October 1984.

Mogul, J. and J. Postel. "Internet Standard Subnetting Procedure." RFC 950, August 1985.

Moy, J. "Multicast Extensions to OSPF." RFC 1584, March 1994.

Perkins, C. et al. "RTP Payload for Redundant Audio Data." RFC 2198, September 1997.

Postel, J. "Internet Protocol." RFC 791, September 1981.

Postel, J. "Transmission Control Protocol." RFC 793, September 1981.

Postel, J. and J. Reynolds. "Domain Requirements." RFC 920, October 1984.

Rekhter, Y. and T. Li. "A Border Gateway Control Protocol 4 (BGP-4)." RFC 1771, March 1995.

Schulzrinne, H. "RTP Profile for Audio and Video Conferences with Minimal Control." RFC 1890, January 1996.

Schulzrinne, H. "RTP: A Transport Protocol for Real-Time Applications." Internet-draft, work in progress, November 2001.

Schulzrinne, H. et al. "RTP: A Transport Protocol for Real-Time Applications." RFC 1889, January 1996.

Schulzrinne, H. and S. Casner. "RTP Profile for Audio and Video Conferences with Minimal Control." Internet-draft, work in progress, November 2001.

Chapter 3

ITU-T. "Methods for subjective determination of transmission quality." Recommendation P.800, August 1996.

ITU-T. "Objective quality measurement of telephone-band (300–3,400 Hz) speech codecs." Recommendation P.861, February 1998.

ITU-T. "Perceptual evaluation of speech quality (PESQ), an objective method for end-to-end speech quality assessment of narrowband telephone networks and speech codecs." Recommendation P.862, February 2001.

ITU-T. "Artificial voices." Recommendation P.50, September 1999.

ITU-T. "Pulse code modulation (PCM) of voice frequencies." Recommendation G.711, November 1988.

ITU-T. "32kb/s adaptive differential pulse code modulation (ADPCM)." Recommendation G.721, October 1988.

ITU-T. "40, 32, 24, 16 kbit/s Adaptive Differential Pulse Code Modulation (ADPCM)." Recommendation G.726, December 1990.

ITU-T. "Coding of speech at 16 kbit/s using low-delay code excited linear prediction." Recommendation G.728, September 1992.

ITU-T. "Dual rate speech coder for multimedia communications transmitting at 5.3 and 6.3 kbit/s." Recommendation G.723.1, March 1996.

ITU-T. "Silence suppression scheme." Recommendation G.723.1, Annex A, November 1996.

ITU-T. "Coding of speech at 8 kbit/s using Conjugate-Structure Algebraic-Code-Excited Linear Prediction (CS-ACELP)." Recommendation G.729, March 1996.

ITU-T. "Reduced complexity 8 kbit/s CS-ACELP speech codec." Recommendation G.729, Annex A, November 1996.

ITU-T. "A silence compression scheme for G.729 optimized for terminals conforming to Recommendation V.70." Recommendation G.729, Annex B, October 1996.

ITU-T. "6.4 kbit/s CS-ACELP speech coding algorithm." Recommendation G.729, Annex D, September 1998.

ITU-T. "11.8 kbit/s CS-ACELP speech coding algorithm." Recommendation G.729, Annex E, September 1998.

ETSI. "Enhanced Full Rate (EFR) speech transcoding." EN 300 726 (GSM 06.60), December 1999.

ETSI. "Adaptive Multi-Rate (AMR) speech transcoding." EN 301 704 (GSM 06.90), April 2000.

Schulzrinne, H. and S. Petrack. "RTP Payload for DTMF Digits, Telephony Tones, and Telephony Signals." RFC 2833, May 2000.

TIA/EIA. "High Rate Speech Service Option 17 for Wideband Spread Spectrum Communications Systems." IS-733, March 1998.

The complete volume of the ITU material, from which the texts reproduced are extracted, can be obtained from

International Telecommunication Union

Sales and Marketing Service

Place des Nations

CH-1211 Geneva 20 (Switzerland)

Telephone: +41 22 730 6141 (English) / +41 22 730 61 42 (French) / +41 22 730 61 43 (Spanish)

Telex: 421 000 uit ch / Fax: +41 22 730 51 94

X.400: S = Sales; P = itu; A = 400net; C = ch

E-mail: sales@itu.int

Web address: www.itu.int/publications

Chapter 4

ITU-T. "Packet-based multimedia communications systems." Recommendation H.323 version 4, November 2000.

ITU-T. "Call signalling protocols and media stream packetization for packet-based multimedia communication systems." Recommendation H.225.0 version 4, November 2000.

ITU-T. "Control protocol for multimedia communication." Recommendation H.245 version 8, July 2001.

ITU-T. "ISDN user-network interface layer 3 specification for basic call control." Recommendation Q.931, May 1998.

ITU-T. "Video codec for audiovisual services at p×64 kbit/s." Recommendation H.261, March 1993.

ITU-T. "Generic procedures for the control of ISDN supplementary services." Recommendation Q.932, May 1998.

Chapter 5

Agrawal, H. et al. "SIP-H.323 Interworking Requirements." work in progress.

Camarillo, G. et al. "Integration of Resource Management and SIP." work in progress.

Campbell, B. et al. "Session Initiation Protocol Extension for Instant Messaging." work in progress.

Donovan, S. "The SIP INFO Method." RFC 2976, October 2000.

Handley, M. et al. "SIP: Session Initiation Protocol." RFC 2543, March 1999.

Handley, M. et al. "Session Announcement Protocol." RFC 2974, October 2000.

Handley, M. and V. Jacobson. "SDP: Session Description Protocol." RFC 2327, April 1998.

Roach, A.B. "Session Initiation Protocol (SIP)-Specific Event Notification." RFC 3265, June 2002.

Rosenberg, J. "The Session Initiation Protocol UPDATE Method." work in progress.

Rosenberg, J. and H. Schulzrinne. "Reliability of Provisional Responses in SIP." RFC 3262, June 2002.

Rosenberg, J. and H. Schulzrinne. "An Offer/Answer Model with the Session Description Protocol (SDP)." RFC 3264, June 2002.

Rosenberg, J. et al. "Session Initiation Protocol (SIP) Extensions for Presence." work in progress.

Rosenberg, J. et al. "SIP: Session Initiation Protocol." RFC 3261, June 2002.

Schulzrinne, H. et al. "Real Time Streaming Protocol (RTSP)." RFC 2326, April 1998.

Sparks, R. "The SIP REFER Method." work in progress.

Sparks, R. "Internet Media Type message/sipfrag." work in progress.

Zimmerer, E. et al. "MIME media types for ISUP and QSIG Objects." RFC 3204, December 2001.

Chapter 6

Arango, M. et al. "Media Gateway Control Protocol (MGCP) Version 1.0." RFC 2705, October 1999.

Cuervo, F. et al. "MEGACO Protocol version 1.0." RFC 3015, November 2000.

Greene, N. et al. "Media Gateway Control Protocol Architecture and Requirements." RFC 2805, April 2000.

Pantaleo, M. "The Megaco/H.248 Gateway Control Protocol, version 2." work in progress.

Chapter 7

ANSI. "Telecommunications-Signalling System Number 7 (SS7)-Message Transfer Part (MTP)." T1.111, 1996.

ANSI. "Telecommunications-Signalling System Number 7 (SS7)-Integrated Services Digital Network (ISDN) User Part." T1.113, 1995.

ANSI. "Telecommunications-Signalling System Number 7 (SS7)-Signaling Connection Control Part (SCCP)." T1.112, 1996.

ANSI. "Telecommunications-Signalling System Number 7 (SS7)-Transaction Capability Application Part (TCAP)." T1.114, 1996.

Freed, N. and N. Borenstein. "Multipurpose Internet Mail Extensions (MIME) Part Two: Media Types." RFC 2046, November 1996.

George, T. et al. "SS7 MTP2-User Peer-to-Peer Adaptation Layer." work in progress.

ITU-T. "Functional Description of the Message Transfer Part (MTP) of Signalling System No. 7." Recommendation Q.701, March 1993.

ITU-T. "Signalling System No. 7—ISDN User Part Functional Description." Recommendation Q.761, December 1999.

ITU-T. "Functional description of the Signalling Connection Control Part." Recommendation Q.711, March 2001.

ITU-T. "Functional description of Transaction Capabilities." Recommendation Q.771, June 1997.

Loughney, J. "Signalling Connection Control Part User Adaptation Layer (SUA)." work in progress.

Morneault, K. et al. "ISDN Q.921-User Adaptation Layer (IUA)." RFC 3057, February 2001.

Morneault, K. et al. "Signaling System 7 (SS7) Message Transfer Part (MTP) 2 User Adaptation Layer." work in progress.

Ong, L. et al. "Framework Architecture for Signaling Transport." RFC 2719, October 1999.

Sidebottom, G. et al. "SS7 MTP3-User Adaptation Layer (M3UA)." work in progress.

Stewart, R. et al. "Stream Control Transmission Protocol." RFC 2960, October 2000.

Telcordia Technologies. "Specification of Signaling System Number 7." GR-246-CORE, December 1999.

Weilandt, E. et al. "V5.2-User Adaptation Layer (V5UA)." work in progress.

Chapter 8

Andersson, L. et al. "LDP Specification." RFC 3036, January 2001.

Awduche, D. "Requirements for Traffic Engineering over MPLS." RFC 2702, September 1999.

Awduche, D. et al. "RSVP-TE: Extensions to RSVP for LSP Tunnels." RFC 3209, December 2001.

Blake, S. et al. "An Architecture for Differentiated Services." RFC 2475, December 1998.

Braden, R. et al. "Resource ReSerVation Protocol (RSVP)—Version 1 Functional Specification." RFC 2205, September 1997.

Davie, B. et al. "An Expedited Forwarding Per-Hop Behavior (PHB)." RFC 3246, March 2002.

Durham, D. et al. "The COPS (Common Open Policy Service) Protocol." RFC 2748, January 2000.

Grossman, D. "New Terminology and Clarifications for DiffServ." RFC 3260, April 2002.

Heinanen, J. et al. "Assured Forwarding PHB." RFC 2597, June 1999.

Jamoussi, B. et al. "Constraint-Based LSP Setup Using LDP." RFC 3212, January 2002.

Ramakrishnan, K. et al. "The Addition of Explicit Congestion Notification (ECN) to IP." RFC 3168, September 2001.

Rosen, E. et al. "Multiprotocol Label Switching Architecture." RFC 3031, January 2001.

Shenker, S. and J. Wroclawski. "General Characterization Parameters for Integrated Service Network Elements." RFC 2215, September 1997.

Shenker, S., C. Partridge, and R. Guerin. "Specification of Guaranteed Quality of Service." RFC 2212, September 1997.

Wroclawski, J. "Specification of Controlled-Load Network Element Service." RFC 2211, September 1997.

INDEX

ABOUT THE AUTHOR

Daniel Collins has worked in the telecommunications industry for 15 years. He spent approximately nine years with Ericsson in various countries, including Ireland, Australia, the United Kingdom and the United States. During that time he worked extensively with both wireline and wireless network technologies. He helped to develop and deploy 2G wireless systems in Europe; he played a major role in the adaptation of GSM standards for use in the United States; and he was a major contributor to the launch of some of the earliest PCS networks in North America.

Since leaving Ericsson, Daniel has worked for a new telecommunications carrier and, more recently as a consultant. In a consultancy capacity, he has provided wireless and VoIP engineering expertise to numerous network operators, consultancy companies and infrastructure vendors. Daniel's clients include PrimeCo Personal Communications (now part of Verizon Wireless), Synacom Technology, AT&T Wireless, Alcatel USA and several other companies.

Collins is co-author of *3G Wireless Networks*, published by McGraw-Hill. He holds a degree in Electrical and Electronic engineering from the National University of Ireland.